D1260105

TESTING OF COMMUNICATING SYSTEMS

IFIP - The International Federation for Information Processing

IFIP was founded in 1960 under the auspices of UNESCO, following the First World Computer Congress held in Paris the previous year. An umbrella organization for societies working in information processing, IFIP's aim is two-fold: to support information processing within its member countries and to encourage technology transfer to developing nations. As its mission statement clearly states,

IFIP's mission is to be the leading, truly international, apolitical organization which encourages and assists in the development, exploitation and application of information technology for the benefit of all people.

IFIP is a non-profitmaking organization, run almost solely by 2500 volunteers. It operates through a number of technical committees, which organize events and publications. IFIP's events range from an international congress to local seminars, but the most important are:

- The IFIP World Computer Congress, held every second year;
- open conferences;
- working conferences.

The flagship event is the IFIP World Computer Congress, at which both invited and contributed papers are presented. Contributed papers are rigorously refereed and the rejection rate is high.

As with the Congress, participation in the open conferences is open to all and papers may be invited or submitted. Again, submitted papers are stringently refereed.

The working conferences are structured differently. They are usually run by a working group and attendance is small and by invitation only. Their purpose is to create an atmosphere conducive to innovation and development. Refereeing is less rigorous and papers are subjected to extensive group discussion.

Publications arising from IFIP events vary. The papers presented at the IFIP World Computer Congress and at open conferences are published as conference proceedings, while the results of the working conferences are often published as collections of selected and edited papers.

Any national society whose primary activity is in information may apply to become a full member of IFIP, although full membership is restricted to one society per country. Full members are entitled to vote at the annual General Assembly, National societies preferring a less committed involvement may apply for associate or corresponding membership. Associate members enjoy the same benefits as full members, but without voting rights. Corresponding members are not represented in IFIP bodies. Affiliated membership is open to non-national societies, and individual and honorary membership schemes are also offered.

TESTING OF COMMUNICATING SYSTEMS

Tools and Techniques

*IFIP TC6/WG6.1 13th International Conference
on Testing of Communicating Systems (TestCom 2000),
August 29–September 1, 2000, Ottawa, Canada*

Edited by

Hasan Ural
Robert L. Probert
Gregor v. Bochmann
University of Ottawa
Canada

KLUWER ACADEMIC PUBLISHERS
BOSTON / DORDRECHT / LONDON

Distributors for North, Central and South America:
Kluwer Academic Publishers
101 Philip Drive
Assinippi Park
Norwell, Massachusetts 02061 USA
Telephone (781) 871-6600
Fax (781) 871-6528
E-Mail < kluwer@wkap.com >

Distributors for all other countries:
Kluwer Academic Publishers Group
Distribution Centre
Post Office Box 322
3300 AH Dordrecht, THE NETHERLANDS
Telephone 31 78 6392 392
Fax 31 78 6546 474
E-Mail < services@wkap.nl >

 Electronic Services < http://www.wkap.nl >

Library of Congress Cataloging-in-Publication Data

IFIP TC6/WG6.1 International Conference on Testing of Communicating Systems
(13th : 2000 : Ottawa, Ont.)
  Testing of communicating systems : tools and techniques : IFIP TC6/WG6.1 13th
International Conference on Testing of Communicating Systems (TestCom 2000),
August 29-September 1, 2000, Ottawa, Canada / edited by Hasan Ural, Robert L.
Probert, Gregor v. Bochmann.
    p. cm. — (International Federation for Information Processing ; 48)
  Includes bibliographical references and index.
  ISBN 0-7923-7921-7
  1. Telecommunication systems—Testing—Congresses. I. Ural, Hasan. II.
Probert, Robert L. III. Bochmann, Gregor von, 1941– IV. Title. V. International
Federation for Information Processing (Series) ; 48.

TK5101.A1 I48292 2000
004.6'2'0287—dc21   00-056159

Printed on acid-free paper.

Printed in the United States of America.

Preface

This volume contains the contributed papers which were accepted and presented at TestCom 2000, the IFIP WG 6.1 International Conference on Testing of Communicating Systems (formerly IWTCS), held in Ottawa, Canada, on August 29th through September 1st, 2000. This conference is the thirteenth in a series of annual meetings sponsored by IFIP WG 6.1. The twelve previous meetings were held in Vancouver, Canada (1988), Berlin, Germany (1989), Mclean, U.S.A. (1990), Leidschendam, the Netherlands (1991), Montreal, Canada (1992), Pau, France (1993), Tokyo, Japan (1994), Evry, France (1995), Darmstadt, Germany (1996), Cheju Island, Korea (1997), Tomsk, Russia (1998), and Budapest, Hungary (1999).

This conference aims to draw together researchers and industrial practitioners in order to promote the fruitful exchange of views, ideas, and experiences. The conference contains nineteen refereed papers which are contained in this volume, as well as two invited keynote addresses, two panel discussions, and ongoing test tool demonstrations, all of which are preceded by a tutorial/workshop day involving a number of tutorials on such topics as interoperability and performance test methods, testing of electronic commerce systems, applications, and frameworks, and the relation of one test description formalism to the evolving TTCN standard. TestCom 2000 was organized under the auspices of IFIP WG 6.1 by S.I.T.E., the School of Information Technology and Engineering, at the University of Ottawa. It was financially supported by a number of corporations including Nortel Networks, Telelogic, Mitel Electronics, and the research organization CITO, Communications and Information Technology Ontario, a center of excellence established by the government of Ontario and supported by a consortium of communication and information technology corporations.

The refereeing process for this conference involved having each paper assessed by three referees, most of whom were technical program committee members, and others who are listed in the proceedings as additional referees. We are certainly grateful for their assistance in helping us to select a high quality program for this conference. Unfortunately, a number of strong contributions could not be accepted given the limited number of sessions available for contributed papers. Also, we thank the steering committee for IFIP WG 6.1, in particular the TestCom steering committee at the time of the preparation of the conference. We would also wish to thank the distinguished keynote speakers.

We would like to express our gratitude to everyone who has contributed to the success of TestCom 2000. In particular, we are grateful to the authors for writing and presenting their papers, the many referees for assessing and making well informed suggestions for improvements on these papers, the members of the technical program committee some of whom traveled a considerable distance to attend the paper selection meetings, the local arrangements committee members for many hours of dedicated service, our graduate students and OCRI (The Ottawa Center for Research and Innovation), for their attention to detail in support of the conference, and particularly our indefatigable secretarial assistant Mrs. Louise Desrochers for many dedicated hours of support and many contributions.

Hasan Ural, Robert L. Probert, Gregor v. Bochmann, Ottawa, Canada, August 2000

Committee Members and Reviewers

STEERING COMMITTEE

Gregor v. Bochmann (University of Ottawa, Canada)
Ed Brinksma (University of Twente, The Netherlands)
Stanislaw Budkowski (INT, France)
Guy Leduc, Chairman (University of Liege, Belgium)
Elie Najm, (ENST, France)
Richard Tenney, (University of Massachusetts, USA)
Ken Turner, (University of Stirling, UK)

CONFERENCE CO-CHAIRS

Hasan Ural, Robert L. Probert and Gregor v. Bochmann
(University of Ottawa, Canada)

TECHNICAL PROGRAM COMMITTEE

B. Baumgarten (GMD, Germany)
A. R. Cavalli (INT, France)
S. T. Chanson (Hong Kong Univ. of S. & T., Hong Kong)
S. Dibuz (Ericsson, Hungary)
R. Dssouli (Univ. of Montreal, Canada)
R. Groz (France Telecom, France)
T. Higashino (Osaka Univ., Japan)

D. Hogrefe (Univ. of Luebeck, Germany)
C. Jard (IRISA, France)
M. Kim (Information and Communications Univ., Korea)
H. Koenig (BTU Cottbus, Germany)
R. Lai (La Trobe Univ., Australia)
D. Lee (Bell   Labs, USA)
M. T. Liu (Ohio State Univ., USA)
G. Luo (Nortel Networks, Canada)
J. de Meer (GMD FOKUS, Germany)
O. Monkewich (Nortel Networks, Canada)
K. Parker (Telstra Research Labs, Australia)
A.   Petrenko (CRIM, Canada)
O. Rafiq (Univ. of Pau, France)
I.   Schieferdecker (GMD FOKUS, Germany)
K. Suzuki (KDD, Japan)
K. C. Tai (North Carolina State Univ., USA)
J. Tretmans (Univ. of Twente, The Netherlands)
A.   Ulrich (Siemens, Germany)
M. U. Uyar (CUNY, USA)
M. R. Woodward (Univ. of Liverpool, UK)
J. Wu (Tsinghua Univ., China)
N. Yevtushenko (Tomsk Univ., Russia)

ADDITIONAL REVIEWERS

Axel Belinfante, University of Twente, The Netherlands
Sergiy Boroday, CRIM, Canada
Leo Cacciari,University of Pau, France
Tibor Csondes, Ericsson, Hungary
Ali Y. Duale, CUNY, USA
Zaidi Fatiha, INT, France
Mariusz Fecko, University of Delaware, USA
Jan Feenstra, University of Twente, The Netherlands
Toru Hasegawa, KDD R&D Labs., Japan
Jens Grabowski, University of Luebeck, Germany
Akira Idoue, KDD R&D Labs., Japan
Sungwon Kang, Korea Telecom, Korea
Beat Koch, University of Luebeck, Germany
Ousmane Kone, LORIA/INRIA, France
Peter Kremer, Ericsson, Hungary
Mang Li ,GMD FOKUS, Germany

PART I

TEST SUITE COVERAGE AND VERIFICATION

1

VERIFICATION OF TEST SUITES

Claude Jard, Thierry Jéron and Pierre Morel *
IRISA, Campus de Beaulieu, F-35042 Rennes, France
{ Claude.Jard, Thierry.Jeron, Pierre.Morel } @irisa.fr

Abstract   We present a formal approach to check the correctness and to propose corrections of hand-written test suites with respect to a formal specification of the protocol implementations to test. It is shown that this requires in general a complex algorithmic comparable to model-checking. The principles of a prototype tool, called VTS, and based on the synthesis algorithms of TGV, are presented. We then prove the usefulness of the technique by checking a significant part of the ATM Forum test suite for the SSCOP protocol.

Keywords: Conformance testing, TTCN, Verification, SDL, SSCOP

1.   INTRODUCTION

The simple idea developed in this article is that, as soon as one has a formal specification, one can check the correctness of test suites written by hand. It is a useful function since many errors of various types remain in manual test cases. It is in particular the case when designing tests in context or distributed tests. One can also go a little further trying to automatically correct test cases.

Test case verification appears easier than the synthesis problem. This latter, already well studied, must still face problems of state explosion or handling complex symbolic systems. However, test case verification is not a commonplace algorithmic problem. Indeed the test cases do not only arise as sequences of interactions which a simulator can reproduce. Test cases are often real reactive programs which can be abstracted by general graphs.

* respectively researcher at CNRS, at Inria and PhD Student of the University of Rennes. The authors would like to acknowledge reviewers for helpful comments. They also acknowledge Amr Hashem, student of the Information Technology Institute in Cairo, who did his stage on the application of our VTS tool to the SSCOP test suite. This work was part of the FORMA French national project on formal methods.

Existing approaches of the problem are based on a co-simulation of the test case with the specification. This is the case with TTCN-Link [12] for SDL specifications and TTCN test cases [7]. The same principle was also used in [4] using the Tetra tool for Lotos specifications and test cases translated from TTCN. Co-simulation is useful for the early detection of errors in test cases but is not sufficient in general for exhaustively checking test cases. The reason is that this technique only allows to look at particular sequences. The first problem is due to possible loops in test cases which may be unfolded only a bounded number of times. It is also not sound for non-deterministic specifications (due to hiding of internal actions for example), as correctness involves the comparison of possible outputs of the specification with the inputs of the test case after the same trace. This problem was mentionned but not solved in [4]. Thus correctness of test cases with respect to the behaviors contained in a formal specification requires the installation of a complex algorithmic comparable with model-checking. We precisely have this experience in the development of the test generation tool TGV [8][1].

Based on some basic blocks of TGV, we have developed a toolset called VTS (for Verification of Test Suites), specialized for the verification of tests. In this case, the test case plays the role of a complete test purpose[2] which strongly guides the traversal of the specification state graph. As the algorithm works on-the-fly by building in a lazy way only the part of the specification state graph (and its observable behavior) corresponding to the test case, the performances are very satisfactory. The principal limitations that one met relate to the necessary abstraction of the test cases expressed in TTCN in the general format of graphs we have in VTS. The suggested technique is illustrated by the validation of the test suite of the SSCOP protocol proposed by the ATM Forum. We found several errors and proposed some corrections.

The continuation of the article is structured as follows: Section 2 presents the testing theory which constitutes the basis of the test-checking toolset. Section 3 is devoted to the principles of test-checking. Section 4 presents the application of VTS to the verification of part of the SSCOP test suite. The article ends in a conclusion and some prospects.

2.   FORMAL CONFORMANCE TESTING

In this section we introduce the models used to describe specifications, implementations and test cases. We then define a conformance relation that precisely states which implementations conform to a given specification. We report to [13] for a precise definition of the testing theory used.

---

[1]TGV generates test cases from specifications in SDL and LOTOS and formalized test purposes.
[2]This means that all observable actions are present in the test purpose, while TGV allows more abstraction.

2.1   Models

The model used for specifications, implementations and test cases is based on the classical model of labelled transition systems with distinguished inputs and outputs.

Definition 1 *An IOLTS is an LTS* $M = (Q^M, A^M, \rightarrow_M, q_0^M)$ *with* Q^M *a finite set of states,* A^M *a finite alphabet partitioned into three distinct sets* $A^M = A_I^M \cup A_O^M \cup I^M$ *where* A_I^M *and* A_O^M *are respectively inputs and outputs alphabets and* I^M *is an alphabet of unobservable, internal actions,* $\rightarrow_M \subset Q^M \times A^M \times Q^M$ *is the transition relation and* q_0^M *is the initial state.*

We use the classical following notations of LTS for IOLTS.
Let $q, q', q_{(i)} \in Q^M, Q \subseteq Q^M, a_{(i)} \in A_I^M \cup A_O^M, \tau_{(i)} \in I^M$, and $\sigma \in (A_I^M \cup A_O^M)^*$.

- $q \overset{\epsilon}{\Rightarrow}_M q' \equiv (q = q' \vee q \overset{\tau_1 \ldots \tau_n}{\rightarrow}_M q')$
- $q \overset{a}{\Rightarrow}_M q' \equiv \exists q_1, q_2 : q \overset{\epsilon}{\Rightarrow}_M q_1 \overset{a}{\rightarrow}_M q_2 \overset{\epsilon}{\Rightarrow}_M q'$
- $q \overset{a_1 \ldots a_n}{\Rightarrow}_M q' \equiv \exists q_0, \ldots q_n : q = q_0 \overset{a_1}{\Rightarrow}_M q_1 \ldots \overset{a_n}{\Rightarrow}_M q_n = q'$.
- $traces_M(q) = \{\sigma | q \overset{\sigma}{\Rightarrow}_M\}$ and $traces_M(M) = traces_M(q_0^M)$.
- $q \text{ after}_M \sigma = \{q' | q \overset{\sigma}{\Rightarrow}_M q'\}$ and $Q \text{ after}_M \sigma = \cup_{q \in Q} q \text{ after}_M \sigma$.
  For an IOLTS M, we sometimes use $M \text{ after } \sigma$ for $q_0^M \text{ after}_M \sigma$.
- $out_M(q) = \{a \in A_O^M | q \overset{a}{\Rightarrow}_M\}$ and $out_M(Q) = \{out_M(q) | q \in Q\}$.

Specifications.   A specification is modelled by an IOLTS $S = (Q^s, A^s, \rightarrow_s, q_0^s)$. This IOLTS describes the complete behavior of the specification, including internal actions. We consider quiescence (livelock and output quiescence) as observable (by timeouts). So we need to model possible quiescence in the specification. Formally, a state q of S is quiescent if $\neg(\exists a \in A_O^s \cup I^s, q \overset{a}{\rightarrow}_s)$ (output quiescence)[3] or $\exists \alpha \in I^{s*}, q \overset{\alpha}{\rightarrow}_s^* q$ (livelock)[4]. The *suspension automaton* S^δ of S is then obtained by considering the special label δ as an output and adding self loops labelled by δ in all quiescent states. This corresponds to [13] except that we also consider livelocks. In practice, S^δ is not build but its construction is mixed with τ-closure (see below).

Now, as testing only deals with observable events (including quiescence), we define a deterministic IOLTS S_{VIS} with same observable behavior as S^δ. $S_{\text{VIS}} = (Q^{\text{VIS}}, A^{\text{VIS}}, \rightarrow_{\text{VIS}}, q_0^{\text{VIS}})$ where $Q^{\text{VIS}} \subseteq 2^{Q^s}$, $A^{\text{VIS}} = A_I^{\text{VIS}} \cup A_O^{\text{VIS}}$ with $A_O^{\text{VIS}} = A_O^s \cup \{\delta\}$ and $A_I^{\text{VIS}} = A_I^s$, $q_0^{\text{VIS}} = q_0^s \text{ after}_s \epsilon, \forall a \in A^{\text{VIS}}, \forall P, P' \in Q^{\text{VIS}}$, $P \overset{a}{\rightarrow}_{\text{VIS}} P' \iff P' = P \text{ after}_s a$.

---

[3] A deadlock ($\neg(\exists a \in A^s, q \overset{a}{\rightarrow}_s)$) is a particular case of output quiescence.
[4] As we consider finite state IOLTS, a livelock is a loop of internal actions. A livelock in the specification is not necessarily an error as it may occur due to abstraction.

S^δ does not need to be built because transitions labelled with δ can be added directly to S_{VIS} during τ-closure. This is easy for deadlocks and output quiescence but involves the computation of strongly connected components (SCCs) of τ actions for livelocks (δ are only synthesized on SCC roots). This is done on-the-fly by a part of TGV called FERMDET [9, 8] which adapts Tarjan's algorithm [11].

Implementations.   We assume (usual test hypothesis) that an implementation can be modelled by an IOLTS $Imp = (Q^{\text{Imp}}, A^{\text{Imp}}, \rightarrow_{\text{Imp}}, q_0^{\text{Imp}})$ with $A^{\text{Imp}} = A_I^{\text{Imp}} \cup A_O^{\text{Imp}} \cup I^{\text{Imp}}$ and $A_I^{\text{s}} \subseteq A_I^{\text{Imp}}$ and $A_O^{\text{s}} \subseteq A_O^{\text{Imp}}$. As usually, we assume that implementations can never refuse an input. We note $\overline{IOLTS}$ the set of input-complete IOLTS. For the definition of conformance, we also need to consider the suspension automaton Imp^δ of Imp.

Test cases.   A test case is modelled by a deterministic IOLTS $TC = (Q^{\text{TC}}, A^{\text{TC}}, \rightarrow_{\text{TC}}, q_0^{\text{TC}})$ where $A^{\text{TC}} = A_I^{\text{TC}} \cup A_O^{\text{TC}}$ with $A_O^{\text{TC}} \subseteq A_I^{\text{s}}$ and $A_I^{\text{TC}} \subseteq A_O^{\text{Imp}}$. Two disjoint subsets of states $Pass \subseteq Q^{\text{TC}}$ and $Inconc \subseteq Q^{\text{TC}}$ and a state $Fail \in Q^{\text{TC}}(Fail \notin Pass \cup Inconc)$ are associated to TC. They correspond to arrival states of transitions carrying verdicts as in TTCN. We assume that a test case is complete for inputs in A_I^{TC} in each non controllable state (state where no output is possible). This is in general the case also for TTCN test cases with the special label $?otherwise$. We restrict ourselves to deterministic test cases without internal actions. This restriction could be avoided to deal with more general test cases including internal actions such as distributed tests. In this case, τ-reduction and determinization should be applied to test cases with FERMDET.

2.2   Conformance Testing

In order to speak about correctness of test cases, we need to define the conformance relation that the test cases are supposed to check. As in [8], we consider the **ioco** relation [13]. Note that it is in fact an extension of **ioco** as we consider livelocks. It says that conformant implementations are IOLTS that allow only outputs of the specification (including δ) after any trace of S^δ (also called suspension trace of S in [13]). It is defined as follows:

Definition 2 *Let Imp (implementation) and S (specification) be two IOLTS,*
   *Imp **ioco** $S \equiv \forall \sigma \in traces(S^\delta), out(Imp^\delta$ **after** $\sigma) \subseteq out(S^\delta$ **after** $\sigma)$.*

3.   VERIFICATION PRINCIPLE

Different properties can be checked on test cases. First some static properties can be checked such as syntactical correctness, existence of verdicts, input completeness, controllability, timer management. These properties can be checked using test cases only. We are more interested in dynamic properties which involve the observable behavior of the specification. We do not pretend to check all properties but only some of them in particular those involving the specification. We could also check some properties involving the test purpose (are Pass verdicts correctly assigned) but this neccessitates also to formalize test purpose which are often very informal.

First, we tackle the problems of laxness and unsoundness. A test is lax if it accepts non conformant implementations which it could be able to reject. Almost conversely, a test is unsound if it rejects conformant implementations. Then in a second part, we deal with the problems of controllability.

This separation corresponds to a difference in algorithmic design. The first problems are solved by a forward traversal of state graphs, while some controllability conflicts are only corrigible by a backward traversal. Moreover, the problems of laxness and unsoundness are strongly dependent on the specification. This is not the case for controllability.

3.1   Test Case against Specification

In this part, we define the types of errors of a test case which are detectable by comparing the behavior of the specification with that of the test.

The concept of comparison leads us to define the synchronous product denoted by PS_{VTS} between a test case TC and the observable behavior of the specification S_{VIS}. Let $TC = (Q^{TC}, A^{TC}, \rightarrow_{TC}, q_0^{TC})$ provided with two sets of states $Pass^{TC}$ and $Inconc^{TC}$ a Fail state and let $S_{VIS} = (Q^{Svis}, A^{Svis}, \rightarrow_{Svis}, q_0^{Svis})$ be the τ-reduced and determinized specification.

Definition 3 *The synchronous product is an IOLTS*
$PS_{VTS} = (Q^{VTS}, A^{VTS}, \rightarrow_{VTS}, q_0^{VTS})$ *where*

- $A^{VTS} = A_0^{VTS} \cup A_I^{VTS}$, *with* $A_0^{VTS} = A_I^{TC} \cup A_I^{VIS}$, *outputs of the product are the outputs of the test case and the inputs of the specification;*
  $A_I^{VTS} = A_I^{TC} \cup A_0^{VIS}$, *inputs of the product are the inputs of the test case and the outputs of the specification,*
- $Q^{VTS} \subseteq (Q^{VIS} \cup \{\perp\}) \times (Q^{TC} \cup \{\perp\}$ *and* $\rightarrow_{VTS}$ *are the smallest sets defined by application of the following rules.*

  - $q_0^{VTS} = (q_0^{VIS}, q_0^{TC})$

  - $\dfrac{(q^{VIS}, q^{TC}) \in Q^{VTS} \wedge q^{VIS} \xrightarrow{a}_{VIS} q'^{VIS} \wedge q^{TC} \xrightarrow{a}_{TC} q'^{TC}}{(q'^{VIS}, q'^{TC}) \in Q^{VTS} \wedge (q^{VIS}, q^{TC}) \xrightarrow{a}_{VTS} (q'^{VIS}, q'^{TC})}$

  - $\dfrac{(q^{VIS}, q^{TC}) \in Q^{VTS} \wedge q^{VIS} \xnrightarrow{a}_{VIS} \wedge q^{TC} \xrightarrow{a}_{TC} q'^{TC},}{(\perp, q'^{TC}) \in Q^{VTS} \wedge (q^{VIS}, q^{TC}) \xrightarrow{a}_{VTS} (\perp, q'^{TC})}$

  - $\dfrac{(q^{VIS}, q^{TC}) \in Q^{VTS} \wedge q^{VIS} \xrightarrow{a}_{VIS} q'^{VIS} \wedge q^{TC} \xnrightarrow{a}_{TC}}{(q'^{VIS}, \perp) \in Q^{VTS} \wedge (q^{VIS}, q^{TC}) \xrightarrow{a}_{VTS} (q'^{VIS}, \perp)}$

The two last rules say that the traces of the specification (resp. of the test)
which do not exist in the test (resp. in the specification) end in particular
states noted $\perp$ in the synchronous product.

Verification and correction of laxness.   A test case is lax if it could reject
a non-conformant implementation but does not. More precisely, TC is lax if
there exists an implementation Imp which does not conform to S because after
a trace σ it allows an output a that S does not, TC can perform the trace $\sigma.a$
but does not produce a *Fail* verdict. Formally:

Definition 4 *Let S be a specification and TC a test case. TC is lax w.r.t S for*
ioco *iff* $\exists Imp \in \overline{IOLTS}, \exists \sigma \in traces(S^\delta) \cap traces(Imp^\delta) \cap traces(TC)$,
$\exists a \in A_o^s$ *such that* $\sigma.a \in traces(TC)$, $a \in out(Imp^\delta$ **after** $\sigma) \wedge$
$a \notin out(S^\delta$ **after** $\sigma) \wedge TC$ **after** $\sigma.a \neq Fail$.

If we notice that Imp^δ is characterized by $\sigma.a$, it is easy to see that the
existential quantification on Imp can be eliminated. Thus the laxness property
can be reformulated while using only the traces of TC and S^δ, thus the product
PS_{VTS}.

Proposition 1 *A test case is lax iff*
$\exists (q^{\text{vis}}, q^{\text{TC}}) \in Q^{\text{VTS}}, a \in A_I^{\text{vts}}, q'^{\text{TC}} \neq Fail \in Q^{\text{TC}} : (q^{\text{vis}}, q^{\text{TC}}) \xrightarrow{a}_{\text{VTS}} (\perp, q'^{\text{TC}})$

We propose a correction which eliminates any laxness from a given test case.
Each time an input of the test case TC not leading to *Fail* does not correspond to
an output of the specification, this transition is replaced by a transition leading
to *Fail* in the corrected test case TC'. We thus obtain the inclusion of the
outputs of the specification in the inputs of the test in each state of the test
where an input is possible and accessible by a trace of the specification. This
is formalized by the following transformation rule:

$$\frac{(q^{\text{vis}}, q^{\text{TC}}) \xrightarrow{a}_{\text{vts}} (\perp, q'^{\text{TC}}) \wedge a \in A_I^{\text{vts}} \wedge q'^{\text{TC}} \neq Fail}{q^{\text{TC}} \not\xrightarrow{a}_{\text{TC}} q'^{\text{TC}} \wedge q^{\text{TC}} \xrightarrow{a}_{\text{TC}} Fail}$$

Verification and correction of unsoundness.   A test is sound if it rejects
only non conformant implementations. Conversely, it is unsound if there exists
a conformant implementation which can be rejected by the test case. Formally:

Definition 5 *Let S be a specification and TC a test case.*
TC is unsound w.r.t. S for **ioco** *iff*
$\exists Imp \in \overline{IOLTS}, Imp$ **ioco** $S \wedge \exists \sigma \in traces(S^\delta) \cap traces(Imp^\delta) \cap traces(TC)$,
$\exists a \in A_O^s, TC$ **after** $\sigma.a = Fail$

Again, the existential quantification on Imp can be suppressed and unsoundness can be expressed on PS_{VTS}.

Proposition 2 *A test case TC is unsound iff*
$$\exists (q^{vis}, q^{rc}) \in Q^{vis}, \ a \in A_I^{vis}, \ q'^{vis} \neq \bot \in Q^{vis}: (q^{vis}, q^{rc}) \xrightarrow{a}_{vis} (q'^{vis}, Fail)$$

In this case, the correction consists in replacing the incorrect transition leading to *Fail* in TC with a new transition in the corrected test case TC' with same label and leading to a new state in the INCONCLUSIVE set. This correction is reflected by the following rule:

$$\frac{(q^{vis}, q^{rc}) \xrightarrow{a}_{vis} (q'^{vis}, \bot) \wedge q'^{vis} \neq \bot \wedge a \in A_I^{vis} \wedge q'^{rc} \notin Q^{rc}}{q'^{rc} \in Q^{rc} \wedge q'^{rc} \in Inconc^{rc} \wedge q^{rc} \xrightarrow{a}_{rc} q'^{rc}}$$

It is easy to see that corrections of laxness and unsoundness do not interfere (correction of laxness cannot produce unsoundness and vice versa). Correction of laxness replaces lax inputs by sound Fail verdicts while correction of unsoundness remove unsound Fail verdicts by unlax inputs leading to Inconc.

Verification and correction of controllability conflicts. Test cases should be controllable in the sense that they should never have the choice between an output and another output or input.

Definition 6 *A test case has a controllability conflict if:*
$$\exists q^{rc} \in Q^{rc}, \ \exists a \in A_o^{rc}, \ \exists x \in A_I^{rc} \cup A_o^{rc} \backslash \{a\} : \ q^{rc} \xrightarrow{a}_{rc} \ and \ q^{rc} \xrightarrow{x}_{rc}$$

Detection of controllability conflict can be done by any forward search in the test case. But correction is more difficult in the general case where test cases have loops. While pruning a test case, accessibility to *Pass* states must be preserved. The solution is then to perform a search (breadth-first or depth first) on the reverse transition relation, to prune other transitions in case of conflict and to forget parts of the test case that become unreachable. This algorithm is detailed in [8] as it is also part of TGV.

Implementation in VTS. The algorithm takes as input a test case in Aldébaran format (general purpose graph format) and a specification described in LOTOS, SDL, BCG (compressed format for graphs) or Aldébaran. In the case of testing in context, the specification should include this context as conformance is defined for the specification in its context. It checks the correctness of the test case (laxness and unsoundness) with respect to the specification behavior. For SDL and LOTOS this behavior is given by simulators (respectively ObjectGéode [14] and the OPEN/CÆSAR interface [5]) which are driven by VTS. VTS implements these verifications by a breadth-first traversal of the synchronous product between the τ-reduced and determinized specification and the test case. This τ-reduction and determinization are performed on-the-fly by the FERMDET tool only on the common traces of the test case and the specification (in fact traces of PS_{vis}). The traces of the test case not leading to *Fail* must be included in those of the specification. Thus for any trace of the test case, we check two aspects. On the one hand, if inputs of the test case are possible, the algorithm checks the equality between these possible inputs not leading to *Fail* and the possible outputs of the specification after the same trace. In addition, if outputs are possible in the test case, then they should be possible inputs of the specification (the equality is not required in this case). The controllability conflicts are detected when a state of the test case has several possible outputs or an output and inputs. Correction is performed by a backward traversal.

4.   APPLICATION TO THE ATM SSCOP TEST SUITE

We have decided to experiment VTS with a real case study.   We chose the B-ISDN ATM Adaptation Layer-Service Specific Connection Oriented Protocol (SSCOP) from the ITU Q.2110 document [10].   It presents several advantages:

- this protocol has been studied for test generation with various tools such as Samstag [6], TVeda [3] and TestGen [2],
- we have a formal SDL specification, which has already been validated and used for automatic test generation [1];
- there is a complete conformance test suite, standardized by the ATM Forum, publicly available at http://www.atmforum.com

4.1   The SSCOP Protocol

The Service Specific Connection Oriented Protocol resides in the Service Specific Convergence Sublayer (SSCS) of the ATM Adaptation Layer (AAL) (see figure 1). SSCOP is used to transfer variable length Service Data Units (SDUs) between SSCOP users. SSCOP provides its service to a Service Specific Coordination Function (SSCF). The SSCF maps the service of SSCOP to the needs of the AAL user. SSCOP uses the service of the CPCS (Common Part Convergence Sublayer) and SAR protocols which provide an un-assured information transfer and a mechanism for detecting corruption of SSCOP Protocol Data Units (PDUs). One currently defined use of SSCOP is within the signaling AAL (SAAL).

SSCOP performs the following functions:

- Sequence integrity: this function preserves the order of SSCOP SDUs that were submitted for transfer by SSCOP.
- Error correction by selective retransmission: through a sequencing mechanism, the receiving SSCOP entity can detect missing SDUs. This function corrects sequence errors through retransmission.
- Flow control.
- Error reporting to layer management.

- Keep alive: this function verifies that the two peer SSCOP entities participating in a connection are remaining in a link connection established state even in the case of a prolonged absence of data transfer.

- Local data retrieval: this function allows the local SSCOP user to retrieve in-sequence SDUs which have not yet been released by the SSCOP entity.

- Connection control: this function performs the establishment, release, and re-synchronization of an SSCOP connection. It also allows the transmission of variable length user-to-user information without a guarantee of delivery.

- Transfer of user data: SSCOP supports both assured and un-assured data transfer.

- Protocol error detection and recovery.

- Status reporting.

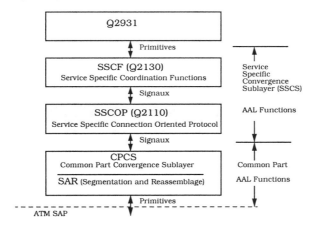

Figure 1.   Situation of SSCOP in the ATM stack

The SDL executable specification of SSCOP was written by Serge Gauthier from CNET (the research center for France-Telecom). The specification was written in 1995 using SDL based on the SDL description given in the final draft document XI/Q2210 of ITU-T study group. It consists in approximately 5000 lines of textual SDL code. The specification was dedicated to test generation, thus it makes some simplifications which do not comply with all the aspects of the standard. Later on, the formal specification has been slightly corrected during the verification works of the FORMA project [1].

4.2   The ATM Test Suite

We considered the conformance abstract test suite for SSCOP, which was published by the ATM Forum Technical Committee on September 1996 under the title "Conformance Abstract Test Suite for the SSCOP for UNI 3.1.". This test suite aligns with the principles defined in the OSI conformance testing methodology and framework ISO 9646 Parts 1-2 [7]. The test scripts are written in TTCN.

The testing architecture considered is the remote testing architecture (see figure 2) with only one lower tester (and PCO). The asynchronous communication on this PCO is reflected in the formal specification by the intercalation of a retransmission process between the SSCOP entity and the environment. This palliates the absence of queue between the SDL model and its environment in the ObjectGeode simulator.

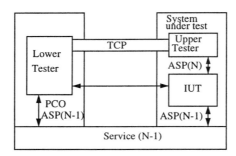

Figure 2.   Remote testing architecture

While the test suite consists of 317 test cases (approximately 500 pages of TTCN), the test cases viable for verification based on our formal specification of the protocol were found to be 110 test cases. The abstraction resulted in the following:

- the test cases testing invalid PDUs (INV) were not considered as the specification does not describe the behavior of the SSCOP on reception of invalid PDUs. INV test cases arise to 186 cases.
- UD and MD PDUs were not considered in the SDL specification. Thus these PDUs have been abstracted in TTCN test cases. Moreover the 12 test cases related to valid UD and MD PDUs were not considered.
- the test cases testing the behavior of clocks and timeouts were abstracted as timers are treated as internal unobservable actions. Timer test cases arise to 9 cases.

Most of the test cases have the same structure: a preamble tries to drive the implementation under test (IUT) to a particular control state of the SSCOP entity as defined in the specification; then a test body is applied to check that the IUT behaves correctly; then follows a state identification behavior which checks the arrival control state; finally a postamble drives the IUT to the initial state.

4.3   Our Approach to Use VTS

We proceeded through the successive following steps.

1. The development of a compiler for the generation of test cases from the test suite. The input is a test suite written in TTCN machine processable (TTCN.MP) format. The output is a set of automata (one for each test case) in the Aldébaran format that the VTS program can process. The compiler also maintains a set of state variables similar to the set maintained by SSCOP. The values of the state variables are used to generate the values of the constraints that VTS can deal with. The compiler has to simulate the operations that are to be executed on the contents of the state variables of the SSCOP protocol mentioned by the test case. This simulation is essential for the sake of generation of PDUs and ASPs to be fired by the tester. It also simulates the changes in the state variables associated with firing transitions that are normally not explicitly mentioned in the test case. This is essential for the sake of generation of PDUs and ASPs expected from the specification.

2. The development of a program for the automatic generation of the supplementary files needed as inputs to the VTS program, namely the "feed" and the "hide" files. The feed file is used during the construction of the transition system of the specification as an input to the ObjectGeode API. The feed file contains the various signals that to be feed into the SSCOP specification from the external environment. The environment refers to both upper and lower layers. The hide file is used to hide the unobservable transitions that are generated but are not observable at the point of control and observation (PCO) due to the test architecture. Unobservable transitions in the test suite may be 1/ inputs or outputs of the queue of the retransmission process representing the asynchronous channel, 2/ signals between the user and the SSCOP specification that cannot be observed form the lower tester, and thus are not mentioned in the test script, 3/ actions on timers of the specification.

3. The use by VTS of the test cases generated by the compiler from the TTCN test suite and of the supplementary files in order to check and correct test cases.

4. The analysis of the errors found by the VTS tool and the variance between the hand written test cases of the ATM Forum test suite and the corrected test cases generated by VTS.

5. The proposal for corrections of the ATM test suite by providing alternatives to incorrect test cases with respect to the SSCOP specification.

4.4   Results

Most test cases failed because of forgotten signals due the classical problem of message crossing inherent to asynchronous communication. In fact these signals are observable due to the expiration of timers before the reception by SSCOP of a signal sent by the tester. One can imagine that those are not real errors but are due to implicit assumptions on the transmission delay between the tester and the IUT. Nevertheless, as the remote testing architecture is considered this assumption should not be done or should at least be documented.

Out of the 110 test cases, 16 test cases failed for other reasons. They all fail in the state identification step: 2 tests in control state 4, 1 in control state 5, 4 in control state 7 and 9 in control state 10.

Out of the 16 defective tests, the following have been corrected:

1. 2 tests of state 4, 1 test of state 5, 2 tests of state 10 by adding a step to correct the value of the state counters. For example, in state 4, VR_SQ is not changed on receiving RS or ER PDUs. The S4_VERIFY test step fails if the last transition performs an RS or ER PDUs as VR_SQ is not incremented. Thus the BGN PDU of the test step is not detected as a retransmission. As VTS does not manipulate symbolic variables but only values, it does not provide a useful correction. Once the problem identified, it can nevertheless be easily corrected by hand by decrementing the value of VT_SQ in the test case before conducting the state verification step.

2. 4 tests of state 7 by adding a step to initialize the values of the state counters. This is detected by VTS as incorrect values of the $USTAT$ PDU in the S10_VERIFY step (state 10 is the arrival state of these test cases). The SSCOP specification resets its state variables after sending a $BGAK$ PDU at state 3. The initialization is not reflected in the test cases. This can be manually corrected by inserting the initialization procedure before state verification.

3. 4 tests of state 10 by introducing an alternative sequence for verification of state 10. S10_VERIFY assumes the reception of an $USTAT$ PDU before entering in the verification step. VTS has shown that there exist situations in which the SSCOP entity transfer directly to state 10 without the generation of an $USTAT$ PDU. The only solution is to change completely the S10_VERIFY procedure.

4. 2 tests of state 10 by changing the values of the USTAT PDU in the procedure for verification of state 10.

4.5   Example of Verification/Correction

We present here an example of the experimentation of the VTS tool on the test case numbered S10_V_P17 in the ATM Forum Test Suite. This test case and its following checking sequence (S10_VERIFY) are presented below in TTCN.GR format. They are exact copies of the ATM Forum test case and test step except that UD and MD PDUs have been abstracted. According to the informal test purpose, this test case verifies that the IUT, in control state 10 (DataTransferReady), saves an SD PDU that sequence number is between the sequence number of the next in sequence and the next highest expected SD PDUs.

Nr	Label	Behaviour Description	Constraint Ref	Verdict
			S10_V_P17	
1		+S10_PREAMBLE		
2		LT_PCO!SD	SD_S_N_S(VT_S+2)	
3		START T_Wait		
4	LB1	LT_PCO?USTAT(VT_MS:= BIT_TO_INT(USTAT.N_MR))	USTAT_R_LIST(VT_S, VT_S+2, VT_S)	
5		LT_PCO!SD	SD_S_N_S (VT_S+1)	
6		LT_PCO!SD	SD_S_N_S (VT_S)	
7		LT_PCO!POLL(VT_PS:= INC_MOD_24 (VT_PS,1))	POLL_S_N_S(VT_S+3)	
8		START T_Wait		
9	LB2	LT_PCO?STAT[CHECK_N_PS (VT_PA, BIT_2_INT(STAT.N_PS), VT_PS)](VT_MS:=BIT_TO_INT (STAT.N_MR))	STAT_R_N_R(VT_S+3)	(P)
10		+S10_VERIFY		
11		+POSTAMBLE		
12		LT_PCO?POLL	POLL_R_GEN	
13		GOTO LB2		
14		+TS_Wait		
15		LT_PCO?POLL	POLL_R_GEN	
16		GOTO LB1		
17		+TS_Wait		

Nr	Label	Behaviour Description	Constraint Ref	Verdict
			S10_VERIFY	
1		LT_PCO!SD	SD_S_N_S(VT_MS+3)	
2		START T_Wait		
3	LB1	LT_PCO?USTAT(VT_MS:= BIT_TO_INT(USTAT.N_MR))	USTAT_R_LIST(VT_S, VT_MS, VT_S)	(P)
4		LT_PCO?POLL	POLL_R_GEN	
4		GOTO LB1		
5		+TS_Wait		

The graph drawn on the left hand side of Figure 3 represents the test case (body) with its different test steps (preamble, checking sequence and postamble) in a graph format as processed by our TTCN compiler. For the sake of clarity we did not represent the transitions ?*otherwise* producing *Inconclusive* verdicts in states 2 and 4 and *Fail* verdicts in states 6, 10, 12 and 15. In the SDL specification and the test suite we have set the parameter VT_MS to 20 and Max_CC to 1 in order to shorten the preambles. The variable VT_S is initially set to 0.

As for most test cases, the VTS tool detects unsoundness (forgotten inputs of END PDU in states 6, 10 and 12) because of a bad treatment of asynchronism. According to the specification, once in control state DataTransferReady an END PDU can be sent if TimerNoResponse expires. The implicit hypothesis made in this test is that the PDUs SD (lines 2 and 5 or transitions $5 \rightarrow 6$ and $11 \rightarrow 12$) and POLL (line 7, transition $8 \rightarrow 9$) sent by the tester are received before the timer expires.

Formally speaking, the test case is also lax as ?*otherwise* in states 2 and 4 should produce a *Fail* verdict and not an *Inconclusive* verdict. In fact it is common practice to deliver only *Inconclusive* verdicts in preambles but the VTS tool does not make this distinction.

The test case is really incorrect in the S10_VERIFY step as the value of the first and last parameters of the USTAT PDU (line 3, transition $12 \rightarrow 13$) are not correct. The VTS tool found this error (which is both laxness and

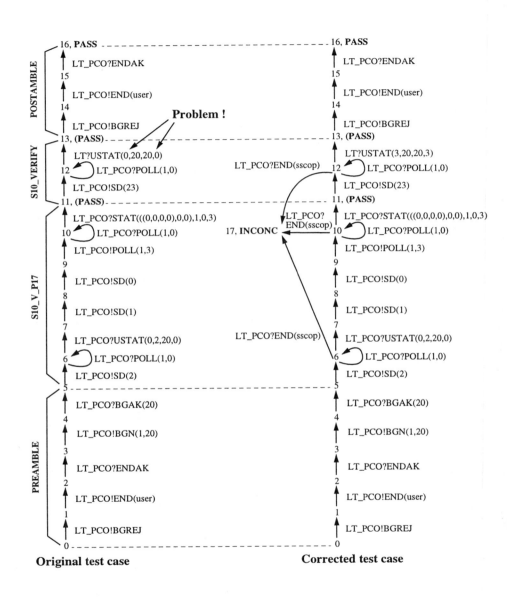

Figure 3.   Initial and corrected test cases

unsoundness) and proposed to correct the test case. The right hand side of Figure 3 corresponds to the corrected test case produced by the VTS tool. New transitions lead to the *Inconclusive* state 17 for missing END PDUs and the transition labelled by USTAT is corrected by changing the first and last parameters values from 0 to 3.

This error was further analyzed by a close look to the TTCN test case. The real reason is that the variable VT_S of the test case is not updated when sending an SD PDU (lines 2, 5, 6 in S10_V_P17) but this variable is used in USTAT PDU (line 3 of S10_VERIFY). Thus it still has value 0 which is sent in USTAT while in the specification the corresponding variable is incremented and the resulting parameter has value 3.

As the VTS tool manipulates instantiated test cases, it was only possible to find the correct values. But finding that incrementations were missing was done by hand. Finding this kind of correction automatically is not possible in general. But some improvements can be made by a symbolic treatment of variables, the help of provers, abstractions and static analysis.

5.   CONCLUSION AND PERSPECTIVES

VTS was originally conceived for testing the TGV tool. Test cases produced by TGV were checked by VTS in order to track bugs in the main algorithm of TGV. But the main interest of VTS is to check manual test cases. This was demonstrated here by its application on an industrial size specification and a significant part of a test suite in which some errors have been found.

The interest of such a tool is evident for complex systems. This is in particular the case for distributed systems because of the difficulty to foresee all behaviors of these systems due to asynchronism, hiding of internal actions and non-determinism. This leads directly to checking distributed test cases. VTS can be easily extended to check the correctness of this kind of test cases by using FERMDET as a front end. Nevertheless correction is more problematic as observed errors can be caused by internal actions.

But VTS suffers from the limitation inherent to enumerative tools. Parameters of specifications and test cases have to be fixed thus correctness cannot be guaranteed for all values of these parameters. Moreover, some of the errors detected by VTS on the SSCOP test suite are errors on values of message parameters. Analyzing the errors and correcting them would be easier with a symbolic treatment of data.

References

[1] M. Bozga, J.-C. Fernandez, L. Ghirvu, C. Jard, T. Jéron, A. Kerbrat, P. Morel, and L. Mounier. Verification and test generation for the SSCOP protocol. *Journal of Science of Computer Programming*, 36(1), 2000.

[2] A. Cavalli, B. Lee, and T. Macavei. Test generation for SSCOP-ATM networks protocol. In *SDL Forum, (INT, Evry)*. Elsevier, September 1997.

[3] I. Disenmayer, S. Gauthier, and L. Boullier. L' outil tveda dans une chaˆne de production de tests d' un protocole de télécommunication. In *CFIP'97 : Ingénierie des Protocoles*, pages 271–286. Hermes, September 1997.

[4] M. Dubuc, G. Bochmann, O. Bellal, and F. Saba. Translation from ttcn to lotos and the validation of test cases. Technical Report PUB 732, Université de Montréal, may 1990.

[5] Hubert Garavel. Open/cæsar: An open software architecture for verification, simulation, and testing. In *Proceedings of TACAS'98 (Lisbon, Portugal)*, volume 1384 of *LNCS*, pages 68–84, Berlin, March 1998. Springer Verlag.

[6] J. Grabowski, R. Scheurer, and D. Hogrefe. Applying SAMSTAG to the B-ISDN Protocol SSCOP. Technical Report A-97-01, part I, University of Lübeck, January 1997.

[7] OSI-Open Systems Interconnection, Information Technology - Open Systems Interconnection Conformance Testing Methodology and Framework. ISO/IEC International Standard 9646-1/2/3, 1992.

[8] Thierry Jéron and Pierre Morel. Test generation derived from model-checking. In Nicolas Halbwachs and Doron Peled, editors, *CAV'99, Trento, Italy*, pages 108–122. Springer, LNCS 1633, July 1999.

[9] P. Morel. *Une algorithmique efficace pour la génération automatique de tests de conformité*. PhD thesis, Rennes I Univ., France, February 2000.

[10] ITU Q.2110: B-ISDN ATM Adaptation Layer - Service Specifi c Connection Oriented Protocol (SSCOP), 1994.

[11] R. Tarjan. Depth-first search and linear graph algorithms. *SIAM Journal Computing*, 1(2):146–160, June 1972.

[12] Telelogic. The SDT TTCN Link Reference Manual, 1997.

[13] J. Tretmans. Test generation with inputs, outputs and repetitive quiescence. *Software—Concepts and Tools*, 17(3):103–120, 1996.

[14] Verilog. ObjectGeode SDL Simulator Reference Manual, 1996.

2

STRUCTURAL COVERAGE FOR LOTOS
A Probe Insertion Technique

Daniel Amyot and Luigi Logrippo
SITE, University of Ottawa, Canada. {damyot,luigi}@site.uottawa.ca

Abstract   Coverage analysis of programs and specifications is a common approach to measure the quality and the adequacy of a test suite. This paper presents a probe insertion technique for measuring the structural coverage of LOTOS specifications against validation test suites. Coverage results can help detecting incomplete test suites, a discrepancy between a specification and its tests, and unreachable parts of a given specification. Such results are provided for several examples, taken from real-life and hypothetical communicating systems for which a LOTOS specification was constructed and validated.

Keywords: Coverage, LOTOS, probes, specification, validation testing.

1.   INTRODUCTION

"When to stop testing?" is and will remain an important problem for communications software validation and verification. Lai [16] mentions that knowing how much of the source code has been covered by a test suite can help estimate the risk of releasing the product to users, and discover new tests necessary to achieve a better coverage. Inexperienced testers tend to execute down the same path of a program, which is not an efficient testing technique.

Coverage measures are considered to be a key element in deciding when to stop testing. Coverage analysis of code is a common approach to measure the quality and the adequacy of a test suite [23]. Coverage criteria can guide the selection of test cases (*a priori*, i.e. before the execution of the tests) and be used as metrics for assessing the quality of an existing test suite (*a posteriori*, i.e. after the execution of the tests). Many methods are available for the measure of different coverage criteria such as statements, branches, data-flow, paths, and so on [9].

This paper covers a different angle of the same question, relating to specification coverage. Specifications, just like programs, can be covered for several reasons and according to several criteria. For example, we could want to cover a specification in the generation of conformance test cases for an implementation, or in order to check whether a specification satisfies abstract requirements. These processes can also gain in quality from the use of coverage measurements. Many formal specification languages already benefit from tool-supported coverage metrics, including SDL with Telelogic's *Tau* [21] and VDM with IFAD's *VDMTools* [13]. Unfortunately, no such tools are currently available for ISO's formal specification language LOTOS [14].

Still, several coverage criteria have been defined for LOTOS. For instance, van der Schoot and Ural developed a technique for static data-flow analysis [20], whereas Cheung and Ren proposed an operational coverage criterion [10]. These two techniques are used mostly for guiding, *a priori*, the generation of test cases from specifications. The first one is based on data usage and the second one is based on the semantics of LOTOS operators.

The availability of a formal specification enables the (automated) generation of test cases based on different coverage criteria [18]. This feature is particularly beneficial in a context of conformance testing, i.e. when the behaviour of an implementation under test is required to conform to its specification [15]. One of the main assumptions behind this use of coverage criteria is that the specification is correct and valid with respect to the system requirements. This validity cannot usually be established formally because initial requirements are often informal. Specifications can however be exercised through different means, including *validation testing* (different from conformance testing), until a sufficient degree of confidence in their validity is reached.

This paper presents a new *a posteriori* coverage technique for LOTOS, based on the specification's *structure*. This technique is intended to be used during the initial validation of the specification against a test suite that captures the main functionalities of the requirements. In this particular context, validation test cases are often generated *manually* rather than automatically. For instance, the *SPEC-VALUE* methodology (Specification-Validation Approach with LOTOS and UCMs) promotes the use of scenarios, which capture informal functional requirements from a behavioural perspective, for the construction of an initial formal specification in LOTOS [1][2][3][4]. These scenarios also guide the generation of a validation test suite to ensure the consistency between the LOTOS specification, which integrates all scenarios, and the requirement scenarios. The specification is considered satisfactory once all the test cases pass successfully *and* once the structural coverage goals are achieved.

To measure this structural coverage, the LOTOS specification is instrumented with probes, which are visited by validation test cases during their execution. Section 2 illustrates a probe insertion technique for sequential programs.

This idea is adapted to the LOTOS context in Section 3. Section 4 provides coverage results coming from experiments with specifications of communicating systems of various natures and complexity. Conclusions follow in Section 5.

2.   PROBES FOR SEQUENTIAL PROGRAMS

Probe insertion is a well-known white-box technique for monitoring software in order to identify portions of code that have not been yet exercised, or to collect information for performance analysis. A program is instrumented with probes (generally counters) without modification of its functionality. When executed, test cases trigger these probes, and counters are incremented accordingly. Probes that have not been "visited" indicate that part of the code is not reachable with the tests in consideration. Obvious reasons could be that the test suite is incomplete, or that this part of the code is unreachable.

Section 2.1 raises several issues related to probe instrumentation, and Section 2.2 gives an illustrative overview of an existing probe insertion technique for well-delimited sequential programs.

2.1   Issues With Probe Instrumentation

The following four points are notable software engineering issues related to approaches based on probe instrumentation of implementation code or of executable specifications. They will be discussed further in the next sections.

1. *Preservation of the original behaviour.* New instructions shall not interfere with the intended functionalities of the original program or specification, otherwise tests that ran successfully on the original behaviour may no longer do so.

2. *Type of coverage.* Because probes are generally implemented as counters, it is easier to measure the coverage in terms of control flow rather than in terms of data flow or in terms of faults. Other techniques, summarized by Charles in [9], are more suitable for the two last types.

3. *Optimization.* In order to minimize the performance and behavioural impact of the instrumentation, the number of probes shall be kept to a minimum, and the probes need to be inserted at the most appropriate locations in the specification or in the program.

4. *Assessment.* What is assessable from the data collected during the coverage measurement represents another issue that needs to be addressed. Questions such as "Are there redundant test cases?" and "Why hasn't this probe been visited by the test suite?" are especially relevant in the context of SPEC-VALUE.

2.2   Probe Insertion Technique

For well-delimited sequential programs, Probert suggests a technique for inserting the minimal number of *statement probes* necessary to cover all branches [19]. Table 1 illustrates this concept with a short Pascal program (a) and an array of counters named Probe[]. The counters indicate the number of times each probe has been reached. Intuitively, (b) shows three statement probes inserted on the three branches of the program. In (c), the same result can be achieved with two probes only. Using control flow information, the number of times that statement3 is executed is computed from Probe[1]- Probe[2]. After the execution of the test suite, if Probe[2] is equal to Probe[1], then the conclusion is that the 'else' branch that includes statement3 has not been covered.

Table 1. Example of probe insertion in Pascal

a) Original Pascal code	b) Three probes inserted	c) Optimal number of probes

statement1;
if (*condition*)
then
  begin
    statement2
  end
else
  begin
    statement3
  end
{end if};
	statement1;
	inc(Probe[1]);
	if (*condition*)
	then
	  begin
	    inc(Probe[2]);
	    statement2
	  end
	else
	  begin
	    inc(Probe[3]);
	    statement3
	  end
	{end if};
		statement1;
		inc(Probe[1]);
		if (*condition*)
		then
		  begin
		    inc(Probe[2]);
		    statement2
		  end
		else
		  begin
		    {No probe here!}
		    statement3
		  end
		{end if};

It has been proven in [19] that the optimal number of statement probes necessary to cover all branches in a well-delimited sequential program is $E-V+2$, where E and V are respectively the number of edges and of vertices of the underlying extended delimited Böhm-Jacopini flowgraph of the program.

The four issues raised in Section 2.1 are covered as follow:

1. *Preservation of the original behaviour*: if the probe counters are variables that do not already exist in the program, then the original functionalities are preserved.

2. *Type of coverage*: the coverage is related to the program control flow.

3. *Optimization*: there exists a way to minimize the number of statement probes so it can be smaller than the number of statements.

4. *Assessment*: this technique covers all branches in a well-delimited sequential program.

3.   PROBES FOR LOTOS SPECIFICATIONS

Test cases extracted (manually) from requirements are often used to establish the validity of a specification. A posteriori measurements help to assess the coverage of the specification structure by the validation test suite. This section presents a structural coverage technique for LOTOS specifications. Similarly to probe insertion for sequential programs, LOTOS constructs can be used to instrument a specification at precise locations while preserving its general structure and its externally observational behaviour. Because the measurement of the structural coverage is performed during the execution of test cases, LOTOS testing theory is briefly discussed in Section 3.1. Then, Section 3.2 introduces a simple insertion strategy, which is improved in Section 3.3. The interpretation of coverage results is discussed in Section 3.4.

3.1   LOTOS Testing

The LOTOS testing theory assumes that the specification, modelled as a labelled transition system (LTS), communicates in a symmetric and synchronous way with external observers, the *test processes* [5]. There is no notion of initiative of actions, and no direction can be associated to a communication.

To verify the successful execution of a test case, the corresponding test process and the specification under test (*SpecUT*) are composed in parallel. They synchronize on all gates but one, the *Success* event, which is added at the end of each test case. If the composed behaviour expression deadlocks prematurely, i.e. if *Success* is not always reached at the end of each branch of the LTS resulting from this composition, then the *SpecUT* failed this test.

In the real world, test cases are often executed more than once when there is non-determinism in either the test or the implementation. Things are simpler at the LOTOS level. LOLA, a tool used to test LOTOS specifications rather than to generate tests, avoids this problem altogether. It determines the response of the *SpecUT* to a test by a complete state exploration of their composition [17]. For each test case, one of the three following verdicts is output by LOLA:

- *Must pass*: all the possible executions (called *test runs*) were successful.
- *May pass*: some test runs were successful, some unsuccessful.
- *Reject*: all test runs failed as they deadlocked prematurely.

3.2   Simple Probe Insertion Strategy

Among the LOTOS constructs, the most interesting candidate for representing a probe is an internal event with a unique identifier. Such event can be composed of a hidden gate name that is not part of any original process in the specification (e.g. *Probe*), followed by a unique value of some new enumerated abstract data type (ADT) (e.g. *P_0*, *P_1*, *P_2*, *P_3,* etc.).

In LOTOS, a *basic behaviour expression* (BBE) is either the inaction **stop**, the successful termination **exit**, or a process instantiation (P[...]). A *behaviour expression* (BE) can be one of the following:

- A BBE.
- A BE prefixed by a unary operator, such as the action prefix (;), a hide, a let, or a guard ([predicate]->).
- Two BEs composed through a binary operator, such as a choice ([]), an enable (>>), a disable ([>), or one of the parallel composition operators (|[...]|, ||, or |||).
- A BE within parentheses.

In this paper, a *sequence* is defined as a BBE preceded by one or more events, separated by the action prefix operator ($e_1; e_2; ... e_n; BBE$). A BBE that is not preceded by any event is called a *single BBE*.

Probes enable the measure of the coverage of every event in a behaviour expression, and therefore in a whole specification. The simplest and most straightforward strategy consists in adding a probe after each event at the syntactic level. For each event *e* and each behaviour expression *B*, the expression *e; B* is transformed into *e; Probe!P_id; B* where *Probe* is a hidden gate and *P_id* a unique identifier. A probe that is visited guarantees, by the action prefix inference rule, that the prefixed event has been executed. In this case, if all the probes are visited by at least one test case in the validation test suite, then the test suite has achieved total *event coverage*, i.e. the coverage of all the events in the specification (modulo the value parameters attached to these events).

Table 2 illustrates this strategy on a very simple specification *S1* (a). Since there are three occurrences of events in the behaviour, three probes, implemented as hidden gates with unique value identifiers, are added to *S1* to form *S2* (b). The validation test suite is somehow derived from scenarios or requirements according to some test plan or functional coverage criteria not discussed here. In this example, it is composed of two test cases (Test1 and Test2), which remain unchanged during the transformation. The third specification (c) will be discussed in Section 3.3.

Probe insertion is a syntactic transformation that also has an impact on the underlying semantic model, i.e. the specification's labelled transition system (LTS). Table 3 shows the LTSs resulting from the expansion of *S1* and *S2*. A LOTOS **exit** is represented by δ at the LTS level. When a test case ending by **exit** is checked (e.g. Test1), LOLA automatically transforms such δ into **i** followed by *Success*. Although the LTSs (a) and (b) are not equal as trees, they are observationally equivalent [14]. Therefore, as shown by Brinksma in [5], the tests that are accepted and refused by *S1* will be the same as those of *S2*.

Table 2. Simple probe insertion in LOTOS

a) Original Lotos specification (S1)	b) Simple probe insertion strategy (S2)	c) Improved probe insertion strategy (S3)

`specification S1[…]: exit`
`   … (* ADTs *)`
`behaviour`
`   a; exit`
`   []`
`   b; c; stop`
`where`
`   process Test1[a]:exit:=`
`      a; exit`
`   endproc (* Test1 *)`
`   process Test2[…]:noexit:=`
`      b; c; Success; stop`
`   endproc (* Test2 *)`
`endspec (* S1 *)`
	`specification S2[…]: exit`
	`   … (* ADTs *)`
	`behaviour`
	`   hide Probe in`
	`   (`
	`      a; Probe!P_1; exit`
	`      []`
	`      b; Probe!P_2;`
	`         c; Probe!P_3; stop`
	`   )`
	`where`
	`   … (* Test1 and Test2 *)`
	`endspec (* S2 *)`
		`specification S3[…]: exit`
		`   … (* ADTs *)`
		`behaviour`
		`   hide Probe in`
		`   (`
		`      a; Probe!P_1; exit`
		`      []`
		`      b; c; Probe!P_2; stop`
		`   )`
		`where`
		`   … (* Test1 and Test2 *)`
		`endspec (* S3 *)`

Table 3(c) presents two traces, resulting from the composition of each test process found in Table 2(a) with *S2*, that cover the events and probes of *S2*. `Test1` covers P_1 in the left branch of (c) whereas `Test2` covers P_2 and P_3 in the right branch. Neither of these tests covers all probes, but together they cover all three probes, and therefore the event coverage is achieved, as expected from such a validation test suite. The fact that the entire LTS is covered here is purely coincidental, as it is usually not the case for complex specifications.

Table 3. Underlying LTSs

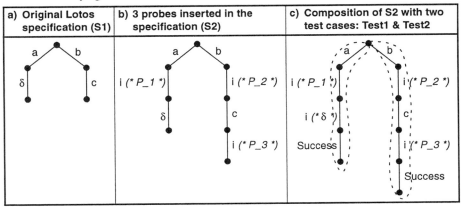

a) Original Lotos specification (S1)	b) 3 probes inserted in the specification (S2)	c) Composition of S2 with two test cases: Test1 & Test2

Going back to the four issues enumerated in Section 2.1, the following observations are made:

1. *Preservation of the original behaviour*: probes are unique internal events inserted after each event (internal or observable) of a sequence. They do not affect the observable behaviour of the specification. This insertion can be summarized by Proposition 1, which coincides with one of the LOTOS congruence rules found in the standard [14] (congruence rules preserve observational and testing equivalences in any context):

$$e; B \approx_c \textbf{hide } \textit{Probe } \textbf{in } (e; \textit{Probe!P\_id; B}) = e; \textbf{i}; B \qquad \text{(Prop. 1)}$$

2. *Type of coverage*: this coverage is concerned with the structure of the specification, not with its data flow or with fault models. The resulting *event coverage* makes abstraction of the semantic values in the events (e.g. the expression `dial?n:nat` abstracts from any natural number n).

3. *Optimization*: none; the total number of probes equals the number of occurrences of events in the specification. Reducing the number of probes is the focus of the next section.

4. *Assessment*: this strategy covers all events syntactically present in a specification. Single basic behaviour expressions are not covered.

3.3   Improved Probe Insertion Strategy

The simple insertion strategy leads to interesting results, yet two problems remain. First, the number of probes required can be very high. The composition of a test case and a specification where multiple probes were inserted (and transformed into internal events) can easily result in a state explosion problem. Second, this approach does not cover single BBEs as such, because they are not prefixed by any event. Single BBEs may represent a sensible portion of the structure of a specification that needs to be covered as well. This section presents four optimizations that help solving these two problems.

In a sequence of events, the number of probes can be reduced to one probe, which is inserted just before the ending BBE. If such a probe is visited, then LOTOS' action prefix inference rule leads to the conclusion that all the events preceding the probe in the sequence were performed. The longer the sequence, the better this first optimization becomes. Table 2(c) shows specification *S3* where two probes are necessary instead of three as in *S2*. This *sequence coverage* implies the coverage of events with fewer probes or the same number of probes in the worst case.

The second optimization concerns the use of parenthesis in *e; (B)*, where *B* is not a single BBE. In this case, no probe is required before *(B)*. The behaviour expression *B* will most certainly contain probes itself, and a visit to any of these probes ensures that event *e* is covered (by the prefix inference rule).

The third optimization is concerned with the structural coverage of single BBEs (without any action prefix), where some subtle issues first need to be explored. Suppose that * is one of the LOTOS binary operators enumerated at the beginning of Section 3.2 ([ ], >>, [>, | [...] |, | |, | | |). If a single BBE is prefixed with a probe in the generic patterns BBE * BE and BE * BBE, then care is required in order not to introduce any new non-determinism. Additional non-determinism could result in some test cases to fail. A probe can safely be inserted before the BBE unless one of the following situations occurs:

- BBE is **stop**: this is inaction. No probe is required on that side of the binary operator (*) simply because this inaction cannot be covered. This syntactical pattern is useless and should be avoided in the specification.

- BBE is a process instantiation P[...]: a probe before the BBE can be safely used except when * is the choice operator ([]), or when * is the disable operator with the BBE on its right side (BE [> P[...]). In these cases, a probe would introduce undesirable non-determinism that might cause some test cases to fail partially: LOLA would return a *may pass* verdict instead of a *must pass*. A solution would be to guard the process instantiation. One way of doing so in many cases would be to partially expand process P with the expansion theorem so an action prefix would appear. Another solution is presented below, in the fourth optimization.

- BBE is **exit**: the constraints are the same as for the process instantiation. The solution is also to prefix this **exit** with some event.

The fourth optimization is concerned with BBEs that are process instantiations. When a process P is not defined as a single BBE, then the necessary number of probes can be further reduced when P is instantiated in only one place in the specification (except for recursion in P itself). In this case, a probe before P is not necessary because probes inserted within P will ensure that the single instantiation of P is covered. This is especially useful when facing a process instantiation as a single BBE. For example, suppose a process Q that instantiates P in one place only, where P is not a BBE and P is not instantiated in any process other than Q and P itself:

$$Q[...] := e1; e2; e3; \textbf{stop} [] P[...]$$

A probe inserted before P would make the choice non-deterministic, which could lead to undesirable verdicts during the testing. However, if P is not a single BBE and if it is not instantiated anywhere else, then no probe is required before P in this expression. Any probe covered in P would ensure that the BBE on the right of the choice operator in Q has been covered. This situation often occurs in processes that act as containers for aggregating and handling other process instances, a common pattern in communicating systems.

Regarding the four issues enumerated in Section 2.1, the improved strategy achieves a larger coverage of the specification than the simple strategy of Section 3.2, and it requires fewer probes to do so.

1. *Preservation of the original behaviour*: probes are unique internal events inserted before each BBE. When such BBE is prefixed by an event, then the probe does not affect the observable behaviour (Proposition 1). When the BBE is not prefixed, a case not addressed by the simple strategy, then special care must be taken in order not to introduce new non-determinism.

2. *Type of coverage*: the *sequence and single BBE coverage* is concerned with the specification structure (it implies the event coverage, Section 2.2).

3. *Optimization*: the total number of probes is less than or equal to the total number of sequences and BBEs in the specification.
4. *Assessment*: this strategy covers all events syntactically present in a specification, as well as single BBEs other than **stop** (which should not be found in the specification anyway).

3.4   Interpretation of Coverage Results

Several problem sources can be associated to probes that are not visited by a test suite. They usually fall into one of the following categories:

- *Incorrect specification*. In particular, the specification could include unreachable code caused by processes that cannot synchronize properly or by guards that can never be satisfied.
- *Incorrect test case*. This is usually detected before probes are inserted, during the verification of the functional coverage of the specification.
- *Incomplete test suite*. Caused by an untested part (an event or a single BBE) of the specification (e.g. a feature of the specification that is not part of the original requirements).
- *Discrepancy*. Due to the manual nature of the construction of the specification from the scenarios or requirements, there could be some discrepancy between a test and the specification caused by ADTs, guards, the choice ([ ]) operator, or other such constructs.

Code inspection and simulation of the specification can help diagnosing the source of the problem highlighted by a missing probe. Several LOTOS tools, including LOLA, also offer reachability and expansion mechanisms that can be helpful in determining whether a specific probe can be reached at all.

4.   EXPERIMENTATION

The structural coverage technique was applied to various specifications and validation test suites developed using the SPEC-VALUE methodology (Section 4.2) and a self-coverage experiment (Section 4.3). But first, current tool support is briefly presented in Section 4.1.

4.1   Tool Support for Structural Coverage Measurement

A filter tool called LOT2PROBE was built for the automated translation of special comments manually inserted in the original specification (e.g. *(*_PROBE_*)*) into internal probe events with unique identifiers (e.g. Probe!P_0;). A new data type that enumerates all the unique identifiers for the probes is also added to the specification. Care was taken not to add any new line to the original specification, in order to preserve two-way traceability between the transformed specification and the original one. Though full automation of probe insertion is possible, the solution developed so far is still semiautomatic because of some special cases (e.g. single BBEs) that are not trivial to handle. However, the manual insertion of these probe comments has the benefit of being more flexible, and it can be done at specification time or after the initial validation.

Batch testing under LOLA can then be used for the execution of the validation test suite against the transformed specification. Several scripts compute probe counts for each test and then output textual and HTML summaries of the probes visited, with a highlight on probes that are not covered by any test.

4.2   Scenario-Based Validation Experiments

The SPEC-VALUE methodology focuses on the construction of a LOTOS specification from a collection of scenarios described with the Use Case Map (UCM) notation [6][7]. UCMs have proven to be useful for the high-level description of communicating systems as they visually describe scenarios in terms of causal relationships between responsibilities, the latter being bound to system components. The specification integrates all UCM scenarios into a component-based description, which is then validated against black-box test cases derived from those same UCMs, which capture functional requirements.

For the sake of simplicity in this paper, the functional coverage goals are considered to be achieved once the test suite is successfully executed. At this point, the specification can be considered from another perspective, namely from the structural coverage viewpoint. The techniques and tools discussed so far are hence applied to obtain results indicating whether or not the validation test suite has covered the entire specification structure. If so, then the confidence in the validity and completeness of the specification and its test suite is increased. If not, then appropriate measures (inclusion of test cases, correction of the specification or of the tests, etc.) can be applied at a very early stage of the design process.

SPEC-VALUE was applied to the following communicating systems:

- *Group Communication Server (GCS)* [1]: an academic example that describes a server with different functionalities for group-based multicast.
- *GPRS Group Call* [2]: a real-life mobile communication feature of the General Packet Radio Service (GPRS), based on GSM. This work was done during the first standardization stage of GPRS [12].
- *Feature Interaction Example (FI)* [3]: an academic case study oriented towards the avoidance and the detection of undesirable interactions

between a collection of telephony features described in the 1998 Feature Interaction Contest.

- *Agent-Based Simplified Basic Call (SBC)*: real-life system developed during a feasibility study for the application of a functional testing process to industrial telephony applications based on agents and the Internet protocol (IP). This work was extended to include several features in [4].
- *Tiny Telephone System (TTS)*: an academic example (basic call plus two telephony features) used as a tutorial for the SPEC-VALUE methodology.

For each of these systems, which are significantly diverse in nature and in complexity, Table 4 summarizes the main characteristics of the LOTOS specification and the test suite constructed from the UCM scenarios. Then, characteristics and results related to the structural coverage are provided. The last column (MAP protocol) will be discussed in Section 4.3.

Table 4. Summary of structural coverage experiments

	System	GCS	GPRS	FI	SBC	TTS	MAP
LOTOS	a) # Process definitions	19	30	13	9	11	14
	b) # Lines of behaviour	750	1400	800	750	375	850
	c) # Abstract data types	29	53	39	8	19	22
	d) # Lines of ADTs	800	1125	750	200	400	375
	e) # Lines of test processes	1600	800	1325	300	375	7725
	f) Total number of lines	3150	3325	2875	1250	1050	8950
Tests	g) # Functional test cases	109	36	37	11	33	603
	h) # Unexpected verdicts	0	0	1	3	0	6
	i) Test time (in seconds)	5	120	11	64	5	16
Coverage	j) # LOTOS events	57	126	94	204	25	156
	k) # LOTOS BBEs	35	86	27	20	22	46
	l) # Sequences	40	74	49	60	18	67
	m) # Probes inserted	54	99	55	64	26	83
	n) Optimization reduction	28%	38%	28%	20%	35%	27%
	o) Overall reduction	41%	53%	55%	71%	47%	59%
	p) # Missed probes	3	11	4	17	0	17
	q) Time, with TestExpand	235	-	165	-	140	-
	r) Time, with OneExpand	31	81	37	18	9	1000
	s) Why probes missed	❸	❶, ❸	❸	❷, ❸	-	❶

The legend for row *s* respects the interpretation of coverage results discussed in Section 3.4:

❶ Unreachable code or error in the LOTOS specification.

❷ Incomplete test suite.

❸ Discrepancies between the LOTOS specification and either the test suite itself or the scenarios from which these tests were derived.

Row *n* shows the reduction obtained using the optimizations on sequences $(n = (k+l-m)/(k+l))$, whereas row *o* represents the reduction relative to the number of events and BBEs in the specification $(o = (j+k-m)/(j+k))$. These measures show the effectiveness of the optimizations discussed in Section 3.3.

LOLA was used in two different ways to generate the coverage results. Row q indicates the number of seconds (on a 300MHz Celeron) taken by LOLA's *TestExpand* command, which does a full exploration of the state space resulting from the synchronization of each test process with the specification. This command was not used on the GPRS, SBC, and MAP specifications and test suites because of their complexity. Row i uses the same command on the specification without probes, but with an option which applies equivalence rules on the fly to reduce the state space (hence resulting in faster executions).

Row r shows the time taken by LOLA's *OneExpand* command to measure the structural coverage. This command performs a partial coverage of the composition through random executions (five executions per test in the above examples). This pragmatic solution handles large state spaces and provides quick and effective coverage results for complex specifications. However, unlike *TestExpand*, the use of *OneExpand* does not guarantee that reachable probes will be covered by random test runs.

As for the missed probes (row p), the reasons and resulting actions are:

- GCS: Additional feature not in the original set of UCM scenarios (two probes). This resulted in the addition of the feature to the UCMs, which in turn led to two new test cases. One other UCM scenario was specified as two alternatives in LOTOS (one probe), so one test case was added.

- GPRS: Unreachable code (one probe). This part of the specification and its probe were removed. This GPRS specification includes robustness conditions that are unreachable when correct client and server processes are composed (ten probes). No system-level test was added, but we checked that these probes were manually reachable by simulating the client and server processes taken individually.

- FI: Partial specification of the whole collection scenarios (4 features out of 13 in the UCMs). The specification structure contains placeholders for scenarios still to be specified (four probes). There was no action taken because this situation was expected. However 1 unexpected interaction was detected between two features (this was fixed in a recent version).

- SBC: Failure conditions were handled by the specification and the scenarios but were not tested as the test suite focused on correct user-level interactions (17 probes). New test cases are required, but none was added to the test suite. This specification and its test suite were intended to be part of a proof of concept rather than to be complete. This also explains the 3 unexpected verdicts. A more complete version is discussed in [4].

Many bugs and inconsistencies were detected and fixed during the validation testing. However, the structural coverage helped fine-tuning these specifications and test suites by detecting several non-trivial problems at a very low cost (from a few seconds to a few minutes).

4.3   Self-Coverage Experiment

The structural coverage technique was also applied to a specification developed in a rather different context. The system under study was GSM's *Mobile Application Part* (MAP) protocol [11], which maintains consistency among databases frequently modified by mobile telephone users.

In this experiment, the MAP specification was derived manually from the standard and validated (also manually) through simulations. Then, an abstract conformance test suite was generated automatically from this specification via TESTGEN, a tool that covers all transitions of the underlying state machine by using Cavalli's unique event sequences [8]. This test suite was converted back to LOTOS test processes in order to check whether or not the structural coverage of the MAP specification was achieved (hence the name *self-coverage*). A 100% coverage was of course expected.

Three major iterations were needed to achieve a satisfactory coverage. In the first one, less than half of the probes were visited by the test suite (417 tests) because of a problem with the data types and guards which caused about half of the specification not to be reachable. The second iteration fixed this bug and resulted in a new test suite (603 tests), whose results are shown in Table 4. Several verdicts were wrong because of remaining non-determinism in the specification. This also caused problems when generating the test suite, which couldn't cover 17 probes. A third version (not shown here) fixed this problem and led to the generation of 684 test cases, with a full structural coverage.

The use of this structural coverage technique helped preventing the generation of a faulty conformance test suite from an incorrect specification. Such a self-coverage approach to testing is an interesting by-product of the technique. It shows it can be useful even in the absence of validation test cases.

5.   DISCUSSION AND CONCLUSIONS

This paper presents a probe insertion technique for measuring the structural coverage of initial LOTOS specifications against validation test suites. This coverage can improve the quality and consistency of both the specification and the tests, hence resulting in a higher degree of confidence in the system's description. The paper describes how probes can be inserted in a specification without affecting its observable behaviour. Different optimizations for reducing the number of probes while preserving the coverage of single BBEs and event sequences are also discussed.

Through experimentation with several communicating system specifications of various sizes and complexity, it is shown that coverage results can help detecting incomplete test suites, discrepancies between specifications and their tests, and unreachable parts of specifications. Results can also be output quickly and at low cost. This technique is valuable not only for scenario-based approaches such as SPEC-VALUE or Yi's [24], but also for checking, through self-coverage, the quality of conformance test suites generated (by other means) from LOTOS specifications.

Complex specifications can be handled by this technique because the structural coverage can also be measured *compositionally*. Probes can be covered independently, even one at a time, through multiple executions of the same test suite. The LOT2PROBE filter allows different variations of the probe comment in the specification, which represent different groups of probes. Partial results only need to be put together at the end. Having fewer probes converted at once reduces the number of internal actions and helps avoiding state explosion.

This technique provides an assessment of how well a given test suite has covered a LOTOS specification rather than providing a guideline on how the specification is to be covered for testing purpose. Validation tests are not intended to replace conformance tests; their respective goals are quite different. Validation tests can be reused throughout the evolution of a specification, but they may not be adequate to ensure conformance of an implementation to a specification. Coverage of the implementation code is still necessary at a later development stage, and it can be measured through conventional techniques.

The structural coverage technique opens the door to other research issues. Coverage measurements could be used as a potential guide for test case management. A test which covers probes already all visited by another test may be a sign of redundancy. Test cases could also be sorted in the test suite according to, for instance, the number of probes they cover. Our structural coverage criterion could also be complemented by a coverage of the abstract data type definitions. Finally, two equivalent specifications written using different styles might lead to different coverage results for the same test suite.

Acknowledgment. We are indebted to the U. of Ottawa LOTOS Group for their collaboration, and in particular to H. Ben Fredj, L. Charfi, P. Forhan, N. Gorse, D. Petriu and J. Sincennes for their work on several UCMs and specifications studied here. We kindly acknowledge NSERC, CITO, FCAR, the U. of Ottawa, Mitel Corp., Nortel , and Motorola for their support.

REFERENCES

[1] Amyot, D., Logrippo, L., and Buhr, R.J.A. (1997) "Spécification et conception de systèmes communicants : une approche rigoureuse basée sur des scénarios d'usage". In: G. Leduc (Ed.), *CFIP 97, Ingénierie des protocoles*, Liège. Hermès, 159-174.

[2] Amyot, D., Buhr, R.J.A., Gray, T., and Logrippo, L. (1999) "Use Case Maps for the Capture and Validation of Distributed Systems Requirements". In: *RE'99, Fourth IEEE International Symposium on Requirements Engineering*, Limerick, Ireland, June 1999, 44-53.

[3] Amyot, D. and Logrippo, L. (2000) "Use Case Maps and LOTOS for the Prototyping and

Validation of a Mobile Group Call System". In: *Computer Communications*, 23(12), 1135-1157. http://www.UseCaseMaps.org/UseCaseMaps/pub/cc99.pdf

[4]   Amyot, D., Charfi, L., Gorse, N., Gray, T., Logrippo, L., Sincennes, J., Stepien, B., and Ware, T. (2000) "Feature description and feature interaction analysis with Use Case Maps and LOTOS". In: *Sixth International Workshop on Feature Interactions in Telecommunications and Software Systems (FIW'00)*, Glasgow, Scotland, UK, May 2000, 274-289.

[5]   Brinksma, E. (1988) "A theory for the derivation of tests". In: S. Aggarwal and K. Sabnani (Eds), *Protocol Specification, Testing and Verification VIII*, North-Holland, 63-74.

[6]   Buhr, R.J.A. and Casselman, R.S. (1996) *Use Case Maps for Object-Oriented Systems*, Prentice-Hall, USA. http://www.UseCaseMaps.org/pub/UCM_book95.pdf

[7]   Buhr, R.J.A. (1998) "Use Case Maps as Architectural Entities for Complex Systems". In: *IEEE Transactions on Software Engineering, Special Issue on Scenario Management*. Vol. 24, No. 12, December 1998, 1131-1155. http://www.UseCaseMaps.org/pub/tse98final.pdf

[8]   Cavalli, A., Kim, S., and Maigron, P. (1993) "Improving Conformance Testing for LOTOS". In: R.L. Tenney, P.D. Amer and M.Ü. Uyar (Eds), *FORTE VI, 6th International Conference on Formal Description Techniques*, North-Holland, 367-381.

[9]   Charles, Olivier. (1997) *Application des hypothèses de test à une définition de la couverture*. Ph.D. thesis, Université Henri Poincaré — Nancy 1, Nancy, France, October 1997.

[10] Cheung, T. Y. and Ren, S. (1992) *Operational Coverage and Selective Test Sequence Generation for LOTOS Specification*. TR-92-07, SITE, U. of Ottawa, Canada, January 1992.

[11] ETSI (1992) Digital Cellular Telecommunication System (Phase 2); *Mobility Application Part (GSM 09.02), Version 4.0.0* (June 1992).

[12] ETSI (1996) Digital Cellular Telecommunications System (Phase 2+); *General Packet Radio Service (GPRS); Service Description Stage 1 (GEM 02.60), Version 2.0.0* (Nov.).

[13] IFAD (1999) *VDMTools*. http://www.ifad.dk/Products/VDMTools

[14] ISO (1989), Information Processing Systems, Open Systems Interconnection, *LOTOS — A Formal Description Technique Based on the Temporal Ordering of Observational Behaviour*, IS 8807, Geneva.

[15] ISO/EIC (1996) *Proposed ITU-T Z.500 and Committee Draft on "Formal Methods in Conformance Testing" (FMCT)*. ISO/EIC JTC1/SC21/WG7, ITU-T SG 10/Q.8, CD-13245-1, Geneva.

[16] Lai, R. (1996) "How could research on testing of communicating systems become more industrially relevant?". In: *9th International Workshop on Testing of Communicating Systems (IWTCS'96)*, Darmstadt, Germany, 3-13.

[17] Pavón, S. and Llamas, M. (1991) "The testing Functionalities of LOLA". In: J. Quemada, J.A. Mañas, and E. Vázquez (Eds), *FORTE III*, IFIP/North-Holland, 559-562.

[18] Poston, R.M. (1996) *Automating specification-based software testing*. IEEE Computer Society Press, Los Alamitos, CA, USA.

[19] Probert, R.L. (1982) "Optimal Insertion of Software Probes in Well-Delimited Programs", *IEEE Transactions on Software Engineering*, Vol 8, No 1, January 1982, 34-42.

[20] van der Schoot, H. and Ural, H. (1997) "Data Flow Analysis of System Specifications in LOTOS". In: *Int. Journ. of Software Eng. and Knowledge Eng.*, Vol.7, No. 1, 43-68.

[21] Telelogic (1999) *Tau Tool*. http://www.telelogic.com/solution/tools/tau.asp

[22] Ural, H. (1992) "Formal methods for test sequence generation". In: *Computer Communications*, 15, 311-325.

[23] Zhu, H., Hall, P.A.V., and May, J.H.R. (1997) "Software unit test coverage and adequacy". In: *ACM Computing Surveys*, 29(4), December, 366-427

[24] Yi, Z. (2000) *Specification and Validation of a Mobile Telephony Feature Using Use Case Maps and LOTOS*. Masters thesis, SITE, University of Ottawa, Canada.

3

FAULT DETECTION POWER OF A WIDELY USED TEST SUITE FOR A SYSTEM OF COMMUNICATING FSMS

Ana Cavalli*, Svetlana Prokopenko°, Nina Yevtushenko°
*Institut National des Télécommunications
9 rue Charles Fourier
F-91011 Evry Cedex
France
Email: Ana.Cavalli@int-evry.fr
° Tomsk State University
36 Lenin av.
634050 Tomsk
Russia
Email: prokopenko.rff@elefot.tsu.ru, qel@asd.iao.tsc.ru

Abstract   This paper studies the fault detection power of a widely used test suite, i.e., a test suite that traverses each possible transition of each component FSM (Finite State Machine) in the reference system. It is shown that such a test suite is complete, with respect to single output faults of a component under test, if the output of the component is accessible during a testing mode. Experiments have been performed showing that a test suite detecting single outputs faults of each component is good: 92 % of transfer and output faults of the composite FSM are detected.

Keywords: conformance testing, embedded testing, fault detection, communicatings FSMs

1.   INTRODUCTION

One important aspect of automatic test generation is the derivation of tests for a complex system of a distinguished structure. The problem is known as gray-box testing [7]. A number of approaches have been developed for testing a system under an assumption that at most one component can be faulty. In this case, the system is divided into two parts: the context that is assumed to be fault-free and the component that should be tested. This problem is well known as embedded testing [7] or testing in context [17]. Some of the approaches proposed to solve the problem for a system of communicating Finite State Machines (FSMs) are heuristic and do not guarantee a complete fault coverage [9]. Other approaches [11,12,13,17,18,19,25,26] deliver complete test suites with respect to various fault domains, i.e. sets of possible implementations of the component under test.

In this paper, we consider a system of communicating FSMs and study fault detection power of a widely used test suite that should traverse each involved transition of each component FSM in the reference system. We formally show that the test suite detects all single output faults of a component when the output of the component under test is accessible during a testing mode. The performed experiments clearly show that a complete test suite w.r.t. single output faults of each component FSM usually also detects "almost all" transfer and output faults of the corresponding composite FSM, i.e. is good enough. There exist a number of methods for such test suite derivation [1,9]. Similar to the paper [19], our results are based on a so-called embedded equivalent that represents all faulty traces of the component FSM that can be detected at points of observation. However, in our case, component FSMs can be partially specified. To illustrate our approach we present the testing of a system composed by telephone services.

The rest of the paper is structured as follows. Section 2 presents some basic notions. Section 3 is devoted to the problem statement and explains how an embedded equivalent for partial FSMs can be derived. This section also the results of the performed experiments. We then present, in Section 4, a set of external input sequences that traverse each transition of a component under test in the reference system. We show that this set of external inputs is a complete test suite w.r.t. single output faults of the component FSM if the component's output is accessible during a testing mode. A procedure is proposed to perform the fault coverage evaluation of a given test suite.

Section 5 comprises an example of testing a system composed by telephone services. Finally, section 6 gives the conclusions of this work.

2. PRELIMINARIES

2.1 I/O Finite State Machines

An I/O finite state machine (often simply called an FSM or a machine throughout this paper) is initialized, possibly partially specified and a finite set of states with $s_0 \in S$ as the initial state, X is a finite nonempty set of inputs, Y is a finite nonempty set of outputs, and h is a behavior function mapping a specification domain $D_A \subseteq S \times X$ into $P(S \times Y)$ where $P(S \times Y)$ is the set of all nonempty subsets of the set $S \times Y$. The machine A is *deterministic* if $|h(s,x)|=1$ for all $(s,x) \in D_A$. In a deterministic FSM, the function h usually is replaced with two functions: next state function δ: $D_A \rightarrow S$ and output function λ: $D_A \rightarrow Y$. Deterministic I/O finite stste machine is denoted by 7-tuple $(S,X,Y,\delta,\lambda,D_A,s_0)$. The machine A is *complete* if $D_A=S \times X$; otherwise, the machine is *partial*. The machine $B=(T,X,Y,g,D_B,s_0)$ is called a *submachine* of A if $T \subseteq S$, $D_B=D_A$ and for all $(s,x) \in D_A$, $g(s,x) \subseteq h(s,x)$.

In the usual way, the function h is extended to a so-called *output function* h^y over an appropriate subset I_A of input sequences with results in the set of nonempty subsets of output sequences. The set I_A is the set of all input sequences where a behavior of the A is defined. The set $h^y(\alpha)$ comprises each output sequence that can be produced by the FSM when the sequence α is submitted. As usual, given input sequences $x_1...x_k$ and output sequence $y_1...y_k$, the sequence $x_1/y_1,...,x_k/y_k$ is called *a trace* of the A if $h^y(x_1...x_k)=y_1...y_k$.

Given two FSMs $A=(S,X,Y,h,D_A,s_0)$ and $B=(T,X,Y,g,D_B,t_0)$, FSM A is called a *trace reduction* of the FSM B, or simply a *reduction* of B, denoted $A \leq B$, if the set of traces of A is a subset of that of the FSM B, i.e. $I_A \subseteq I_B$ and for any input sequence $\alpha \in I_A$, $h^y(s_0, \alpha) \subseteq g^y(t_0, \alpha)$. If for some input sequence $\alpha \in I_A \cap I_B$ it holds that $h^y(s_0, \alpha) \not\subseteq g^y(t_0, \alpha)$ then the sequence α is said to *distinguish* the FSM A from B.

For the class of deterministic FSMs, the reduction relation coincides with a so-called quasi-equivalence relation [5]. Deterministic FSM $A=(S,X,Y,\delta,\lambda,D_A,s_0)$ is a reduction of deterministic FSM $B=(T,X,Y,\Delta,\Lambda,D_B,t_0)$ if and only if B is quasi-equivalent to A, i.e. $I_A \subseteq I_B$ and for any input sequence $\alpha \in I_A$, the output sequences of A and B to α coincide.

2.2   Fault Domain

One important aspect of high quality test generation is to specify an appropriate fault model. We further assume that a reference system and a system under test are modeled by deterministic FSMs specified over the same input and output alphabets. We refer to an FSM modeling the reference system as to a *reference* FSM while referring a *fault domain* to the set $\mathfrak{I}$ of FSMs modeling all possible systems under test. A FSM $B \in \mathfrak{I}$ is called an *implementation*. The B is called a *faulty* or a *nonconforming* implementation if the reference FSM is not a reduction of B; otherwise it is a *conforming* implementation.

A finite set of finite input sequences of the reference machine RM is a *test suite* (w.r.t. $\mathfrak{I}$) if it detects at least a single nonconforming implementation. A test suite which detects each nonconforming implementation is called a *complete* test suite w.r.t. the fault domain $\mathfrak{I}$. Formally, given a reference FSM RM and a fault domain $\mathfrak{I}$, a test suite TS is complete w.r.t. $\mathfrak{I}$ if for each FSM $B \in \mathfrak{I}$ such that RM is not a reduction of B, the TS has a sequence distinguishing RM from B.

Thus, if a reference FSM is deterministic, a system under test is modeled by an FSM of the set $\mathfrak{I}$ and is not detected by a complete test suite w.r.t. $\mathfrak{I}$ then one concludes that the reference system is a reduction of the system under test, i.e. a behavior of the system under test coincides with that of the reference system under any input sequence where the behavior of the reference system is specified.

As it is claimed in [14,15,24], a complete test suite derived from a partial reference FSM RM is different from that derived from the reference FSM where each undefined transition of the RM is augmented as a looping transition with the *Null* output. In the latter case, during a testing mode one also checks the looping transitions that are never traversed during a working mode. By this reason, in this paper, we derive a test suite without augmenting partial FSMs. We now consider two widely used fault domains.

Given a deterministic reference FSM $RM=(S,X,Y,\delta,\lambda,D_A,s_0)$ with n states, a fault domain $\mathfrak{I}_{RM}$ has each FSM $B=(S,X,Y,\Delta,\Lambda,D_B,s_0)$ such that $D_B \supseteq D_{RM}$ and for each $(s,x) \in D_{RM}$, $\delta(s,x)=\Delta(s,x)$, i.e. only output faults may occur in an implementation. The fault domain $\mathfrak{I}_n$ comprises each complete FSM over alphabets X and Y with at most n states. In the former case, a transition tour of the reference FSM is a complete test suite [20]. In the latter case, also transfer faults are involved. Procedures for a complete test suite derivation w.r.t. the fault domain $\mathfrak{I}_n$ are well developed [2,4,10,14,16,21,22,23,24].

However, most of them only deal with a reduced reference FSM and deliver tests with length proportional to the product of number of transitions and number of states of the reference machine.

2.3   Composition of FSMs

We consider a system composed by two FSMs, as shown in Figure 1. We refer to symbols of alphabet X_1 and X_2 as to external inputs, to symbols of alphabets Y_1 and Y_2 as to external outputs while referring to symbols of alphabets Z and U as to internal actions. In fact, one of the alphabets X_1 or X_2 (Y_1 or Y_2) can be empty.

In order to define the composition of FSMs we do the following assumptions. The system has always a single message in transit, i.e. the environment submits the next input only when the system has produced an output to the previous input. Moreover, a component machine accepting an input may produce either an external or an internal output. If the component machines fall into infinite internal dialog when an appropriate external input sequence is submitted we say the system falls into live-lock under the input sequence and its behavior is not specified under the input sequence. If the behavior of one of component FSMs is not specified under a submitted input we also assume the system behavior is not specified.

Under the above assumptions we can derive the composite FSM *Context◊Spec* of the context FSM *Context* and the component FSM *Spec* using various algorithms [11,17]. Below we briefly sketch the algorithm from the paper [11].

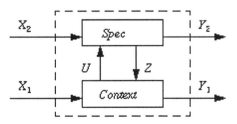

Figure 1. System model

Let the component FSMs *Context* and *Spec* have the sets Q and T of states, respectively. Then the composite FSM is a FSM over alphabets $X_1 \cup X_2$ and $Y_1 \cup Y_2$ with the state space that is a subset of $Q \times T$. States of the set $Q \times T$ are divided into stable and transient states. By definition, the initial state q_0t_0 is stable. Otherwise, the state is stable if it is reached after the system has produced an external output. We start from the initial state q_0t_0..

Given a stable state qt and input x, there is a transition labeled with x/u (x/z) from the initial state qt to a transient state $q't$ (qt') if x/u (x/z) takes the FSM *Context* (*Spec*) from the state q (t) to the state q' (t'). There is a transition labeled with x/y from the state qt to a stable state $q't$ (qt') if x/y takes the FSM *Context* (*Spec*) from the state q (t) to the state q' (t').

Given a transient state $q't'$ with an incoming transition labeled by a/z (a/u), a pair of internal actions z/u (u/z), and a transient state $q''t'$ ($q't''$), there is a transition labeled with z/u (u/z) from the state $q't'$ to the state $q''t'$ (from the state $q't'$ to $q't''$) if there is an outgoing transition from the state q' to the state q'' under z/u in the *Context* (under u/z in the *Spec*). There is a transition labeled with z/y (u/y) from the state $q't'$ to a stable state $q''t'$ (from the state $q't'$ to $q't''$) if there is a transition from the state q' to the state q'' under z/y in the *Context* (from the state t' to the state t'' under u/y in the *Spec*). The stable states cannot be merged with transient states. Two transient states with the same names are merged if they have an incoming transition labeled with a pair with the same output part.

If no stable state is reachable then a transition from the state qt under x is undefined. Otherwise, given the final stable state $q't'$ with an incoming transition x/y, z/y or u/y, the composite FSM has a transition from the state qt to $q't'$ under x/y. As mentioned above, no stable state is reachable if one of the following conditions holds: a) a transition of the context (or the component) at a current state is undefined under a submitted input; b) the system falls into live-lock, i.e. has a cycle labeled with internal actions.

The procedure is repeated unless all reachable stable states with possible external inputs are considered.

3.   TEST SUITE DERIVATION FOR AN EMBEDDED COMPONENT W.R.T. OUTPUT FAULTS

We consider a reference system composed by two deterministic FSMs, as shown in Figure 1.

3.1   Discussion about Faults

Our first step is to specify a type of faults that should be detected with a derived test suite, i.e. to define an appropriate fault domain. Usually two types of faults are considered, namely output and transfer faults. It is well known that a test suite detecting all transfer and output faults is a high-quality one. However, it can hardly be used in practical situations, for its length is proportional to the product of number of transitions and number of states of the reference FSM.

On the other hand, there are publications [see, for example, 3] where the authors claim that for a proper kind of FSMs length of a test suite detecting "almost all" faults is proportional to the number of states of the reference FSM, i.e. tests can be used in practical situations. The result is also confirmed with a widely distributed opinion that long tests are needed to detect a very low percent of proper faulty implementations which can be treated separately (if necessary). By this reason, we performed some experiments before selecting a fault domain.

Given an FSM, experiments have been performed in order to evaluate the fault coverage of a transition tour of the FSM w.r.t. transfer faults. Given a complete reference FSM $RM=(S,X,Y,\delta,\lambda,D_A,s_0)$ and a test suite, fault coverage of the test suite is calculated as ratio $(m/n)\cdot100\%$. Here n is number of FSMs over alphabets X and Y with the state set S which are not equivalent to RM, while m is number of such FSMs which are detected with the given test suite, i.e. have an unexpected output sequence to an appropriate test case. Reference FSMs derived in pseudo-random way with up to 10 states have been studied. The average fault coverage of a transition tour is equal to 96%. In order to perform experiments with more complex reference FSMs we evaluated fault coverage of a transition tour w.r.t. 100 implementation FSMs also derived in pseudo-random way. The same result has been obtained.

We then considered a composition of two complete communicating FSMs. Given a complete test suite w.r.t. single output faults of the context *Context* and of the component *Spec*, we have evaluated its fault coverage w.r.t. output faults of the reference composite FSM *Context◊Spec*. Our experiments show it is about 95%. In other words, given a test suite complete w.r.t. single output faults of the context *Context* and of the component *Spec*, the fault coverage w.r.t. transfer and output faults of the composite FSM is expected to be about 92%, i.e. such test is of a practical use.

In the following sections, we show a set of external input sequences which traverse each transition of a component under test in the reference composition is a complete test suite w.r.t. single output faults of the component when the component's output is accessible during a testing mode. Length of such test suite is proportional to number of transitions of the component FSM involved with a given context and there exist a number of methods [1,9] for such test suite derivation without constructing a composite FSM that usually has huge number of states.

3.2   Fault Assumptions and Test Architecture

1. We further assume that the context is correctly implemented.

2. Only single output faults are possible in the component under test. In other words, the next state function of a faulty component implementation *Imp* is an extension of that of the reference component *Spec*. Thus, the set $\mathfrak{R}_{Spec}$ of all possible component implementations is the set of possible extensions of all deterministic sub-machines of the FSM *Spec'* [8] that has the same next state function as *Spec* with each output for each transition.

3. An implementation *Imp* of the component under test is conforming if *Context◊Spec* is a reduction of *Context◊Imp* while the *Context* (the *Imp*) only produces internal output sequences where a behavior of the *Spec* (*Context*) is specified.

4. We have an access to the internal output of the implementation component *Imp*, i.e. can observe the internal output the component has been produced (Figure 2). However, we cannot control its internal inputs.

Formally speaking, fault domain of our interest is the set $Context◊\,\mathfrak{R}_{Spec}$ of all composite FSMs of the reference context FSM and a component FSM, possibly with a single output fault. However, the reference FSM now is slightly different from the *Context◊Spec*, because of the point of observation at the internal output of the component FSM during a testing mode. This new situation is illustrated by Figure 2.

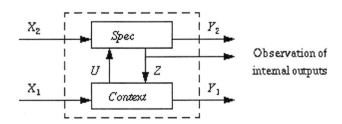

Figure 2. Test architecture

It is well known that if the reference FSM *Spec* is a reduction of an implementation component FSM *Imp* then the reference FSM *Context◊Spec* also is a reduction of the *Context◊Imp*. However, generally, a converse is not true. There can exist FSM *Imp* such that the machine *Context◊Spec* is a reduction of *Context◊Imp* while the reference component *Spec* is not a reduction of *Imp*. In the case, when machines *Context*, *Spec* and *Context◊Spec* are complete such machines *Spec* and *Imp* are called *equivalent in the context* while *Imp* also is called a *conforming implementation in the context* [18]. Thus, if one requires to have only reference internal outputs for the machine *Imp*, some conforming implementations can be rejected. The following example illustrates the situation.

Example. Consider FSMs *Context*, *Spec* and *Context◊Spec* in Figure 3 when the external input and output sets X_2 and Y_2 of the component *Spec* are empty. By direct inspection, one can assure that *Context◊Spec* is equivalent to the FSM *Context◊Imp* where the FSM *Imp* is shown in Figure 3d, i.e. the *Context◊Spec* is a reduction of the FSM *Context◊Imp*. Therefore, a conforming implementation system at its observation points can produce an unexpected output sequence $z_2z_1y_1$ when x_1 is submitted, instead of the reference z_1y_1.

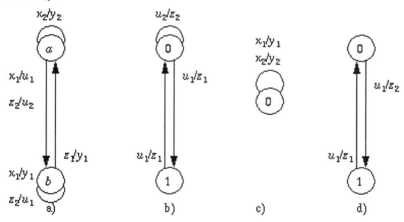

Figure 3. FSMs *Context* (a), *Spec* (b), *Context◊Spec* (c) and FSM *Imp* (d)

More rigorous analysis is necessary to determine which internal outputs are allowed for an implementation component FSM when an external input is submitted. In the case when FSMs *Context*, *Spec* and *Context◊Spec* are complete, this analysis can be performed based on the embedded equivalent [19] of the component FSM *Spec* in the given context. An FSM *Imp* is a conforming implementation of the *Spec* in the given context if and only if the FSM *Context◊Imp* is complete and the FSM *Imp* is a reduction of the embedded equivalent.

A complete test suite w.r.t. the fault domain *Context*◊$\mathfrak{R}_{Spec}$ can be derived by a procedure proposed in [19]. We first derive a complete test suite *TS* w.r.t. the fault domain $\mathfrak{R}_{Spec}$ from the embedded equivalent. An internal test is then translated into a external test suite. We cannot use this approach directly for FSMs *Context* and *Spec* that can be partially specified. Moreover, we are interested only in single output faults; thus, a simpler procedure for a complete test suite derivation can be expected.

3.3   Trace Detecting FSM

Given an external input sequence α such that the behavior of the reference FSM *Context*◊*Spec* is defined under α, a trace over alphabets $X_2 \cup U$ and $Y_2 \cup Z$ is said to be *detectable with* α [25] if any implementation system with the component *Imp* having this trace has an unexpected behavior when α is submitted, i.e. either the output response of the implementation system to α is different from that of the reference system or the behavior of the implementation system is not defined under α. A trace is said to be *detectable* if it is detectable with an appropriate external input sequence. Due to definition, the behavior of the implementation system is not defined if the system falls into live-lock when α is submitted or the behavior of the context is undefined under a current input. An observation point at the internal output of the component enables to detect these faults.

In this section, we derive a so-called *Trace Detecting FSM (TDF)*. We show that the set of traces of a nonconforming implementation component FSM intersects the set of traces of the labeling paths which have no cycles and go from the initial state to a designated *Fail* state in the *TDF*. As usual, we further refer to a path that has no cycles as to a *simple* path.

In fact, the *TDF* is a particular case of an embedded equivalent when the FSMs *Context*, *Spec* and *Context*◊*Spec* can be partial, and the test architecture allows to observe the internal output of the component under test, and we are interested only in output faults. By this reason, to derive the FSM *TDF*, we use the machine *Spec'* where *Spec'* has the same next state function as the *Spec* and all possible outputs for each defined transition, instead of the chaos machine [17,19]. The latter together with observation of the internal outputs of the component under test allows to determine internal traces implying live-locks as well as to simplify a procedure for a complete test suite derivation.

To determine all the detectable traces we first derive a FSM *F* that represents all possible composition traces and recognizes those of them that induce an unexpected behavior of the composition by a designated state *Fail*. The FSM is derived as a composition of the context and the component FSM *Spec'*. We use the FSM *Context◊Spec* to recognize traces with an unexpected external output projection. FSM *F* then is projected onto alphabets of the component *Spec* by a subset construction replacing each subset having the state *Fail* by a designated state *Fail* without outgoing transitions.

Let the FSMs *Context*, *Spec* and *Context◊Spec* have the state sets Q, T and S, respectively. The state space of the FSM F is a subset of $Q \times T \times S$. States of the set $Q \times T \times S$ are divided into stable and transient states. By definition, the initial state $q_0 t_0 s_0$ is stable. Otherwise, the state is stable if it is the fail-state or has an incoming transition with an external output. The stable states cannot be merged with transient states. Two transient states with the same names are merged if they have an incoming transition labeled with a pair with the same output part. We start from the initial state $q_0 t_0 s_0$. Then we apply the following procedure:

1. Given a stable state *qts* and input *x*, there is a transition labeled with *x/u* (*x/z*) from the initial state *qts* to a transient state *q'ts* (*qt's*) if *x/u* (*x/z*) takes the FSM *Context* (*Spec'*) from the state *q* (*t*) to the state *q'* (*t'*). There is a transition labeled with *x/y* from the stable state *qts* to a stable state *q'ts'* (*qt's'*) if *x/y* takes the FSM *Context* (*Spec'*) from the state *q* (*t*) to the state *q'* (*t'*) while *x/y* takes the FSM *Context◊Spec* from the state *s* to the state *s'*. If the output of *Context◊Spec* at the state *s* to *x* is defined and is different from *y*, we specify a transition from the state *qts* to the designated state *Fail* labeled with *x/y*. The reason is if a component has a trace with the corresponding projection then an unexpected external output will be produced when an appropriate external input sequence is submitted to the implementation system.

2. Given a transient state *qts* with an incoming transition labeled by *a/z* (*a/u*) and a pair *z/u* (*u/z*), there is a transition from the state *qts* to the designated state *Fail* labeled with *z/u* (*u/z*) if one of the following conditions holds:

a) *z/u* (*u/z*) provides a cycle labeled with pairs of internal actions, i.e. when a component implementation FSM has a trace with the corresponding projection the composition falls into live-lock;

b) the context (the component) is undefined at the state *q'* (*t'*) under input *z* (*u*), i.e. the context (the component) does not expect a submitted input at the current state.

If none of the above conditions holds then there is a transition labeled with z/u (u/z) from the state qts to a transient state $q'ts$ (from the state qts to $qt's$) where q' is a successor of q under z/u in the *Context* (t' is a successor of t under u/z in the *Spec*).

3. There is a transition labeled with z/y (u/y) from the transient state qts to a stable state $q'ts'$ (from the state qts to $qt's'$) if there is a transition from the state q to the state q' under z/y in the *Context* (from the state t to the state t' under u/y in the *Spec*) and x/y takes the FSM *Context* (*Spec*) from the state s to the state s'. If the output of the *Context◊Spec* to x at the state s is defined and is different from y, we specify a transition from the state qts under z/y (u/y) as a transition to the designated state *Fail*.

4. The procedure is repeated unless all reachable stable states with possible external inputs are considered. Moreover, since there is an observation point at the internal output of the component under test, given a state of the FSM such that all its outgoing transitions result in *Fail*, we replace the state with the *Fail* state.

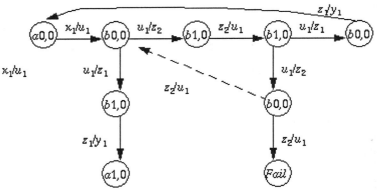

Figure 4. A fragment of the FSM F

Example. Consider FSMs *Context*, *Spec* and *Context◊Spec* shown in Figure 3. We derive the machine *Spec'* by adding each output to each defined transition of the *Spec*. Consider the initial stable state $(a0,0)$. There is a transition under x_2/y_2 from the state to the same stable state and a transition under x_1/u_1 from the state to a transient state $(b0,0)$. There are two transitions from the state $(b0,0)$: a transition under u_1/z_1 to the transient state $(b1,0)$ and a transition under u_1/z_2 to the transient state $(b1,0)$. The latter two transient states cannot be merged since they have incoming transitions labeled by pairs with different output parts. There is a transition from the former state $(b1,0)$ to a stable state $(a1,0)$ under z_1/y_1 while there is a transition from the latter state $(b1,0)$ to a transient state $(b1,0)$ under z_2/u_1. There is a transition under u_1/z_1 from the latter transient state $(b1,0)$ to a transient state $(b0,0)$ where is a transition to a stable state $(a0,0)$ under z_1/y_1. A transition under u_1/z_2 from the latter transient state $(b1,0)$ provides a cycle labeled with internal pairs, i.e. there is a transition to the *Fail* state under u_1/z_2. This fragment of FSM F is shown in Figure 4. In the same way, the fragment of the FSM is derived for a stable state $(a1,0)$.

By construction, the FSM F has complete in about formation traces of the *Spec'* that induce an unexpected behavior of the composition, i.e. the following statement holds.

Proposition 3.1. Given the component *Spec*, let *Spec'* denote the machine with the same next state function and each possible output for each transition. A trace of the machine *Spec'* is detectable if and only if it has a prefix β/γ such that the FSM F has a trace with the corresponding projection β/γ resulting in the *Fail* state.

Thus, our next step is to project the obtained FSM F onto the set of pairs $\{\alpha/\beta: a \in X_2 \cup U, \beta \in Y_2 \cup Z\}$. We perform this using a subset construction [HoU179]. Two states of the FSM are merged into a single state if there exist traces to these states with the same projection over the component alphabets $X_2 \cup U$ and $Y_2 \cup Z$. Each subset comprising a designated state *Fail* is replaced with the state *Fail* without outgoing transitions. If for some state, all its outgoing transitions result in *Fail* then we replace the state with the *Fail* state. Denote *TDF* the obtained FSM and call it *trace detecting FSM*.

Theorem 3.2. Given the trace detecting FSM *TDF* derived by the above procedure and an implementation component FSM *Imp* that is a submachine of *Spec'*, the reference FSM *Context◊Spec* is not a reduction of *Context◊Imp* if and only if the machine *Imp* has a trace that labels a simple path in the *TDF* from the initial state to the *Fail* state.

Proof. If part is a corollary to Proposition 3.1.

Only if part. Suppose now that we have a nonconforming submachine *Imp* of *Spec'*, i.e. the reference FSM *Context◊Spec* is not a reduction of the *Context◊Imp*. Due to Proposition 3.1, the FSM F has a trace resulting in the *Fail* state such that its projection onto alphabets of the FSM is a trace $a_1/b_1...a_k/b_k$ taking the FSM *TDF* from the initial to the *Fail* state. Moreover, suppose that the trace labels some path in the FSM *TDF* from the initial state to the *Fail* state that has a cycle. In other words, we have $S_0 - a_1/b_1 \rightarrow S_1 - a_2/b_2 \rightarrow S_2 ... - a_k/b_k \rightarrow Fail$, along with $S_i = S_j$ for some $i < j$, $i,j = 0,...,k-1$. States of the *TDF* are subsets of state of the FSM F. Each state of the set S_0 is of a kind qt_0s, i.e. has t_0 as the component part of the state. When projected, the component part of each state of the subset S_1 is a successor of the state under the I/O pair a_1/b_1 in the *Spec'* and so on. Thus, the FSM *Imp* has also a trace

$t_0 - a_1/b_1 \to ... t_{i^-} a_{j+1}/b_{j+1} \to t_{j+1} ... - a_k/b_k..$

Therefore, the *TDF* has a trace

$S_0 - a_1/b_1 \to ... S_{i^-} a_{j+1}/b_{j+1} \to S_{j+1} ... - a_k/b_k \to Fail,$

that labels a path resulting in the *Fail* state and has no above cycle, i.e. FSM *Imp* has a trace that labels a simple path resulting in the *Fail* state.

❑

Example. The FSM *TDF* for FSMs *Context* and *Spec* in Figure 3 is shown in Figure 5.

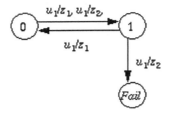

Figure 5. FSM *TDF*

4.   TEST SUITE DERIVATION

4.1   Test Suite Derivation w.r.t. Output Faults

As a corollary to Theorem 3.2, we can establish sufficient conditions for a test suite to be complete w.r.t. the fault domain $Context\Diamond \Re_{Spec.}$

Proposition 4.1. Given a test suite, the test suite is complete w.r.t. the fault domain $Context\Diamond \Re_{Spec}$ if it detects each trace labeling a simple path in the *TDF* from the initial state to the *Fail* state.

Example. Due to proposition 4.1, we obtain a complete test suite $\{x_1x_1\}$ w.r.t. the fault domain $Context\Diamond \Re_{Spec}$ after translation of the set of internal traces $\{u_1/z_1,u_1/z_2; u_1/z_2,u_1/z_2\}$ which label simple paths in the *TDF* (Figure 5) from the initial state to the *Fail* state.

Thus, coming back to procedure proposed in [19] for a complete test suite derivation we now do not need to derive a complete test suite from a nondeterministic FSM *TDF*. An internal test can be derived as the set of all traces of the labeled simple paths in the FSM from the initial state to the *Fail* state. However, we notice that the condition of Proposition 4.1 is not necessary, i.e. a shorter complete internal test may exist. We now establish a necessary and sufficient condition for a test suite detecting all single output faults of the component FSM.

4.2   Test Suite Derivation w.r.t. Single Output Faults

A single output fault t - $a/b \rightarrow t'$ of the component FSM is *detectable* if there exists a simple path in the *TDF* from the initial state to the *Fail* state such that the path traverses the transition t - $a/b \rightarrow t'$, while all other transitions traversed by the path are transitions of the reference component FSM *Spec*. We say that the path, or the trace labeling the path, *captures* the fault t - $a/b \rightarrow t'$. The trace detecting FSM can be essentially simplified if we are interested only in single output faults of the component FSM. To represent such faults we use the FSM $Spec_{sof}$. The next state function of the FSM coincides with that of the *Spec*; output b is in the set of outputs to input a at the state t if and only if the single output fault t - $a/b \rightarrow t'$ is not detectable.

Proposition 4.2. Given an implementation $Imp \in \mathfrak{R}_{Spec}$, *Imp* is a conforming implementation if and only if *Imp* is an extension of some sub-machine of $Spec_{sof}$.

Therefore, to detect an implementation component FSM with a single output fault we need an external case that induces a trace traversing the corresponding transition in the component FSM. A procedure for deriving an internal test suite complete w.r.t. single output faults of the component FSM is almost transparent. A complete test suite is a transition tour traversing each detectable transition of the FSM *Spec'*. The obtained internal test then is translated into external test suite.

Proposition 4.3. An external test suite is complete w.r.t. single output faults of the component FSM if and only if the set of corresponding traces of the component FSM in the reference system traverses each detectable transition.

Proof. Only if part is a corollary of Proposition 4.2.

If part. Let $Imp \in \mathfrak{R}_{Spec}$, t - $a/b \rightarrow t'$ be a faulty detectable transition and αx be the shortest prefix of a test case such that the corresponding traces of the component FSM in the reference system is tailed by t - $a/b \rightarrow t'$. All other transitions of the trace induced by α are reference transitions; by this reason, the global states of the systems $Context \Diamond Imp$ and $Context \Diamond Spec$ after

α coincide. Thus, the external input induces the faulty transition t - $a/b \rightarrow t'$ in the component implementation *Imp*, i.e. an internal (or external) output sequence of the *Imp* is not in the set of output sequences of *Spec$_{sof}$*. Therefore, αx detects the nonconforming implementation *Imp*.

$\Box$

Here we notice that in general case, a test suite traversing each transition of a component under test does not detect its all output faults in the component under test. We can only guarantee that it detects each single output fault. However, if there is the reference output response of the component under test to each test case then there are no output faults in the component.

Given a test suite, an implementation component FSM *Imp*$\in \Re_{Spec}$ is a conforming implementation if and only if the output sequence at the component's outputs to each test case is in the set of corresponding output sequences of the FSM *Spec$_{sof}$*. Thus, a test suite derived by procedure proposed in [9] is complete w.r.t. single output faults of the component FSM when there is an observation point at the internal output of a component under test. The same conclusion can be drawn about the Hit-or-Jump method in [1]. However, both methods do not use a trace detecting FSM; a verdict "pass" is only drawn in the case when the output sequence at the component's outputs coincides with that of the reference component *Spec*. By this reason, some conforming implementations can be rejected by a test suite delivered by the above methods.

4.3   Fault Coverage Evaluation

Given a test suite *TS*, we can calculate its fault coverage w.r.t. single output faults. As usual, the fault coverage is calculated as the ratio m/n multiplied 100% where m is number of single output faults detected with the test suite *TS* while n is number of all such detectable faults. Number n can be calculated as the product $l|Y_2 \cup Z|$-p where l is number of defined transitions of the FSM *Spec* which can be involved with the given context [1] while p is number of single output faults over these transitions which are not detectable. In the case of complete FSMs, an implementation with such fault is equivalent to the specification component FSM *Spec* in the given context [18].Number n is calculated in the same way: $m=k|Y_2 \cup Z|$-q, where k is number of transitions of the component FSM traversed with the test suite in the reference system while q is number of single output faults over these transitions which are not detectable.

Example. Consider the reference component FSM *Spec* (Figure 3). There are two single output faults, namely $0 - u_1/z_2 \to 1$ and $1 - u_1/z_2 \to 0$ over transitions involved with the given context. The fault $0 - u_1/z_2 \to 1$ is not detectable since there is no trace in the FSM *TDF* (Figure 5) from the initial state to the *Fail* state such that the trace traverses this transition while all other transitions traversed by the trace are transitions of the reference component *Spec*. An implementation component FSM *Imp* with the fault is shown in Figure 3d. The system of the context and the FSM *Imp* is a conforming implementation. To detect an implementation component FSM with the fault $1 - u_1/z_2 \to 0$ an external test case should induce an internal trace with a prefix u_1/z_1, u_1/z_2. By direct inspection, one can assure the test case x_1x_1 possesses the feature. We also notice that a test case x_1 does not detect any nonconforming implementation w.r.t. single output faults of the component FSM.

5.   TELEPHONE SERVICES EXAMPLE

We demonstrate our approach by testing a system composed by telephone services: the Basic Call Services (BCS), Call Forward Unconditional (CFU) and Original Call Screening (OCS). The approach is illustrated by testing OCS in the context of the BCS and CFU. The I/O FSMs modeling the services (Figure 6) have been obtained from SDL specifications of the system [13].

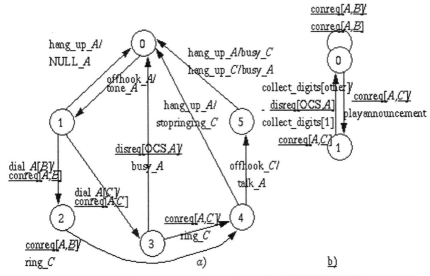

Figure 6. Context FSM (a) and OCS FSM (b)

Figure 7 presents the corresponding composite FSM while Figure 8 shows the trace detecting FSM.

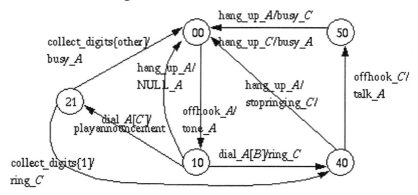

Figure 7. The composite FSM

By direct inspection, one can assure each single output fault is detectable, i.e. there are 12 detectable single output faults. Given a test suite $TS=\{$off_hook_A, dial_$[A,B]$; off_hook_A, dial $[A,C]$, collect_digits[other]$\}$, we determine the set T of all single output faults detected with the test suite TS. In our example, $T=$
$\{0$ - conreq_$[A,B]$/conreq_$[A,C] \rightarrow 0$;
0 - conreq_$[A,B]$/disreq_$[OCS,A] \rightarrow 0$;
0 - conreq_$[A,B]$/play_announcement $\rightarrow 0$;
0 - conreq_$[A,C]$/conreq_$[A,B] \rightarrow 1$;
0 - conreq_$[A,C]$/conreq_$[A,B] \rightarrow 1$;
0 - conreq_$[A,C]$/disreq_$[OCS,A] \rightarrow 1$;
1 - collect_digits[other]/conreq_$[A,B] \rightarrow 0$;
1 - collect_digits[other]/conreq_$[A,B] \rightarrow 0$;
1 - collect_digits[other]/play_announcement $\rightarrow 0\}$.
Thus, three faults remain undetectable with the test suite, i.e. the fault coverage $\Phi(TS)$ of the TS w.r.t. single output faults is equal to $(12/15)100\%=80\%$.

Consider a single output fault 1 - collect_digits[1]/conreq_$[A,B] \rightarrow 0$ undetected with the TS. By direct inspection of Figure 6 (a), one can assure that to induce the trace traversing the transition 1 - collect_digits[1]/conreq_$[A,B] \rightarrow 0$ in an implementation OSC, the context must enter the state 1, i.e. the external sequence off_hook_A, dial [A,C] should be applied. At this point we

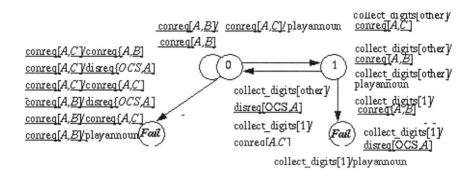

Figure 8. FSM *TDF*

have the context at the state 3, the OCS at the state 1. The external input sequence collect_digits[1] then induces the desired transition if the implementation OSC has such a transition. However, the context does not expect the input conreq_[A,B] at the current state 3, i.e. it is a faulty output. Thus, a test case off_hook_A, dial_[A,C], collect_digits[1] detects the single output fault 1 - collect_digits[1]/conreq_[A,B] → 0.

By direct inspection, one can assure the same test case also detects single output faults 1 - collect_digits[1]/disreq_[OCS,A] → 0 and 1 - collect_digits[1]/play_announcement → 0, i.e. the fault coverage of the test suite with the extra test case is 100%. Therefore, the set {off_hook_A, dial_[A,B]; off_hook_A, dial [A,C], collect_digits[other]; off_hook_A, dial_[A,C], collect_digits[1]} is complete w.r.t. single output faults. Here we notice that a complete test suite not necessary traverses each transition of the reference composed FSM *Context◊Spec*, as it happens in the above example.

Thus, if at observation points of a system under test we observe the reference output sequence tone_A, conreq_[A,B], ring_C to the off_hook_A, dial_[A,B]; the reference output sequence tone_A, play_announcement, disreq_[OCS,A], busy_A to the off_hook_A, dial_[A,C], collect_digits[other]; while the reference output sequence tone_A, play_announcement, conreq_[A,C], ring_C to the off_hook_*A*, dial_[*A*,*C*], collect_digits[1], then the system under test has the reference system as its reduction. Otherwise, the system under test is a nonconforming implementation.

6.   CONCLUSION

In this paper, we studied fault detection power of a test suite widely used in practice for a system of communicating Finite State Machines (FSMs). The test suite traverses each involved transition of each component FSM in the reference system. We have shown that the test suite is complete w.r.t. single output faults of a component FSM when the output of the component is accessible during a testing mode. The performed computer experiments have shown that a test suite w.r.t. single output faults of each component FSM usually also detects "almost all" transfer and output faults in the corresponding composite FSM. We illustrated our approach by testing a system composed by telephone services.

Some of the reviewers did very interesting remarks to the work presented here: for our assumption that the output of the component is accessible during testing, the reviewer signalled that there are real protocols where outputs are not observable, in particular, when an unit implements several layers it is not possible to observe such outputs. Our answer is that if different layers are implemented by an unit, we will consider this unit as a component. On the other hand, our experience on the test of services on a CORBA platform shows that it is possible to observe exchanged messages between software components.

References

[1]   Cavalli, A., Lee, D., Rinderknecht, C., and Zaïdi, F. *Hit-or-Jump: An algorithm for embedded testing with applications to IN services.* Proceedings of Joint Inter. Conf. FORTE/PSTV99, pp: 41-58.

[2]   Chow, T.S. (1978). *Test software design modeled by finite state machines.* IEEE Transactions on Software Engineering, 4(3): 178-187.

[3]   David, R., and Wagner K. *Analysis of detection probability.* IEEE Transactions on Computers, 39(10): 1284-1291.

[4]   Fujiwara, S., v.Bochmann,G., Khendek, F., Amalou, M., and Ghendamsi, A. (1991). *Test selection based on finite state models.* IEEE Transactions on Software Engineering, 17(6): 591-603.

[5]   Gill, A. *Introduction to automata theory.* Mc Graw-Hill, NY, 1962.

[6] Hopkroft, J.E. and Ulman, J.D. (1979). *Introduction to automata theory, languages and computation.* Addison-Welsey, NY.

[7] Information technology. (1991) *Open systems interaction. Conformance testing methodology and framework.* International standard IS-9646.

[8] Koufareva, I., Petrenko, A., Yevtushenko, N. *Test generation driven by user-defined faults.* Proceedings of 12th IWTCS, pp: 215-236.

[9] Lee, D., Sabnani, K.K., Kristol, D.M., and Paul, S. (1996). *Conformance testing of protocols specified as communicating finite state machines - a guided random walk based approach.* IEEE Transactions on Communications, 44(5): 631-640.

[10] Lee, D., and Yannakakis, M. (1996). *Principles and methods of testing finite state machines, a survey.* IEEE Transactions, 84(8):1090-1123

[11] Lima, L.P., and Cavalli, A.R. (1997). *A pragmatic approach to generating test sequences for embedded systems.* Proceedings of 10th IWTCS, pp: 125-140.

[12] Lima, L.P., and Cavalli, A.R. (1997). *Application of embedded testing methods to service validation.* Second IEEE Intern. Conf. on Formal Engineering methods.

[13] Lima, L.P. (1998). *A pragmatic method to generate test sequences for embedded systems.* Ph.D.Thesis. Institute National des telecommunications, Evry, France.

[14] Petrenko, A. *Checking experiments with protocol machines.* Proceedings of 4th IWTCS, 1991, pp: 83-94.

[15] Petrenko A., Yevtushenko, N., and Dssouli, R. (1994). *FSM based strategies for testing communicating FSMs.* Proceedings of 7th IWTCS, pp:181-196.

[16] Petrenko A., Yevtushenko, N. and v.Bochmann, G. (1996). *Testing deterministic implementations from their nondeterministic specifications.* Proceedings of 9th IWTCS, pp: 125-140.

[17] Petrenko A., Yevtushenko, N., and v. Bochmann, G. (1996). *Fault models for testing in context.* Proceedings of Joint Inter. Conf. FORTE/PSTV96, pp: 125-140.

[18] Petrenko, A., Yevtushenko, N., v. Bochmann, G., and Dssouli, R. (1996). *Testing in context: Framework and test derivation.* Computer Communications, 19: 125-140.

[19] Petrenko, A., and Yevtushenko, N. (1997). *Testing faults in embedded components.* Proceedings of 10th IWTCS, pp: 125-140.

[20] B.Sarikaya and G.v.Bochmann. *Synchronization and Specification issues in protocol testing.* IEEE Transactions on Communications, Vol. COM-32, April 1984, pp. 389-395.

[21] Vasilevsky, M.P. (1973). *Failure diagnosis of automata.* Cybernetics, (4): 653-665.

[22] Vuong, S.T., Chan, W.W.L., and Ito, M.R. (1989). *The UIO-method for protocol test sequence generation.* In IFIP TC6 Second Inter. Workshop on Protocol Test Systems, pp. 161-175.

[23] Yannakakis, M., and Lee, D. (1995). *Testing finite state machines: fault detection.* Journal of Computer and System Sciences, (50): 209-237.

[24] Yevtushenko, N., Petrenko, N. *On fault detection power of checking experiments with automata.* Automatic Control and Computer Science. Allerton Press, N.Y., 1989, Vol. 23, No.3, pp: 3-7.

[25] Yevtushenko, N., Cavalli, A.R., and Lima, L.P. (1998). *Test minimization for testing in context.* Proceedings of 11th IWTCS, pp: 127-145.

[26] Yevtushenko, N., Cavalli, A.R., and Anido, R. (1999). *Test suite minimization for embedded nondeterministic finite state machines.* Proceedings of 12th IWTCS, pp: 237-250.

PART II

TESTABILITY AND TEST FEASIBILITY

4

DETERMINATION OF TEST CONFIGURATIONS FOR PAIR-WISE INTERACTION COVERAGE

Alan W. Williams[§]
School of Information Technology
University of Ottawa
Ottawa ON K1N 6N5, Canada
awilliam@site.uottawa.ca

Abstract   Systems constructed from components, including distributed systems, consist of a number of elements that interact with each other. As the number of network elements or interchangeable components for each network element increases, the trade off that the system tester faces is the *thoroughness of test configuration coverage*, versus availability of limited resources (time and budget). An approach to resolving this trade off is to determine a minimal set of test configurations that test each pair-wise combination of components. This goal gives a well-defined, cost-effective level of test coverage, with a reduced number of system configurations. To select such a set of test configurations, we show how to apply the method of covering arrays, and improve on previous results.

Keywords: System testing, Test coverage, Interactions, Component-based testing

1.   INTRODUCTION

A common source of system faults is unexpected interactions between system components. The risk is magnified when there are a number of interchangeable components for each element in a system. A manufacturer of these system components would want to test as many of the potential system

[§]   The author gratefully acknowledges the support of: the IBM Centre for Advanced Studies, Nortel Networks Ltd., Mitel Corporation, ObjecTime Ltd., CITO, NSERC, and the University of Ottawa.

configurations as possible, to reduce the risk of interaction problems. However, the number of potential system configurations grows exponentially. The scenario in Figure 1 has four parameters, each with three values. There are $3^4 = 81$ possible configurations.

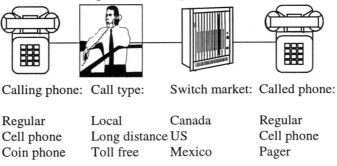

Calling phone:	Call type:		Switch market:	Called phone:
Regular	Local	Canada		Regular
Cell phone	Long distance	US		Cell phone
Coin phone	Toll free	Mexico		Pager

Figure 1.   A System Test Scenario

A system tester faces the constraints of time and money, and all possible configurations cannot be tested within any reasonable allotment. For each system test configuration, a suite of system tests must be executed. Changing between configurations normally requires additional effort. Therefore, reducing the number of system test configurations will produce significant savings. How, then, can the risk of interaction faults be managed in a realistic test plan?

One approach is to at least test for all two-way interactions among various system components. This leads to a reduced set of test configurations. The assumption is that the risk of an interaction among three or more components is balanced against the ability to complete system testing within a reasonable budget. This paper investigates this approach, using covering arrays to determine the set of test configurations that cover all two-way interactions.

This paper starts with an overview of design of statistical experiments, and then shows how to apply these concepts to the testing of software. Orthogonal arrays and covering arrays are defined. An algorithm to generate orthogonal arrays is presented, and these are used to derive building blocks for constructing larger covering arrays. An algorithm, improving on the results from [14], for larger covering arrays is presented, first using an example, and then in general. Algorithm results are provided with respect to the number of configurations produced. The paper concludes with a summary, and opportunities for further work.

2.   APPLYING STATISTICAL EXPERIMENTAL DESIGN TO SOFTWARE TESTING

There is a great deal of work in the literature of statistics, devoted to the subject of experimental design. Except for a few instances ([2], [3], [4], [5], [6] [9], [10]), the results have not generally been applied to the testing of software. Areas where these techniques are often used to detect interactions are in chemical, biological, and medical research. They are increasingly being advocated for use in engineering quality control.

We can make excellent use of these techniques when testing software. The software testing "experiment" that is proposed is to determine if all of the software components are truly free of unwanted interactions. A successful run of the experiments would confirm this independence. A single experiment would consist of running the entire set of test cases for an application, over a specific test configuration. (The prior existence of a suitable "complete" test suite is assumed.)

The application of this method turns out to be simpler than the general situation of experimental design, and this allows some of the constraints to be removed. This means that smaller designs with fewer configurations can be used while still achieving the goals of software testing.

There are two considerations that arise in applying statistical experimental design to software testing. The first is that the values for each system configuration parameter are typically selected from a set of discrete components. The sets of factors and levels are readily apparent, and there is no need to select a set of values from a continuous real-valued range, as would be the case with traditional experimental design.

Second, a statistical analysis of real-valued results is not required. The expected test results are determined directly (e.g. from specifications or customer requirements.) After running each test case, the result is a test verdict: one of "pass," "fail," or perhaps "inconclusive." One experiment corresponds to running the entire test suite, so an overall verdict for the entire test suite is needed to represent the result of the experiment. The assumption is then that if pass verdicts result for all experiments – that is, running the test suite for all configurations in the design – the probability of unwanted interactions is small (depending on the coverage of the test suite used for each configuration). A discrete result for each experiment can be determined on an individual basis. There is no need to analyse a set of real-valued results, with confidence intervals and so on.

Therefore, the requirement for balance can be removed. If every possible pair-wise combination of values is covered, any interactions between those values can be detected. Removing the need for balance provides for a significant reduction in test configurations required.

One possible goal is to determine interactions caused by (up to) a specific number of parameters. In particular, covering of two-way interactions seems to be a reasonable goal that balances the requirement of having a practical number of test configurations, while still finding most of the potential interaction problems in a system. Results obtained in an empirical study [4] show that covering two-way interactions results in excellent code coverage. When required, experiments can be designed to find interactions among three or more specific parameter values, at the expense of an increased number of configurations. This paper will focus on pair-wise parameter value coverage.

3.   ORTHOGONAL AND COVERING ARRAYS

In this section, we first set up the test configuration problem. Next, orthogonal arrays and covering arrays are defined. An algorithm is provided for generating orthogonal arrays. Based on these arrays, we introduce "basic" and "reduced" arrays, which can be used as building blocks to build larger covering arrays.

3.1   The Configuration Problem

Suppose there are k independent system parameters. The property of independence means that the selection of a particular value for one parameter does not effect the selection of any other parameter values. (See [14] for how to handle dependencies among parameters.) Each parameter has a set of discrete values that the parameter can take. Suppose parameter i has n_i values, which are enumerated as $1...n_i$. Without loss of generality, assume that the parameters are ordered so that $n_1 \geq n_2 \geq ... \geq n_k$.

A test configuration consists of a selection of values for each parameter. Therefore, it consists of a k-tuple of parameter values. The goal is to minimise the number of k-tuples that achieve a test coverage criterion.

We also introduce a value $n = f(n_1,...,n_k)$ that is a scalar representation for the "number of values" in various array constructions. Aside from conversion to a scalar value, the reason for introducing a separate variable is that some of the constructions we shall use have additional constraints on n. These constraints can be incorporated into the function f.

3.2 Definitions

A standard construction used for the design of statistical experiments is known as an orthogonal array.

Let $O(c,k,n,t)$ denote an orthogonal array, where:
- c is the number of rows in the array. The k-tuple forming each row is a single test configuration, and thus c is the number of test configurations.
- k is the number of columns, which represents the number of parameters.
- The entries in the array are the values $1,...,n$, where $n = f(n_1,...,n_k)$. Typically, this means that each parameter would have (up to) n values.
- t is the strength of the array (see below).

An orthogonal array has *strength* t if in any $c \times t$ sub-matrix (that is, select any t columns), each of the n^t possible t-tuples (rows) appears the same number of times (>0). In other words, all t-way combinations of parameter values occur the same number of times. It is this balance property that defines the orthogonal array. This property allows the effect of a single parameter to be detected, as well as interactions among (up to) t parameters.

The definition of covering array (see [12] or [13]), is identical to the definition of orthogonal array, except that the strength of the array is redefined as follows:

A *covering array*, denoted $C(c,k,n,t)$, has *strength* t if in any $c \times t$ sub-matrix (that is, selecting t columns), each of the n^t possible t-tuples (rows) appear at least once. In other words, all t-way combinations of parameter values occur at least once. This definition of strength captures the desired t-way interaction property.

This compares with an orthogonal array, where each t-interaction elements appears at least once, and also the same number of times. An orthogonal array satisfies the requirements of a covering array, but a covering array may not be orthogonal. Table 1 shows examples of an orthogonal array (left) and a covering array (right). In the orthogonal array, you can select any two columns, and each ordered pair (1,1), (1,2), (2,1), and (2,2) appears exactly once. In the covering array, each of those ordered pairs will occur at least once, but some may be repeated

Table 1. Examples of an orthogonal array (left) and a covering array (right)

1	1	1		1	1	1	1
1	2	2		2	2	2	1
2	1	2		2	2	1	2
2	2	1		2	1	2	2
				1	2	2	2

Testing at the system level normally occurs in practice after each software component has been tested as an individual unit. If each software component has passed unit testing, it is not necessary to extract the effect of a single parameter. The ability to analyse real-value results is also lost, but this is not required when test verdicts are derived independently. Therefore, the covering array preserves all of the properties needed for software testing, and we can reduce the number of configurations required, as compared with an orthogonal array.

For the rest of the paper, unless explicitly specified, a default value of $t = 2$ is assumed, and the orthogonal and covering arrays will be denoted as $O(c,k,n)$, and $C(c,k,n)$, respectively. This corresponds to the testing goal of pair-wise coverage of parameter values.

3.3   Orthogonal Array Construction

Orthogonal arrays do not necessarily exist for arbitrary values of c, k, and n (see [11] for restrictions). We will use a construction for arrays of strength 2 adapted from Bose [1] for $O(n^2,n+1,n)$. In this construction, n is required to be a prime power. Therefore, we shall define function $n = f(n_1,...,n_k)$ such that $n \geq n_1$ and $n = p^m$ for some prime number p and integer $m \geq 1$. The value of n is the largest number of parameter values, increased to the next highest prime power. In this particular case, the value of f is only dependent on n_1, and not on any of the other parameters.

Let $GF(n) = \{x_0,x_1,...,x_{n-1}\}$ be a Galois (finite) field of order n where $x_0 = 0$ and $x_1 = 1$. If the value of n is a prime number, then the integers modulo n can be used. For values where n is a prime power, see, for example, [7] for an algorithm to generate a Galois field.

To find the element of the array $O_{ij}(n^2,n+1,n)$ for $i = 1,...,n^2$, $j = 1,...,n + 1$, calculate $u = \lfloor (i - 1) / n \rfloor + 1$ and $w = ((i - 1) \bmod n) + 1$. Then, $O_{i1}(n^2,n+1,n) = u$, $O_{i2}(n^2,n+1,n) = w$. When $j > 2$, determine q such that, using the Galois field arithmetic, $x_q = x_{j-2} \cdot x_u + x_w$. and $O_{ij}(n^2,n+1,n) = q$.

This algorithm produces orthogonal arrays that have several useful properties, which will be taken advantage of in subsequent constructions:
- The first row is all ones.
- The first n rows have entries equal to the row index, excepting the first column.
- The first column consists of n ones, n twos, etc.
- Each column of $O(n^2,n+1,n)$, except the first, is comprised of sets of permutations of $(1,...,n)$.

In Figure 1 (in section 1), a scenario was illustrated where there are four system parameters, each with three possible values. There are 81 possible configurations, but pair-wise coverage can be achieved with only 9 configurations.

3.4   Building Blocks for Covering Arrays

Two additional types of arrays are introduced, based on the orthogonal array. These "basic" and "reduced" arrays will be used as building blocks for construction of larger covering arrays.

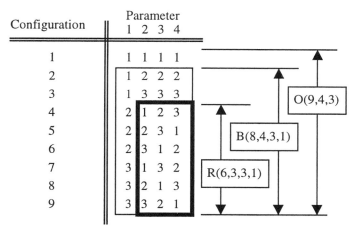

Figure 2.   Orthogonal (O), Basic (B), and Reduced (R) arrays

Define a "basic" array $B(n^2-1,n+1,n,d) = O(n^2,n+1,n)$ with the first row removed, and each column duplicated d times consecutively. The array $B(n^2-1,n+1,n,d)$ has $n^2 - 1$ rows and $(n + 1) \times d$ columns.

Define a "reduced" array $R(n^2-n,n,n,d) = O(n^2,n+1,n)$ with the first n rows removed, the first column removed, and each (remaining) column duplicated d times consecutively. $R(n^2-n,n,n,d)$ has $n^2 - n$ rows and $n \times d$ columns.

Figure 2 shows an example of an orthogonal array $O(9,4,3)$, and shows a corresponding basic array $B(8,4,3,1)$ and reduced array $R(6,3,3,1)$.

The general strategy for using the basic and reduced arrays is that the reduced array has fewer rows, so it contributes fewer test configurations. However, with fewer columns, the reduced array has less parameter capacity. The basic array will be used only when the reduced array cannot hold enough parameters.

We now introduce some building blocks that are not based on an orthogonal array. However, they are useful for constructing larger covering arrays, as we shall see. Let $I(c,d)$ denote an array that contains all ones, and is c rows by d columns. Let $N(n^2-n,n,d)$ denote an array that contains a $n \times d$

block of twos, vertically concatenated with a $n \times d$ block of threes, and so on up to n. $N(c,n,d)$ has c rows and $n \times d$ columns. These constructions are used to add parameter capacity when possible.

For illustrative purposes, the notation in Table 2 will be used to show how building blocks are combined to form larger arrays

Table 2. Notation for large array construction

O(9,4,3)	O(9,4,3)	O(9,4,3)	I(9,1)
R(6,3,3,4)			N(6,3,1)

Table 2 indicates that three copies of $O(9,4,3)$, and then $I(9,1)$ are to be concatenated horizontally. Below the three copies of $O(9,4,3)$, the array $R(6,3,3,4)$ is to be concatenated vertically, and below the array $I(9,1)$, the array $N(6,3,1)$ is to be concatenated. The resulting large array has 13 columns and 15 rows.

4.   CONSTRUCTING LARGER COVERING ARRAYS

In this section, we show how to construct covering arrays that are not subject to the restrictions on the number of parameters that apply to orthogonal arrays.

The orthogonal array $O(n^2,n+1,n)$ is constrained to (a maximum limit of) $k = n + 1$ parameters [11]. When $k > n + 1$, then there are two approaches that can be taken. One is to use a value of $n = k - 1$, instead of setting n as a function of the number of values. This will cause the number of configurations to grow to k^2. The number of configurations is proportional to the square of the number of parameters. However, it is possible to reduce the number of test configurations to be proportional to the logarithm base $n + 1$ of the number of parameters.

A "divide and conquer" approach can be taken. Suppose we want to construct a covering array for twelve parameters, each of which has three values. As a first stage, start by horizontally concatenating three copies of $O(9,4,3)$ (see Table 2). This results in an array with twelve parameters, and three values for each. The next step is to examine how close this array is toward achieving pair-wise coverage.

Consider the first column of the first copy of $O(9,4,3)$. Because of the orthogonal array construction, pair-wise coverage is achieved with the remaining three columns in that first copy of $O(9,4,3)$. Each of the additional copies of $O(9,4,3)$ has duplicates of those same three other columns. The only column where pair-wise coverage has not yet been achieved is the first column in each of the other copies of $O(9,4,3)$. These columns are identical with the first column in the first copy of $O(9,4,3)$. In these identical columns, only the pairs (1,1), (2,2), and (3,3) are covered.

The strategy is then to cover the missing ordered pairs in a second stage. This is where the reduced array can be used. The array $R(6,3,3,1)$ covers the non-(x,x) pairs for three parameters. The columns of R(6,3,3,1) can be inserted under each of the first columns of the O arrays.

It turns out that this process can be repeated for the each column in the first repetition of $O(9,4,3)$. The effect is to duplicate the columns of $R(6,3,3,1)$, effectively forming $R(6,3,3,4)$.

Since we have expanded the array in the vertical direction, there is now capacity for a thirteenth parameter. We can add a column of nine 1's (that is, $I(9,1)$) to the right of the O arrays, and the values 2 to n (repeated n times each vertically) next to the R array (that is, $N(6,3,1)$).

Why does this work? We are making use of the properties of the Bose orthogonal array construction. In the R array, we have $n-1$ permutations of $1,\dots,n$ in each column. Therefore, we can construct a column of n 2's, n 3's, and so on up to n. This covers all combinations in the other columns except for the $(1,x)$ combinations. We can then fill the part of the column next to the O arrays with 1's. Except for the first column of the O arrays (and their identical counterparts), the O array columns are permutations of $1,\dots,n$. The first columns are structured as n 1's followed by n 2's and so on up to n's. Filling the column with 1's will definitely achieve pair-wise coverage.

In fact, we need to fill the column with n 1's at the start, and then in every nth position after that. The remainder of the positions could be left as "don't care" values, where the value of the parameter does not affect the final coverage. It could be set to any legal value to construct a test configuration.

Because this step results in "don't care" values and an unbalanced use of values (the "1" in the example), the tester can judiciously choose the correspondence between the enumerated values $1,2,\dots,n$ and actual system configuration values. For example, if a particular system element has not been as thoroughly tested in the unit testing phase, the system tester could assign this element to correspond to "1" in the extra column, and fill in the "don't care" values with ones as well. This will result in this particular system element being used in more test configurations. For the rest of this section, the don't care values will be arbitrarily filled in as 1's, to form a column of n^2 1's.

Pair-wise coverage of all parameters has now been achieved. A notation for this *two-stage* construction of C(15,13,3) is shown in Table 3.

Table 3. Construction of C(15,13,3)

Stage 1	O(9,4,3)	O(9,4,3)	O(9,4,3)	I(9,1)
Stage 2	R(6,3,3,4)			N(6,3,1)

The selection of the reduced array is by the rule "use $R(n^2-n,n,n,n+1)$ under each group of n repetitions of $O(n^2,n+1,n)$." Up to $n^2 + n$ parameters can be handled using the reduced array.

If there are $n^2 + n < k \le (n + 1)^2$ parameters, the reduced array cannot be used in the second stage, because the first stage will require $n + 1$ copies of $O(n^2,n+1,n)$ instead of n. Instead, the basic array $B(n^2-1,n+1,n,n+1)$ is used in the second stage. This is at the expense of $n - 1$ additional configurations.

Using the reduced array instead of the basic array means that we cannot add an additional column. The basic array does not have the property that its columns are consecutive permutations of $1,...,n$.

For example, if there are sixteen parameters, each with three values, we can construct $C(17,16,3)$ as shown in Table 4.

Table 4. $C(17,16,3)$ in shorthand notation

O(9,4,3)	O(9,4,3)	O(9,4,3)	O(9,4,3)
B(8,4,3,4)			

If there are more than $(n + 1)^2$ parameters, the entire process can be repeated recursively, both horizontally and vertically.

The extra column presents some additional difficulties for additional stages. However, we can replicate the extra columns as we did with the rest of the array, and then cover the missing combinations. It turns out that we can just use the same method of covering the missing combinations of identical columns: add a reduced array underneath the duplicated columns. We can then add a final extra column, again because of the increased vertical dimension. An example of a three-stage construction is shown Table 5.

Table 5. A three-stage construction of covering array $C(21,40,3)$

O	O	O	O	O	O	O	O	O	I(9,3)	I(15,1)
R(6,3,3,4)		R(6,3,3,4)		R(6,3,3,4)		N(6,3,3)				
R(6,3,3,12)									R(6,3,3,1)	N(6,3,1)

For each stage, the number of parameters that can be accommodated is multiplied by $n + 1$ (with an extra column possibly added), so the number of stages required is $s = \lceil \log_{n+1}(k) \rceil$.

4.1 Algorithms for Covering Array Construction

To construct a covering array of strength 2, we are given k, the number of parameters, and that parameter i has n_i values, which are enumerated as $1, \ldots, n_i$. Let n be the least integer where $n \geq n_1$ and $n = p^m$ for some prime number p and integer $m \geq 1$. The number of stages required is $s = \lceil \log_{n+1}(k) \rceil$. If $s = 1$, then a subset of k columns of $O(n^2, n+1, n)$ can be used as $C(c,k,n)$, and it is not necessary to proceed further.

4.1.1 Determining the building blocks to use

Let g_r represent the number of columns to be used in the arrays at stage r. Specifically, $g_r = n + 1$ indicates that at stage r, copies of the basic array are used. If $g_r = n$, then copies of the reduced array are used. The initial number of columns for arrays in each stage is $g_1 = n + 1$ (representing the use of the orthogonal array $O(n^2, n+1, n)$ in the first stage), and $g_r = n$ for $r = 2, \ldots, s$. We are hoping to use reduced arrays for all stages after the first one.

Algorithm 1. Determine parameter capacity
To calculate the values of g_r, an algorithm to determine the parameter capacity will be used as a subroutine:
Parameter capacity $h \leftarrow n + 1$
For $r = 2, \ldots, s$:
$\quad h \leftarrow h \times g_r$
$\quad$ If $g_r = n$ then $h \leftarrow h + 1$ (for extra column)
Algorithm 1 is of order $\log(k)$, since the for loop will execute up to s-1 times.

Algorithm 2. Determine usage of basic versus reduced arrays
$g_1 \leftarrow n + 1$, and $g_r \leftarrow n$ for $r = 2, \ldots, s$
$r \leftarrow 2$.
Determine parameter capacity h.
While $h < k$:
$\quad g_r \leftarrow n + 1$; increment r
$\quad$ Redetermine h
Algorithm 2 is of order $\log(k)$, since the for and while loops will execute up to s-1 times. After using Algorithm 2, $g_r = n + 1$ indicates that at stage r,

copies of the basic array are used. If $g_r = n$, then copies of the reduced array are used.

Algorithm 3.   Determine column duplication values

The next calculation is the values to use for d in the basic or reduced arrays at each stage. Let d_r represent the column duplication value at stage r. Then, $d_1 = 1$, and $d_r = d_{r-1} \times g_{r-1}$ for $r = 2,\ldots,s$. Algorithm 3 is of order $\log(k)$, since there are s values of the d_r.

Algorithm 4.   Values to set up "extra" columns

For each stage that uses reduced arrays, we need to determine the parameters for the extra columns. Let e_r be the number of extra columns in stage r. Thus, at stage r, the arrays $I(c,e_r)$ and $N(n^2-n,n,e_r)$ are used, where c is the current number of rows in the covering array. Then, $e_r = 0$ if $g_r = n + 1$, or $e_r = (d_s\ g_s) / (d_r\ g_r)$ if $g_r = n$

Then, there is the duplication factor for the reduced arrays that are added for the extra columns. The set of extra columns added at each stage is treated differently at each subsequent stage. Let λ_{rq} be the duplication factor for the reduced array added at stage q, for the extra columns added at stage r. For the columns added at any particular stage, the duplication factor starts at 1 in the next stage, and is multiplied by n for subsequent stages. $\lambda_{rq} = 1$ if $q = r + 1$ and $r + 1 \le s$, and $\lambda_{rq} = \lambda_{r,q-1} \times n$ for $q = r + 2,\ldots,s$

Calculation of the values for e_r is of order $\log(k)$, since there are s values of the e_r. Calculation of the λ_{rq} is of order $\log^2(k)$, since there are (up to) s values of r and q.

4.1.2   Building the covering array

The results of the preceding section give us the structure of the building blocks that are needed to construct $C(c,k,n,2)$. This section describes how to assemble the building blocks.

It is assumed that if an array has a duplication factor of 0, it is a null array, and concatenating a null array results in the original array.

Algorithm 5.   Construct $C(c,k,n,2)$

The following algorithm builds $C(c,k,n,2)$:

Start with C as $O(n^2,n+1,n)$ repeated $(d_s\ g_s) / (n + 1)$ times horizontally

For $r = 2,\ldots,s$:

Add to the right of C, $I(c,e_r)$; c is current number of rows in C.

Let A be $R(n^2-n,n,n,d)$ if $g_r = n$, or $B(n^2-1,n+1,n,d)$ if $g_r = n + 1$,

Construct array S as A repeated $(d_s\ g_s) / (d_r\ g_r)$ times horizontally

For $q = 2\ldots r$:

Add to the right of S $e_r / (n\lambda_{rq})$ copies of $R(n^2-n,n,n,\lambda_{rq})$

Horizontally concatenate array $N(n^2-n,n,e_r)$ to the right of S
Vertically concatenate array S below C

Overall, algorithm 5 is of order $k\log^2(k)$. The nested r and q loops are executed on the order of $\log^2(k)$ times. The horizontal concatenations are repeated order k times as the largest number of repetitions is k/n times.

Because the algorithms are executed in sequence, the overall complexity is on the order of $n^2 + k\log^2(k)$. Construction of the O, R, and B arrays is of order n^2. This is done prior to the execution of algorithm 5, so the complexities are added.

5.   ALGORITHM RESULTS

5.1   Results for the Example

For the example in Figure 1 (in section 1), Table 6 shows the test configurations determined from C(9,4,3) that provide pair-wise parameter value coverage for this situation:

Table 6. Test configurations for the scenario in Figure 1

#	Caller	Type	Market	Called
1:	Regular	Local	Canada	Regular
2:	Regular	Long distance	US	Cell phone
3:	Regular	Toll free	Mexico	Pager
4:	Cell phone	Local	US	Pager
5:	Cell phone	Long distance	Mexico	Regular
6:	Cell phone	Toll free	Canada	Cell phone
7:	Coin phone	Local	Mexico	Cell phone
8:	Coin phone	Long distance	Canada	Page
9:	Coin phone	Toll free	US	Regular

5.2   Number of Configurations Generated

With regard to the number of test configurations for k parameters, with values $n_1...n_k$, the best case lower bound on the number of configurations produced is $\lceil \log_{n+1}(k) \rceil (n^2 - n) + n$ where n is the next integer $\geq n_1$ that is a prime power. The worst case upper bound is $\lceil \log_{n+1}(k) \rceil (n^2 - 1) + 1$. In general, the number of configurations is proportional to the square of the number of values, and the logarithm of the number of parameters.

Figure 3 shows the results for the algorithm for various combinations of parameters and values. The numbers of values are in the different lines on the graph, while the x-axis is the number of parameters. If you follow the different numbers of values, you can see the quadratic influence of the number of values, but any specific line grows logarithmically with the number of parameters.

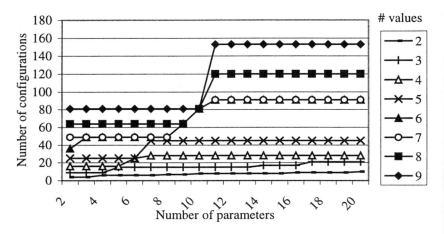

Figure 3.   Results from the algorithm

5.3   Comparison with the IPO Algorithm

The algorithms presented here have been implemented in a tool (versions exist in C++ and Java) called "TConfig". For comparison, the TConfig tool was run using the situations described in the paper by Lei and Tai [8] on their in-parameter-order (IPO) strategy. This is a heuristic strategy that starts with two parameters, and configurations for each combination of the values for those two parameters. The strategy is then to grow the array horizontally by adding a single parameter at a time, and then growing the array vertically by adding test configurations as needed. Table 7 shows the run times and the number of configurations generated by each method. To ensure a fair comparison, the IPO strategy was implemented using the same compiler and class library, and the algorithms were run on the same machine. In fact, the IPO results presented here are slightly better than those reported in [8].

The complexity of the IPO algorithm is $n^5 \times k^3$ [8]. The algorithm presented here has complexity of $n^2 + k\log^2(k)$, which is a considerable improvement.

In all cases, because the algorithm presented here is deterministic, it runs much faster than the IPO method. For situations where the number of parameters are the same for each value, TConfig almost always generated fewer configurations to achieve pair-wise parameter value coverage; the exception is case 6. For situations where the numbers of values for each parameter differ, TConfig generated slightly more configurations.

Overall, the TConfig tool clearly outperforms the IPO method for execution time, and produced fewer test configurations in 13 of the 16 cases.

Table 7. Comparison with the IPO method

Case #	System	IPO		TConfig	
		# config	Time(s)	# config	Time (s)
1	4 parameters, 3 values	10	0.06	9	0.05
2	13 parameters, 3 values	20	0.44	15	0.05
3	61 parms: 15 × 4 values, 17 × 3 values, 29 × 2 vals.	34	2.75	40	0.05
4	75 parms: 1 × 4 values, 39 × 3 values, 35 × 2 values	27	3.90	30	0.05
5	100 parameters, 2 values	15	3.52	14	0.05
6	20 parameters, 10 values	219	47.46	231	0.44
7	10 parameters, 4 values	31	0.11	28	0.05
8	20 parameters, 4 values	34	0.99	28	0.05
9	30 parameters, 4 values	41	2.42	40	0.05
10	40 parameters, 4 values	42	5.00	40	0.05
11	50 parameters, 4 values	47	9.51	40	0.05
12	60 parameters, 4 values	47	15.98	40	0.05
13	70 parameters, 4 values	49	25.43	40	0.05
14	80 parameters, 4 values	49	38.50	40	0.05
15	90 parameters, 4 values	52	53.72	43	0.06
16	100 parameters, 4 values	52	75.02	43	0.11

6. CONCLUSIONS AND FURTHER WORK

This paper has shown that methods from statistical experimental design can be applied to the problem of determining a set of system test configurations that meet a defined goal for coverage of potential 2-way interactions. We presented improvements in algorithms for generating covering arrays. These covering arrays are constructed to ensure that pair-wise coverage of software components is achieved.

Further work needs to be done to address the situation where there are varying numbers of values for each parameter. A potential approach is to combine this approach with a heuristic such as the IPO method to deal with the differing number of values for each parameter.

7.   ACKNOWLEDGMENTS

I would like to thank Dr. Robert Probert, and everyone in the Telecommunications Software Engineering Research Group at the University of Ottawa for their support and feedback on this work.

REFERENCES

[1]   R.C. Bose. On the application of the properties of Galois fields to the construction of hyper Graeco-Latin squares. Sankhya 3:323-338 (1938).

[2]   R. Brownlie, J. Prowse, and M.S. Phadke. Robust Testing of AT&T PMX/StarMail using OATS. AT&T Technical Journal, 71(3): 41-47 (May/June 1992).

[3]   K. Burroughs, A. Jain, and R.L. Erickson. Improved Quality of Protocol Testing Through Techniques of Experimental Design. In Proceedings of Supercomm/ICC '94, 1994, pp. 745-752 1994.

[4]   D.M. Cohen, S.R. Dalal, M.L. Fredman, and G.C. Patton. The AETG System: An Approach to Testing Based on Combinatorial Design. IEEE Transactions on Software Engineering, 23(7): 437-444, (July 1997).

[5]   D.M. Cohen, S.R. Dalal, J. Parelius, and G.C. Patton. The Combinatorial Approach to Automatic Test Generation. IEEE Software, 13(5): 83-88, (September 1996).

[6]   I.S. Dunietz, W.K. Ehrlich, B.D. Szablak, C.L. Mallows, A. Iannino. Applying Design of Experiments to Software Testing. In Proceedings of ICSE '97, pages 205-215, Boston MA USA, (1997).

[7]   W.J. Gilbert. Modern Algebra with Applications. Wiley Interscience, New York NY USA, 1976.

[8]   Y. Lei, K.C. Tai. A test generation strategy for pairwise testing. In Proc. of the 3rd IEEE High-Assurance Systems Engineering Symposium, pages 254-161, Nov. 1998.

[9]   R. Mandl. Orthogonal Latin squares: An application of experiment design to compiler testing. Communications of the ACM, 28(10): 1054-1058 (October 1985).

[10]   W.B. Perkinson. A Methodology for Designing and Executing ISDN Feature Tests Using Automated Test Systems. In Proceedings of IEEE GLOBECOMM '92, 1992.

[11]   C.R. Rao. Factorial Experiments Derivable from Combinatorial Arrangements of Arrays. Journal of the Royal Statistical Society, 9(1): 128-139 (1947).

[12]   B. Stevens, Transversal Covers and Packings, Ph.D. thesis, University of Toronto, 1998.

[13]   B. Stevens, E. Mendelsohn. Efficient Software Testing Protocols. In Proceedings of CASCON '98, pages 279-293, Toronto ON Canada, December 1998.

[14]   A.W. Williams, R.L. Probert, A Practical Strategy for Testing pair-wise Coverage of Network Interfaces. In Proc. of the 7th International Conference on Software Reliability *Engineering (ISSRE '96),* pages 246-254, White Plains NY USA, October 1996.

5

INCREMENTAL TESTING AT SYSTEM REFERENCE POINTS

Ina Schieferdecker, Mang Li, Axel Rennoch
GMD FOKUS
Kaiserin-Augusta-Allee 31, D-10589 Berlin, Germany
phone: +49 30 3463-7000, fax: +49 30 3463-8000
{schieferdecker, m.li, rennoch}@fokus.gmd.de, http://www.fokus.gmd.de/tip

Abstract   RM-ODP (Reference Model of Open Distributed Processing) is a reference model for open information processing systems. Conformance is one of the key aspects of open systems. Conformance ensures the interworking in a multi-vendor information processing environment that may evolve over the time. An essential partitioning concept for open system is that of reference points (RPs). RPs consist of a set of interfaces together with potential interactions at these interfaces. RP specifications define conformance requirements, so that they can be used to determine the conformance of a system. However, the testing of RPs of real systems is restricted as their functionality is often very complex. Refinement of the RP structure is required to ease both their realization and also their testability. In particular, the incremental development of RPs that allows both the evolution of the system architecture and the step-by-step implementation by vendors should be supported. A new concept - that of an RP-facet - has been recently defined to structure RPs and to improve their testability. The focus of this paper is on structuring principles and dependence relations used by the RP-facet concept. They serve as a basis for incremental and efficient testing at RPs. An example taken from use scenarios on service access in open systems is elaborated in the paper.

Keywords: Conformance Testing, Reference Points, Dependence Relations, ODL, MSC, TTCN, ODP

1. INTRODUCTION

Conformance is one of the key aspects of open systems. Conformance ensures the interworking in a multi-vendor information processing environment that may evolve over the time. Conformance evaluation accompanies the whole development process of a system. It is in particular used in the late phase of technology maturity, where it is carried out in form of conformance testing. The consideration of conformance is an essential part of the system architecture and specification. In particular, the design for testability and the specification for testability have to be addressed.

A reference model for open information processing systems is the RM-ODP (Reference Model of Open Distributed Processing) REF[8]. Five viewpoints, the object model and the conformance assessment are basic concepts of RM-ODP. The enterprise, information, computational, engineering and technology viewpoints provide consideration of a system from different concerns. The object model is the basis for the flexibility, modularity and scalability of ODP systems. An object is an entity that contains information and offers services at its interfaces. An ODP system is composed of interacting objects, which may be organized in object groups. As objects or groups of objects may origin from different sources, the reference point concept is used as a basis for conformance assessment. Reference points (RPs) are potential conformance points at which the conformance of an ODP system is determined.

TINA (Telecommunications Information Networking Architecture) [15] is an open system architecture based on RM-ODP. TINA adopts and specializes many concepts of RM-ODP, including those introduced above. TINA is in its maturity phase, where conformance evaluation plays an important role. In TINA, telecommunication stakeholders are generalized by business roles, e.g. consumer, retailer or third-party service provider. Since every stakeholder represents an autonomous administrative domain, the implemented sub-systems used by stakeholders operate in a heterogeneous and unpredictable/uncontrollable environment. Inter-domain RPs are introduced to ensure the interoperability of the various sub-systems.

However, the testability of TINA RPs is restricted. At first, the TINA RP specification template is lacking a behavioral description, so that current TINA RP specifications are inadequate for conformance testing. Secondly, the functionality of an inter-domain RP is often very complex. A refinement of the RP structure would ease their implementation and testing. In particular, an incremental development of RPs that allows both the evolution of the system architecture and the step-by-step implementation by vendors would be beneficial.

The TINA Conformance Testing Framework Request for Proposal (RFP) REF[17] initiated a facet concept to improve the testability of TINA RPs. This paper presents an extension of our contribution to the RFP [4]. The focus of this paper are various *dependence relations* of RP-facets, which serve as a basis for efficient testing of RP-facets. The paper is structured as follows: Section 2 presents an overview on the notion of RPs in ODP and TINA. The structure of an RP is analyzed in Section 3. Based on this analysis, the notion of RP-facets and an approach to the specification of RP-facets are given. Section 4 discusses the basic principles for conformance testing of RP-facets. The main ideas for the definition of an efficient test campaign for RP-facets is given in Section 5. Conclusions finish the paper.

2.   REFERENCE POINTS

In RM-ODP, all interfaces are defined as RPs, while subsets of those RPs to which conformance requirements apply are chosen as conformance points. Four classes of RPs are defined: perceptual, programmatic, interworking and interchange. They correspond to object interfaces to human beings, to other objects or to storage media. Programmatic RPs refer to interfaces allowing logical access to functions. They correspond to intra-domain RPs in TINA. At an interworking RP communication between several systems can be established. It corresponds to an inter-domain RP in TINA. Perceptual RPs referring to interfaces between the system and the outside world, and interchange RPs constituted by interfaces bridging a system and an external physical storage medium, are not in the scope of TINA.

TINA RPs impose conformance requirements on the involved interfaces, so that there is no separation between RPs and conformance points. TINA RPs reside between generalized telecommunication stakeholders, i.e. business roles. The TINA business model (see Figure 1) defines five business roles: *consumer, retailer, third-party service provider, broker*, and *connectivity provider*, and the corresponding domains. Inter-domain RPs and intra-domain RPs are defined. Except that intra-domain RPs are located within administrative domains, they are guided by the same principles as inter-domain RPs.

In this paper, we consider inter-domain TINA RPs. Their functionalities are realized by various objects such as user agent, initial agent, provider agent, etc. Conformance requirements that are imposed on the interfaces of the involved objects are analyzed and a testing method is presented. We consider the Ret-RP between consumer and retailer as an example. In terms of telecommunication services, retailer is the service provider and consumer is the service user.

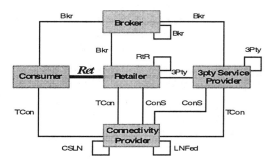

Figure 1.   TINA business model

In general, the computational viewpoint of RPs is defined by object interfaces, which are specified in computational languages. TINA RPs are specified using ODL (Object Definition Language) [9]. ODL provides syntax for structural description of objects and object groups. A formalization of behavioral specification is not prescribed.

The current TINA RP interfaces are operational meaning that the interactions at interfaces occur in form of operation invocations. With an operation invocation, a client requests the execution of some functions by the server object, which are provided by the operational interface of the server object. Typically, an operation invocation returns the termination (results or exceptions) to the client. "Oneway" operations are special case of operations that do not require a response to the client.

The Ret-RP is separated into an access part and a usage part. The access part contains interfaces that are required to establish a contractual relationship between consumer and retailer. It is referred to as the access session. The access session is service-independent. The usage part of Ret-RP captures service session related interfaces. In contrast to the access session, a service session is service specific and is built only upon an access session.

3.   RP FUNCTIONALITIES AND RP-FACETS

The whole Ret-RP is characterized by a number of functionalities being either mandatory or optional. The functionalities are in general realized by various operations at different interfaces, so that functionalities impose a logical structure on a RP: the RP can be partitioned into those operations and interfaces that are used for a selected functionality and into the rest. The logical grouping at an RP has been reflected already in the Ret-RP specification with the concept of feature sets. It has been further refined with the concept of segment[1] in [12]REF.

RP-facets reflect another structuring principle for refining TINA RPs. It has been introduced in [13]. An RP-facet is a meaningful and *self-contained* portion of functionalities. Let us have a first look at the example: the Ret-RP has the functionalities *Login, StartService, Invitation* and *Logout*, which stand in a logical dependency. For example, the *Login* is the precondition for *StartService*, what is the precondition for *Invitation* and *Logout*. Self-contained subsets are {*Login*}, {*Login, Logout*}, {*Login, StartService*}, {*Login, StartService, Logout*}, {*Login, StartService, Invitation*}, and {*Login, StartService, Invitation, Logout*}. Furthermore, *Login, StartService*, and *Logout* are mandatory for Ret-RP as they constitute the basic activities. They are therefore part of the so called core-facet. This results in two facets: the core-facet with {*Login, StartService, Logout*} and an extended facet with {*Login, StartService, Invitation, Logout*}, either of which can be implemented by a system.

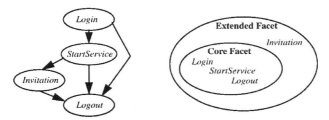

Figure 2.   Selected functionalities and facets of TINA Ret-RP

Each RP is composed of one or more RP-facets. There is a core-facet that provides a minimum set of functionalities. Additional interfaces and interactions can be specified to provide extended functionalities. A RP-facet depends on the presence of the core-facet and may depend on the presence of other RP-facets. Dependent RP-facets form a chain of cohesive sets.

The RP-facet concept facilitates conformance testing by testing the dependent RP-facets according to their order in this chain (see Section 5). Functionalities of a RP-facet are specified by the signature and behavior of operations. In addition, a RP-facet is associated with one of the architectural parts separated by the RP, referred to as *RP-facet role* (e.g. retailer or consumer).

3.1   Structures and Dependencies at a Reference Point

The initial definition of RP-facets [13] includes two relations:
- a *dependence relation* between operations (e.g. an operation can be invoked only after the invocation of another operation), and

- an *order relation* on RP-facets (e.g. a RP-facet is a subset of another one).

There are two additional kinds of dependencies: A RP offers different functionalities that are realized by sets of operations. For each functionality, a number of use scenarios are defined to be the required or forbidden patterns of behavior. The use scenarios refer to preconditional functionalities for the functionality under consideration. A functionality is in this case *causally related* to its preconditional functionalities. For example, *StartService* can be used only after a successful *Login*.

Another kind of relation exists in the case that more than one RP instances are involved in a communication, e.g. in a conference call scenario. It is possible that the involved RP instances belong to the same RP type (e.g. both are of type Ret-RP) or that they are instances of different RP types (e.g. of type Ret and of type 3Pty-RP). For example, an invitation is realized by two Ret-RP instances: one instance at the domain of the inviting consumer and the other at the domain of the invited consumer. We denote this relation between RP instances as *usage relation*. It places requirements on the test execution environment. When testing the global behavior, the consistency of events that occur across RPs (instances) has to be considered.

The structuring principles for an RP are summarized in Figure 3.

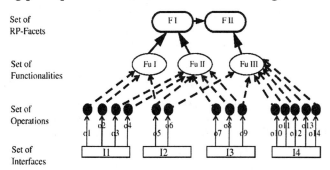

Figure 3.   RP structure

Subsequently, we give a series of definitions to reflect the dependence relations at RPs:

Definition 1:   An <u>interface</u>[2] I has a set of operations SO_I. A <u>reference point</u> R has a set of interfaces SI_R.

Definition 2:   Let o be an operation at an interface I of reference point R, i.e. $o \in SO_I$. Let $o_1..o_n$ be further operations at R, i.e. $o_i \in SO_R$, $i=1..n$. o is <u>dependent</u> on $o_1..o_n$ iff a successful invocation of o requires previous successful invocations of $o_1.. o_n$[3]. o is <u>independent</u> iff there is no operation o_k, $k \in \{1, ...n\}$ on which o is dependent.

Definition 3:   A __functionality__ Fu is a set of operations SO_{Fu} at potentially different interfaces of R. The set of operations covers all those operations that are invoked by the required use scenarios $(o_1,...,o_k)$, ... of this functionality. The set of functionalities of R is denoted by SFu_R.

A RP can be structured into functionalities in different manners. Functionalities reflect the use of a RP by its user, i.e. the service aspects are considered here. For the example in Figure 2, another possibility would be to consider one functionality for the *Login/out* behavior and the operations related to *StartService* and/or *Invitation* may comprise other functionalities.

Definition 4:   Let fu_1, $fu_2 \in SFu_R$ each with its corresponding sets of use scenarios $\{u_1,...\}$, $u_i=(o^1_{i1},..,o^1_{in})$ and $\{r_1, ...\}$, $r_j=(o^2_{j1},..,o^2_{jm})$. fu_2 is __causally related__ to fu_1 iff there is an i and j with o^2_{j1} (i.e. a start operation of a use scenario of fu_2) dependent on o^1_{in} (i.e. a successful end operation of a use scenario of fu_1). Otherwise, fu_1 is __causally unrelated__ to fu_2.

We extend the definition of a RP-facet given in [13]. RP-facets are now composed of functionalities, so that RP-facets define a "logically" complete partitioning of RPs. While functionalities can be defined in a rather flexible way, a RP-facet contains all those functionalities between which dependencies on their operations exist. Therefore, a RP-facet is self-contained.

Definition 5:   A __RP-facet__ F_R is a set of functionalities of a reference point with the following properties:

- *it is non-empty, and*

- *$\forall fu \in SFu_R$ $\forall fu' \in SFu_R$ causally related to fu: if $fu \in F_R$ then $fu' \in F_R$ (the self-containment property).*

The set of all RP-facets is denoted by SF_R.

Definition 6:   The __order relation__ $\leq$ on RP-facets uses the subset relation: $\forall F1$, $F2 \in SF_R$: $F1 \leq F2$ iff $F1 \subseteq F2$.

The relations at a RP are used to direct the conformance testing process for a RP (Section 5).

3.2   RP-Facet Specification

The unambiguous specification of RP-facet including its static and dynamic models is crucial for testability. A formal specification supports in particular automated test generation and the possibility to validate tests for their soundness against the specification. The reuse of specification parts of

RPs under test is desired as it makes test development more efficient and allows a better integration of system development with test development.

Our approach for specifying RP-facets is based on the ODL [9] for signatures of RP-facets in combination with Message Sequence Charts (MSC) [10]. Additions are needed to cover specific aspects of RP-facets according to the concepts introduced in the previous section. The specification template for RP-facets comprises:

- an indication to the related RP and the RP-facet role,
- static specification of the RP-facet in ODL, and
- behavioral specification of the RP-facet in terms of use scenarios, including representations of dependence relations in MSC.

The RP-facet specification and test case generation cycle are presented in Figure 4.

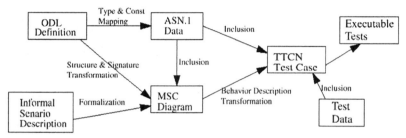

Figure 4.   RP-facet specification and test generation

MSC has been selected as it presents the communication behavior in a very intuitive and transparent manner and can be used for the behavior specification of objects defined in ODL. The standard test notation TTCN (Tree and Tabular Combined Notation) [7] is used to formulate test cases for RP-facets since there is a short step between specification and test execution. Test suites in TTCN can be easily maintained, flexibly modified and extended. ASN.1 [14] is used as a common data representation form for MSC and TTCN. Mappings for data types and constants from ODL to ASN.1 need to be defined.

3.2.1 Example Scenarios

The specification approach is applied to a (simplified) example of the TINA Ret-RP [16], where the following two features are considered:

(fa) The mandatory part of the Ret-RP's access session. It includes login and logout of a consumer with a retailer, using the operation *namedAccess* of the retailer's *i_Initial* interface, as well as start and termination of services at the retailer's *i_Access* interface in case of successful login.

(fb) An optional feature of the Ret-RP's service session part, which supports the invitation of an additional consumer. The invitation is initiated by a consumer that participates already in the multiparty service. The signature of used interactions is given below (see also Section 3.2.2).

```
module TINARet {
  interface i_ProviderInviteReq {
    void inviteUserReq (in UserDetails u, out InvitationReply rep); };
  interface i_PartyInd {
    void inviteUserInd (in UserDetails u); };
  interface i_ProviderVotingReq {
    void voteReq(in ParticipantId myId, in Vote vote); };
  interface i_PartyInfo {
    oneway void inviteUserInfo (in UserDetails u); }; };
```

The inviting consumer invokes *inviteUserReq* on the retailer's *i_ProviderInviteReq* interface with *UserDetails* of the invited consumer as input parameter. The retailer indicates the invitation by calling *inviteUserInd* of the *i_PartyInd* interface of all other consumers in the same service session. The invited consumer identifies himself by the *UserDetails* contained in *inviteUserInd*. He/she uses *voteReq* of the *i_ProviderVotingReq* interface to indicate his/her *ParticipantId* and *Vote* to the invitation. In case the invitation is accepted, all consumers except the inviting one are informed by *inviteUserInfo* on their *i_PartyInfo* interfaces.

Applying the definitions for RP-facets, we can derive a facet *Ret_Retailer_core* from *fa* representing a core RP-facet, and a facet *Ret_Retailer_add1* from *fa+fb*, which is an additional optional RP-facet.

3.2.2 Structural Specification

ODL provides means to specify on type level the system structure and relation between objects. The so called templates are used to define types for interface instances, object instances, and for object groups.

A basic concept in ODL is that of an *object instance*, which might have multiple *interface* instances of multiple interface templates. An object encapsulates its behavior and state. The access to and the visibility of the object functionality are realized in a controlled manner via the objects' interfaces. ODL contains no language features for the formal behavior description. Textual explanations are used instead.

Object instances are represented by *object templates*. Groups of objects are defined in terms of *object group templates*. They enable aggregation of object templates to increase the conceptual level at which programs can be designed. They increase also the modularity of designs.

Current TINA RP specifications use a subset of ODL, that is the OMG-IDL (Interface Definition Language) [11]. The example specification presented in Section 3.2.1 shows that it can be hardly identified which interface is provided/used by which business role. The distinction of *supported* and *required* interfaces is a build-in concept of ODL. The improved structural specification of the example is shown below. The business roles retailer and consumer are represented by two COs (Computational Objects). Each of them contains a list of required interfaces and a list of supported interfaces.

```
module TINARet {
  CO Retailer {
    requires Consumer::i_PartyInd; Consumer::i_PartyInfo;
    supports i_Initial; i_Access; i_ProviderInviteReq;
             i_ProviderVotingReq;
    interface i_Initial {...};
    interface i_Access {...};
    interface i_ProviderInviteReq {...};
    interface i_ProviderVotingReq {...}; };
  CO Consumer {
    requires Retailer::i_Initial; Retailer::i_ProviderInviteReq;
             Retailer::i_Access; Retailer::i_ProviderVotingReq;
    supports i_PartyInd; i_PartyInfo;
    interface i_PartyInd {...};
    interface i_PartyInfo {...}; }; };
```

The transformation of ODL type and constant definitions into ASN.1 representations is rather straight-forward. Please refer to [13] for details.

3.2.3 Behavioral Specification

The behavior of a RP-facet is given as use scenarios and specified by MSCs. The rules how to apply MSC in this context are given in [13] and shortly explained here.

MSCs describe patterns of interaction between a number of independent components of a system. The basic model of interaction is that of *asynchronous* communication by means of *message* passing between the components, which are called *instances*. A MSC describes the order in which interactions and other events take place.

The behavioral specification of a RP-facet is organized by a MSC document. The example MSC document (Figure 5) defines an instance kind for the core RP-facet *Ret_Retailer_core*. It contains the declaration of instances for the business roles separated by the RP, *RetailerCore* and *ConsumerCore*, and interfaces involved in the RP-facet, *i_Initial* and *i_Access*. In addition, the data language ASN.1 is declared.

ODL operations are transformed to asynchronous MSC messages to allow the representation of alternative invocation outcomes under exceptional conditions. A message corresponding to requests (by suffix *_req*) on an operation is defined. If the operation is not a "oneway" operation, a message related to replies (by suffix *_rpl*) on the operation is defined in addition. Each potential exceptional outcome (by suffix *_exp*) of an operation is translated into a separate message. Figure 5 shows examples for the operations *namedAccess* and *startService*.

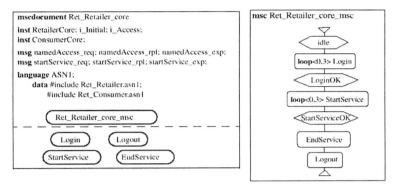

Figure 5.   MSC document and HMSC for Ret_Retailer_core

The behavior of a core RP-facet is represented by a High-level MSC (HMSC) diagram, which refers to utility MSC diagrams. In this example, the diagram *Ret_Retailer_core_msc* is supported by utility diagrams *Login*, *Logout*, *StartService* and *EndService*.

The *dependence relation* and *causal relation* are represented by MSC or HMSC expressions, where the MSC sequential operator is used (see Figure 6). For example, if *o1* needs to be invoked before *o2*, it will be represented by *M1* **seq** *M2*, with *M1* reflecting the invocation of *o1* and *M2* the invocation of *o2*. In the case that several outcomes of *o1* and/or *o2* are possible, the alternative operator **alt** in combination with conditions will be used in addition. More complex behavior definitions for RP-facets will use also parallel (**par**), loop (**loop**) and optional (**opt**) expressions.

When specifying an additional core-based RP-facet, the *order relation* is reflected by the MSC's **inherits** construct, for example *mscdocument Ret_Retailer_add1 inherits Ret_Retailer_core*. This notion allows inclusion of instances and MSCs defined in the inherited instance kind. New definitions can be added in the inheriting instance kind, for example the diagram *Invitation* (Figure 6) for *Ret_Retailer_add1*.

4.   A TEST METHOD FOR RP-FACETS

The test method for RP-facet is based on the Conformance Testing Methodology and Framework (CTMF) [6] for OSI systems and its test notation TTCN [7].

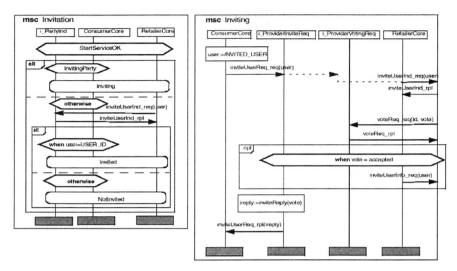

Figure 6.   MSC diagram Invitation and Inviting

REFREFCTMF test architectures are built on an asynchronous communication between *systems under test* (SUTs) and *test systems* (TS). A *point of control and observation* (PCO) is an abstract location, where stimuli are sent to the SUT and reactions of the SUT are observed, either in form of *protocol data units* (PDUs) or *abstract data types* (ASPs). In decentralized test architectures, where typically several *parallel test components* (PTCs) communicate with the SUT, co-ordinated by a *main test component* (MTC), more than one PCOs can be assigned to a PTC.

In [5] and REF[13]we elaborated the analogy of the ODP's object model and the OSI reference model that leads to rules for mapping MSC constructs to TTCN constructs. For example, MSC messages for object operations and attributes are represented by TTCN ASPs. Furthermore, two kinds of PCO are defined due to the distinction of supported and required interfaces. A supported interface of a RP-facet role is interpreted by a *client-PCO* at which request-ASPs are sent to an SUT and reply-ASPs or exception-ASPs from the SUT are observed. Whereas, a required interface of a RP-facet role is interpreted by a *server-PCO* over which request-ASPs from a SUT are received and reply-ASPs or exception-ASPs to the SUT are sent.

A test component may be associated with several PCOs of both kinds. Following the approach of component-based test systems [2], we propose to use one PTC for each emulated entity (e.g. derived from a MSC instance of the corresponding MSC diagram) that interact with the RP-facet role instance to be tested. In the example, three PTCs are derived to represent the alternative behaviors of a consumer, namely *PTC_Inviting*, *PTC_Invited* and *PTC_NotInvited* (Figure 7).

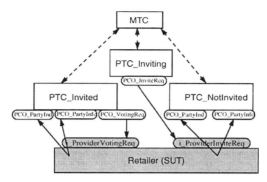

Figure 7.   Test configuration

PTC_Inviting represents the consumer's behavior as an inviting user. It uses the client-PCO *PCO_InviteReq* to access the retailer's *i_ProviderInviteReq* interface. *PTC_Invited* represents the consumer's behavior as an invited user. It uses the client-PCO *PCO_VotingReq* to access the retailer's *i_ProviderVotingReq* interface, and accepts retailer's requests on interfaces *i_PartyInd* and *i_PartyInfo* of an invited consumer at the server-PCOs PCO_PartyInd and *PCO_PartyInfo*. Finally, *PTC_NotInvited* represents a consumer who is not invited but in the same session as the inviting and invited consumers. It accepts retailer's requests on interfaces *i_PartyInd* and *i_PartyInfo* at the server-PCOs *PCO_PartyInd* and *PCO_PartyInfo*.

The PTCs are co-ordinated by the MTC. The behavior of a PTC is specified in a test step that is called when the *CREATE* construct is used to instantiate a PTC. Typically, the test purpose relevant activities are reflected in PTCs test steps as e.g. for *PTC_Invited* shown in Table 1. This test step takes two parameters for PTC instantiation. The first one is an identifier representing a consumer. The second one can be used to trigger either an „invitation is accepted" or an „invitation is refused" case. The PTC's behavior corresponds to the MSCs shown in Figure 6.

5. A TEST CAMPAIGN FOR AN RP-FACET

A test campaign for a RP-facet makes use of the structure at the RP (see also Section 3), i.e. it considers the dependence relation on operations, the causal relation on functionalities and the order relation on RP-facets. This results in a three-level test hierarchy for a RP.

Table 1.   Test step for PTC_Invited

Nr	Behavior Description	Constraints Ref	Verdict
	Test Step Dynamic Behavior		
	Test Step Name: PTC_Invited (id: UserId; vote: Vote)		
	Default: Invited_Default		
1	+SetupAccess(id)		
2	+GetRef(PCO_VotingReq, i_ProviderVotingReq)		
3	+CreateServant(PCO_PartyInd, i_PartyInd)		
4	+CreateServant(PCO_PartyInfo, i_PartyInfo)		
5	PCO_PartyInd ? inviteUserInd_req (uid:=inviteUserInd_req.uid)	inviteUserInd_req_r1	
6	PCO_PartyInd ! inviteUserInd_rpl	inviteUserInd_rpl_s1	
7	[id = uid]		
8	PCO_VotingReq ! voteReq_req	voteReq_req_s1(vote)	
9	PCO_VotingReq ? voteReq_rpl	voteReq_req_r1	(P)
10	[vote = ACCEPTED]		
11	PCO_PartyInfo ? inviteUserInfo_req	InviteUserInfo(id)	(P)
12	+ReleaseSession		
13	[vote = REFUSED]		
14	START testTimer		
15	?TIMEOUT testTimer		(P)
16	+ReleaseSession		
17	[id <> uid]		(F)
18	+ReleaseSession		

RP test groups are defined:

- per operation and will consider the availability and signature for an operation at an interface,
- per functionality and will consider the required and forbidden use scenarios for the functionality, and
- per RP-facet. They will consider the causal order of using the different functionalities of a RP- facets and will take into account the order relation with respect to other RP-facets.

Some test cases for operations are re-used by test cases for RP functionalities, which again are re-used by test cases for RP-facets.

Let us assume two RP facets F1 and F2 with F1≤F2, that F1 has been tested already, and that F2 has to be tested in addition. First, we determine the functionalities of F2 that are not part of F1. The additional functionalities use operations, which are tested first. The dependence relation on operations is considered here: in a first step, the independent operations are tested. Afterwards, those operations are tested whose preconditional operations have been tested successfully. An operation for which one of the preconditional operations failed a test, is assigned an inconclusive test verdict. The tests on dependent operations are repeated until all of them have been tested or assigned an inconclusive test verdict.

In the next phase of testing, the functionalities are tested with respect to the defined use scenarios. Indeed, only those functionalities need to be tested whose operations have passed their tests. In the last phase of testing, the RP-facet as such is considered. Test cases that reflect the sequential and parallel use of functionalities at the RP will be executed. The conformance assessment for the tested RP-facet results from considering the individual test results of operations, functionalities and their combination.

For the basic test configuration given in Figure 7, a test should check the invitation functionality if more than two customers are involved. This is reflected by the creation of several test components of PTC_Inviting, PTC_Invited, and PTC_NotInvited. The parameterization for these test components is handled by the MTC, so that the right mixture of invitations, acceptances and non-acceptances of invitations will be tested. For those types of test, we propose to make use of the upcoming TTCN3 version, which will support dynamic configurations and a flexible handling on the creation and termination of PTCs. As the syntax of TTCN3 is not fixed yet, we do not give an example here.

6. CONCLUSIONS

RPs are an established concept to structure open distributed systems and for the definition of conformance requirements. Current RP specifications e.g. in TINA tend to be too large and complex, what negatively impacts their implementability and testability.

This paper discusses structuring principles of RPs into functionalities and RP-facets. The main concept of an RP-facet is that it is a self-contained portion of an RP, so that stakeholders may implement selected RP-facets only, while preserving a level of completeness for the realized subset of the RP. A combination of existing specification methods is used to support the formal definition of RP-facets. Further, the paper analyses different relations implied by these structures. It considers the dependence relation on operations, the causal relation on functionalities and the order relation on RP-facets. These relations are used to direct an efficient testing of RP-facets.

Indeed, looking at RPs and their structure as a basis for determining test cases for distributed systems opens new ways of handling the conformance testing. In fact, a number of issues are still open. For example, we will investigate the impact of different notions of dependability of operations or the combination of use scenario based test cases in test configurations with several test components.

REFERENCES

[1]   R. V. Binder: Testing Object-Oriented Systems, Models, Patterns and Tools, Addison-Wesley, 1999.

[2]   M. Born, I. Schieferdecker, M. Li: UML Framework for Automated Generation of Component-Based Test Systems, Proc. of SNPD '00, France, May 2000.

[3]   S. Ghosh, A.P. Mathur: Issues in Testing Distributed Component-Based Systems.- In Proc. of the First Intern. ICSE Workshop on Testing Distributed Component-Based Systems, Los Angeles, U.S.A, May 1999.

[4]   GMD FOKUS: Final Subm. to TINA Conformance Testing Framework RFP, Mar 2000.

[5]   M. Li, I. Schieferdecker, A. Rennoch: Testing the TINA Retailer Reference Point, Proc. of ISADS'99, Tokyo, Japan, Mar. 1999.

[6]   ISO/IEC 9646-2: OSI Conformance Testing Methodology and Framework - Part 2: Abstract test suite specification, 1991.

[7]   ISO/IEC 9646-3: OSI Conformance Testing Methodology and Framework - Part 3: The Tree and Tabular Combined Notation (TTCN), edition 2, Dec. 1997.

[8]   ITU-T Rec. X.901 I ISO/IEC 10746-1: Information Technology - Open Distributed Processing - Reference Model: Overview, Aug. 1997.

[9]   ITU-T Z.130: Object Definition Language (ITU-ODL), Mar. 1999.

[10] ITU-T Z.120: Message Sequence Charts (MSC'2000), Nov. 1999.

[11] OMG: Common Object Request Broker Architecture (CORBA), version 2.3, 1999.

[12] OMG: Telecommunications Service Access and Subscription, Joint Revised Subm., telecom/00-02-02, Feb. 2000.

[13] Schieferdecker, M. Li, A. Rennoch: Formalization and Testing of Reference Point Facets, Proc. of FMICS'00, Berlin, Germany, Apr. 2000.

[14] Steedman, D.: Abstract Syntax Notation One (ASN.1), Techn. Appraisals Ltd., 1990.

[15] TINA-C: TINA Reference Points, version 3.1, Jun. 1996.

[16] TINA-C: Ret Retailer Reference Point Specification, version 1.1, 1999.

[17] TINA-C: TINA-CAT WorkGroup Request for Proposals, TINA Conformance Testing Framework, Version 1.0, Jul. 1999.

---

1.   The concept of segments is for specification purposes only. It has been developed in independently and in parallel with the RP-facet concept, which is also dedicated at conformance testing.

2.   An interface denotes here an interface instance of an interface type, i.e. potentially there are a number of interfaces of the same interface type at a reference point.

3.   Please note that we abstract here from the parameters in the operation invocation. Therefore, the dependence relation could be refined to cover further aspects of dependencies. For example, if operation o_2 is only executable when operation o_1 returns x, then o_2 can be defined to be result-dependent on o_1. This is for further study.

4.   For readability reason, the description is simplified at some places, e.g. the usage of timers whenever a RECEIVE event is expected is not shown.

6

GENERATION OF FEASIBLE TEST SEQUENCES FOR EFSM MODELS

Ali Y. Duale and M. Ümit Uyar
Electrical Engineering Department
The City College of the City University of New York
[duale, umit]@ee-mail.engr.ccny.cuny.edu

Abstract   A method that enables the generation of realizable test sequences from a class of EFSMs is presented. If the interdependencies among the variables used in the actions and the conditions of EFSMs are not considered during test generation, the test sequences may be unrealizable in a test laboratory. Algorithms for the detection and elimination of inconsistencies from the EFSM models are presented. Once inconsistencies are eliminated, realizable test sequences can be generated from the resulting consistent EFSM by using the methods available for FSM models.

Keywords: conformance testing, EFSM, Estelle, FSM, VHDL.

1.   INTRODUCTION

The aim of conformance testing is to detect errors and increase the level of confidence on an implementation. A major challenge in conformance testing of complex communication systems is the development of methods that automatically generate test sequences with minimum lengths while maximizing the fault coverage. The automation of test generation from the extended finite-state machines (EFSMs) is mainly impaired by the existence of *inconsistencies* among the variables used in the actions and conditions of the EFSM models.

*Prepared through collaborative participation in the Advanced Telecommunications & Information Distribution Research Program (ATIRP) Consortium sponsored by the U.S Army Research Laboratory under the Federated Laboratory Program, Cooperative Agreement DAAL01-96-2-000-2. The US Government is authorized to reproduce and distribute reprints for Government purposes notwithstanding any copyright notation thereon.

The specifications written in formal description languages such as VHDL and Estelle are typically modeled as EFSMs.

Conformance test generation from the finite-state machine (FSM) or labeled transition system (LTS) models has been an active research area [2, 6, 9, 10, 11, 13, 17, 18, 19]. However, generating feasible test sequences from the EFSMs still remains as an open research problem. Although several researchers made significant contributions towards testing EFSM models [3, 4, 12, 14, 15, 16, 19, 20], the inclusion of infeasible paths in the test sequences may be inevitable as long as the underlying models are EFSMs. Without a proper analysis of the interdependencies among the variables used in the actions and conditions of the EFSMs, considerable effort may be wasted on test generation since infeasible portions of these test sequences will be discarded later. The method presented in this paper enables the generation of only feasible test sequences from a class of EFSMs. It is assumed that the specification consists of a single process with linear actions and conditions. It is also assumed that pointers, recursive functions, and syntactically endless loops are not present in the specification.

The methodology presented here first considers the inconsistencies caused by the actions of the EFSM (if any), and proceeds with the elimination of inconsistencies among the conditions of the EFSM (if any).

The examples presented in this paper are based on VHDL. However, the algorithms are applicable to all EFSMs with the aforementioned properties. Currently, these algorithms are being applied to solve the *conflicting timers problem*, which arises when a protocol has multiple timers running concurrently. Due to the conflicting timers, a test sequence of a protocol such as the Estelle specification of the MIL STD 188-220 [8] may be interrupted by unexpected timeouts. Preliminary results show that generating test sequences for the MIL-STD 188-220 after eliminating the timer inconsistencies significantly improves the test coverage by including more transitions into the test sequences without timer interruptions [7].

The rest of the paper is organized as follows. Section 2. provides background information. Analysis of inconsistencies in EFSM models is presented in Section 3.. An example is worked out in Section 4.. Concluding remarks are given in Section 5..

2.   PRELIMINARIES

The control structure of a communication protocol can be modeled as an FSM whereas the data portion of a protocol is usually modeled as an EFSM.

An $FSM = < S, I, O, \delta, \lambda >$ where S, I, O, δ, and λ are a finite set of states, a finite set of inputs, a finite set of outputs, a state transition function such that $\delta: SxI \rightarrow S$, and an output function such that $\lambda: SxI \rightarrow O$, respectively. An $EFSM = < S, I, O, X, P, B, T >$ where S, I, and O are defined as in an FSM;

X, P, B, and T are a set of variables, a set of parameters associated with the input set, a set of Boolean expressions which reference to a memory location, and a set of actions of the EFSM that consists of spontaneous transitions (ST) and input transitions (IT), respectively.

Both FSMs and EFSMs can be represented as directed graphs. A graph $G(V, E)$ *is a structure defined by two nonempty sets, V and E*. The elements of V and E are called *nodes* (or *vertices*) and *edges*, respectively. A pair of ordered nodes, (v_i, v_j), of a *directed* graph describes an edge $e_k \in E$ which leaves v_i and terminates on v_j. The nodes v_i and v_j are called the *head* and the *tail nodes* of e_k, respectively.

A *path (P)* in $G(V, E)$ is a non-null sequence of consecutive edges. If the starting and ending nodes of a path (i.e., v_1 and v_r) are the same, the path forms a *loop*. For simplicity, throughout this paper, the term *loop* refers to the VHDL *while/for* constructs, rather than the ordinary loops. (The algorithms are applicable to both graphs with the ordinary loops and while/for loop constructs.) A node in G is assumed to be identified as the *initial node* from which all other nodes are reachable, and vice versa. $G'(V', E')$ is called a *subgraph* of $G(V, E)$ (expressed as $G'(V', E') \subseteq G(V, E)$) if:

$$V' \subseteq V \text{ and } E' \subseteq E : \forall e_i' \in E' \ tail(e_i'), \ head(e_i') \in V'$$

3.   INCONSISTENCIES

One of the major differences between the FSMs and EFSMs stems from the memory (i.e., the variables used in conditions) associated with the EFSMs. Unlike FSM graphs, the traversal of an edge $e_k = (v_i, v_j)$ of an EFSM graph mainly depends upon the condition and action variables of the edges in the paths leading to v_i. In an EFSM graph, the traversal of an edge leaving a node v_i may not be possible due to the conflicting conditions of the path(s) from v_0 to v_i, whereas for an FSM graph all paths are feasible. The complexity of testing EFSM models increases when one or more variables used in the edge conditions can assume multiple values at the same node.

A comprehensive analysis of the inconsistencies among the actions and conditions of the EFSM models is given in the subsequent sections. To improve readability, a simple example is chosen in this paper; however, the example is designed to depict the inconsistencies commonly present in real-life communication protocol specifications (e.g., MIL-STD 188-220).

Let us represent the condition of an edge as:

$$a_{00}x_0 + a_{01}x_1 + \ldots + a_{0(m-1)}x_{m-1} < op > d$$

and the action of an edge as:

$$x_i = a_{01}x_0 + a_{11}x_1 + \ldots + a_{1(m-1)}x_{m-1} + d$$

where m, op, and d are the number of variables, an operator, and a constant, respectively.

Definition 1 **Condition inconsistency**: *If there is no solution for the set of equations formed by the accumulated conditions of the edges of a sub-path $e_1 \cdot e_2 \cdot \ldots \cdot e_i$ and an edge e_j, where $head(e_j)$ can be reached from $tail(e_i)$ or $head(e_j) = tail(e_i)$, then e_i and e_j are said to have a condition inconsistency.*

Definition 2 **Action inconsistency**: *If there is no solution for the set of equations formed by the actions of an edge e_i and the condition of another edge e_j, where $head(e_j)$ can be reached from $tail(e_i)$ or $head(e_j) = tail(e_i)$, then the two edges of e_i and e_j are said to have an action inconsistency.*

Definition 3 **Consistent EFSM**: *An EFSM which is free of both action and condition inconsistencies is called a consistent EFSM.*

In a consistent EFSM, the variables used in the conditions and actions do not impose any restrictions over the paths. Hence, all paths of the graph representation of a consistent EFSM are feasible.

In this paper, a depth-first (DF) and a modified breadth-first (MBF) graph traversals are used to detect inconsistencies together with *symbolic execution* and linear programming. Note that symbolic execution is utilized as a tool to detect inconsistencies, not as a test generation tool. For the inconsistency elimination, a graph splitting technique based on symbolic execution and linear programming [1] is introduced.

Once the inconsistencies are eliminated from an EFSM graph, realizable test sequences can be automatically generated from the resulting consistent EFSM by using the test generation methods available for FSM models (see for example [2] - [4], [6], [10] - [20]).

3.1   Definitions

Let us introduce the terms used throughout the paper as:
- $V_{v_i}^{reachable} \subseteq V$: set of nodes reachable from v_i without touching v_0, the initial node.
- $E_{v_i}^{reachable} \subseteq E$: set of outgoing edges of $V_{v_i}^{reachable}$.
- $E_{v_i}^{out} \subseteq E$: set of outgoing edges of v_i.
- $E_{e_i \leadsto e_j}^{out} \subseteq E$: set of edges in the paths between $tail(e_i)$ and $head(e_j)$.
- $E_{v_i}^{in} \subseteq E$: set of incoming edges of v_i.
- $E_{v_i}^{in(T)} \subseteq E_{v_i}^{in}$: set of incoming edges of v_i that have already been traversed.
- $E_{v_i}^{in(NT)} = E_{v_i}^{in} - E_{v_i}^{in(T)}$: set of incoming edges of v_i that have not traversed.
- $Loop_{v_i} \in V$: node v_i is a loop entry/exit node.

- $V^{Loop_{v_i}} \subset V$: set of nodes that constitute the loop body whose entry/exit node is v_i.
- $E^{exit}_{Loop_{v_i}} \subset E$: set of outgoing edges of $Loop_{v_i}$ whose *tail* nodes are in the set $V^{reachable}_{Loop_{v_i}} - V^{Loop_{v_i}}$.
- $VAR = \{var_1, var_2, \cdots, var_m\}$: set of variables used in the edge conditions and actions of $G(V, E)$.
- $VAR^{con-used}_{v_i} \subseteq VAR$: set of variables used in the conditions of outgoing edges of $V^{reachable}_{v_i} \cup \{v_i\}$.
- $VAR^{dif-modified}_{v_i} \subseteq VAR$: set of variables modified differently in the paths leading to v_i.

- $\text{Cons}(e_k, e_r, v_i) = \begin{cases} 1 & \text{if traversing } e_k \text{ or where } e_r, tail(e_k) = \\ & tail(e_r), \text{ does not make } e_j \in E^{out}_{v_i} \\ & \text{infeasible} \\ 0 & \text{otherwise} \end{cases}$

For simplicity, the notations $v_{i(s)}$ and v_i as well as $e_{i(s)}$ and e_i will be used interchangeably where appropriate.

3.2   Detection and Removal of Inconsistencies

The algorithms presented in this paper eliminate the inconsistencies by creating new nodes and edges, which increases the size of the original graph. However, these new nodes and edges are created only when necessary. Therefore, although the well-known state explosion cannot be avoided for all EFSMs, the unnecessary growth of the state space is prevented. For the cases where the state explosion is unavoidable, the size of the new graph is constantly monitored as the algorithms eliminate the inconsistencies.

3.2.1   Action Inconsistencies. Variables modified in the paths leading to a node v_i may cause action inconsistencies with the condition of another edge e_r, where $head(e_r) \in V^{reachable}_{v_i}$. Therefore, the effects of the variables modified differently by the actions of the paths leading to a node v_i on the conditions of the edges reachable from v_i need to be analyzed.

In general, the effects of edge actions on variables (i.e., variable modifications) can be represented as matrices. For an EFSM graph with m variables, $var_1, var_2, \ldots, var_m$, a pair of matrices $A(mxm)$ and $\tilde{B}(mx1)$ called *the modification matrix* and *the modification vector*, respectively, are defined.

The *accumulated effects* of the actions in the paths leading to a node v_i can be represented by a set of *Action Update Matrix* pairs defined as:

$$\text{AUM}(v_i, J) = \{A_{v_i,0}, \tilde{B}_{v_i,0}, A_{v_i,1}, \tilde{B}_{v_i,1}, \cdots, A_{v_i,J-1}, \tilde{B}_{v_i,J-1}\}$$

where $A_{v_i,k}$, $\tilde{B}_{v_i,k}$, and J are the kth modification matrix, k^{th} modification vector ($0 \leq k < J$), and the number of AUM pairs associated with v_i, respectively. The symbolic values of a variable var_r are represented in the rth rows in AUM(v_i, J). Only one AUM pair, where A and $\tilde{B}$ are initialized to the identity matrix and to a zero vector, respectively, is created for the initial node.

The number of AUM pairs associated with v_i solely depends on the number of different ways in which the actions of the edges leading to v_i modify variables. If the overall variable modifications of the actions of any two paths leading to v_i are the same, only one AUM pair is sufficient to account for the effects of the actions in the two paths. Therefore, only unique AUM pairs are associated with v_i.

In general, after traversing an edge $e_k = (v_i, v_j)$, new AUM pairs are formed for v_j by applying the actions of e_k to the AUM(v_i, J). Depending upon the net modification on AUM(v_i, J) by the actions of e_k, the number of new AUM pairs formed for v_j may be less than or equal to that of v_i.

3.2.1.1   Detection of Action Inconsistencies. In this paper, a two-phase MBF graph traversal is designed to handle the detection of the action inconsistencies. The two phases of the MBF graph traversal will be referred to as P1-MBF and P2-MBF. Phase one of the MBF graph traversal, P1-MBF, can be viewed as the main graph traversal from which the P2-MBF may be invoked multiple times.

In the P1-MBF graph traversal, certain restrictions are imposed on the traversal of some edges as described below:

- The traversal of the edges of $E_{Loop_{v_i}}^{exit}$ is postponed until the edges of $(E_{Loop_{v_i}}^{out} - E_{Loop_{v_i}}^{exit})$ are traversed.
- The traversal of the outgoing edges of $v_j \in V^{Loop_{v_i}}$ is postponed until the edges of $(E_{Loop_{v_i}}^{in} - \{e_m = (v_x, v_y) : v_x \in V^{Loop_{v_i}}\})$ are traversed.
- The analysis of a loop (except for nested loops) whose entry/exit node is $loop_{v_i}$ is avoided if there is a non-traversed edge $e_m = (v_x, v_y)$, where $v_x \notin V^{Loop_{v_i}}$ and $Loop_{v_i} \in V_{tail(e_m)}^{reachable}$. Upon traversing an edge $e_k = (v_x, Loop_{v_i})$, it is checked if the following condition is true:

$$\exists \, e_r = (v_y, v_w) \in E_{v_w}^{in(NT)} \text{ such that}$$
$$(Loop_{v_i} \in V_{v_w}^{reachable} \wedge Loop_{v_i} \notin V^{Loop_{v_w}} \wedge head(e_r) \notin V^{Loop_{v_w}})$$

$$(1)$$

If statement (1) is true, the traversal of the edges of $E_{Loop_{v_j}}^{out}$ is postponed. As can be seen from (1), for nested loops, an inner loop is analyzed before traversing certain incoming edges of the outer loop entry/exit node.

In the P1-MBF graph traversal, upon traversing an edge $e_k = (v_x, v_y)$, it is checked if an action inconsistency between the edges in the paths leading to v_y and an edge $e_r = (v_w, v_z)$ reachable from v_y exists. The detection of action inconsistencies starts with checking the number of AUM pairs associated with v_y. If there is more than one, the effects of AUM(v_y, J) on the conditions of $E_{v_y}^{reachable}$ must be analyzed. Such analysis becomes complicated when the paths between v_y and v_w contain loops.

When a node with multiple AUM pairs is visited by using P1-MBF, the P2-MBF graph traversal is initiated. If there is a loop in the path(s) between v_y and $tail(e_i)$, P2-MBF aims to postpone the analysis of the effects of the AUM(v_y, J) on the conditions of $e_i \in E_{v_y}^{reachable}$ until this loop is completely analyzed. To cope with the difficulty of analyzing the effects of the AUM(v_y, J) on the conditions of $E_{v_y}^{reachable}$, P2-MBF allows the traversal of each edge of a loop body at most once. In the P2-MBF graph traversal, edges of $E_{Loop_{v_i}}^{exit}$ are not traversed until the loop whose entry/exit node is $Loop_{v_i}$ is completely analyzed by P1-MBF. In addition, unlike P1-MBF, the loop iterations are not advanced during the P2-MBF graph traversal.

In P2-MBF, the traversal of a given path terminates if at least one of the following statements is true:
- an action inconsistency is detected
- the path contains a loop entry/exit node $Loop_{v_i}$ where
  - all edges in $(E_{Loop_{v_i}}^{out} - E_{Loop_{v_i}}^{exit})$ are traversed or
  - another entry/exit node $Loop_{v_j} \in V^{Loop_{v_i}}$ is visited by traversing an edge $e_k = (v_x, Loop_{v_j})$

The action inconsistency detection algorithm is given in Figure 1, where the first *while* loop is P1-MBF and the *if* statement within the second *while* loop implicitly represents P2-MBF. The algorithm stops after finding the first action inconsistency. If there is no action inconsistency, the algorithm stops upon completing the traversal of the graph in the MBF manner.

3.2.1.2   Elimination of Action Inconsistencies.

The algorithm Figure 2 eliminates an action inconsistency between $e_k = (v_x, v_y)$ and $e_r = (v_w, v_z)$ of a loop-free graph by placing them into two separate subgraphs to prevent these two edges from being included in the same path. The two subgraphs are formed by splitting nodes and edges of $(\{v_y\} \cup V_{v_y}^{reachable})$ and $(E_{v_y}^{out} \cup E_{v_y}^{reachable})$ such that each subgraph contains either e_k or e_r but not both. When a node $v_{i(s)}$ is split, the new duplicate of v_i is denoted as $v_{i(s^*)}$, where s^* represents the number of times that $v_{i(s)}$ is split. Similarly, the duplicate of an edge e_i is denoted in the same manner (i.e., $e_{i(s^*)}$).

Only the edges of $E_{v_y}^{in} = (E_{v_y}^T \cup E_{v_y}^{NT})$ that do not conflict with the edges in $E_{v_w}^{out}$ are duplicated during the splitting. However, it is not apparent if placing

---

Action Inconsistency Detection
begin
    input: $G(V, E)$
    output: E^{conf}
    goal: to detect an action inconsistency
    Let $Conf(e_i, e_j) = 1$ if by traversing e_i, e_j becomes infeasible;
    $E^{conf} = \emptyset$;
    $v_x = v_0$;
    $V = V - \{v_x\}$;
    DONE_MBF = False;
    while $((NOT\ DONE\_MBF) \wedge (V \neq \emptyset))\{$
        while $((E_{v_x}^{out} \neq \emptyset) \wedge (NOT\ DONE\_MBF))\{$
            traverse $e_k = (v_x, v_y) \in E_{v_x}^{out}$;
            $E_{v_x}^{out} = E_{v_x}^{out} - \{e_k\}$;
            if $((VAR_{v_y}^{dif-modified} \cap VAR_{v_y}^{con-used} \neq \emptyset) \wedge$
                $(\exists\ e_r = (v_w, v_z) \in E_{v_y}^{reachable} : Conf(e_k, e_r) = 1))\{$
                $E^{conf} = \{e_k\} \cup \{e_r\}$;
                DONE_MBF = True;
            $\}$
        $\}$
        select a new node $v_x \in V$ as determined by the MBF;
        $V = V - \{v_x\}$;
    $\}$
    return E^{conf};
end

Figure 1.   Action Inconsistency detection algorithm.

copies of the edges in $E_{v_y}^{NT}$ in the same subgraph with $e_k = (v_x, v_y)$ will cause action inconsistencies. Copies of such edges are temporarily included in the same subgraph with e_k.

The effects of the actions of $E_{v_w}^{depend}$ on the outgoing edges of $v_w \in V_{tail(e_k)}^{reachable}$ must be analyzed later when each edge in $E_{v_w}^{depend}$ is traversed, where $E_{v_w}^{depend}$ is the set of edges temporarily included in the subgraph containing e_k. When an edge $e'_{i(s*)} \in E_{v'_{w(s*)}}^{depend}$ is traversed, it is checked if the actions of $e'_{i(s*)}$ are inconsistent with the conditions of the outgoing edges of $v'_{w(s*)}$. If they are, $e'_{i(s*)}$ is removed from the graph. Recall that $v'_{w(s*)}$ is one of the copies of $v_{w(s)}$ whose outgoing edges conflicted with $e_{k(s)}$.

The graph splitting is slightly different when an action inconsistent is detected by traversing $e_k = (v_x, v_y)$, where v_y is a loop entry/exit node, as will be described later.

Since the graph topology changes with the creation of the new edges and nodes, the action inconsistency detection algorithm should be re-invoked after each graph split. As defined in Section 3.1, a path from v_i to v_j cannot include v_0. Hence, the inconsistency removal algorithm does not split v_0.

As for the loop-free graphs, the action inconsistency detection is performed by using the two-phase MBF graph traversal with P1-MBF and P2-MBF. For

Action Inconsistency Elimination

input: $e_k = (v_x, v_y)$ and $e_r = (v_w, v_z)$, where an action inconsistency exists between e_k and e_r

goal: to eliminate the inconsistency between e_k and e_r by creating two subgraphs such that each subgraph contains either e_k or e_r, but not both

begin

Split $v_{y(s)}$ into two nodes as: $v'_{y(s)}$ and $v'_{y(s*)}$;

Split $\forall v_{i(s)} \in V^{reachable}_{v_{y(s)}}$ into two nodes as:

$$v'_{i(s)} \in V^{reachable}_{v'_{y(s)}} \text{ and } v'_{i(s*)} \in V^{reachable}_{v'_{y(s*)}};$$

Split $\forall e_{m(s)} = (v_{i(s)}, v_{j(s)}; L_{e_{m(s)}}) \in E^{reachable}_{v_{y(s)}}$ into two edges as:

$$e'_{m(s)} = (v'_{i(s)}, v'_{j(s)}; L_{e_{m(s)}}) \in E^{reachable}_{v'_{y(s)}} \text{ and}$$

$$e'_{r(s*)} = (v'_{i(s*)}, v'_{j(s*)}; L_{e_{m(s)}}) \in E^{reachable}_{v'_{y(s*)}};$$

Split $\forall e_{m(s)} = (v_{i(s)}, v_{y(s)}; L_{e_{m(s)}}) \in E^{in(T)}_{v_{y(s)}}$ such that $\text{Cons}(e_{m(s)}, e_{k(s)}, v_{w(s)}) = 1$, into two edges as:

$$e'_{m(s)} = (v_{i(s)}, v'_{y(s)}; L_{e_{m(s)}}) \in E^{in}_{v'_{y(s)}} \quad \text{and}$$

$$e'_{m(s*)} = (v_{i(s)}, v'_{y(s*)}; L_{e_{m(s)}}) \in E^{in}_{v'_{y(s*)}} \quad \text{where } v_{w(s)} \text{ is the node}$$

whose outgoing edges are inconsistent with $e_{k(s)}$;

Create a duplicate of $\forall e_{m(s)} = (v_{i(s)}, v_{y(s)}; L_{e_{m(s)}}) \in E^{in(T)}_{v_{y(s)}}$ such that $\text{Cons}(e_{m(s)}, e_{k(s)}, v_{w(s)}) = 0$ as:

$$e'_{m(s)} = (v_{i(s)}, v'_{y(s)}; L_{e_{m(s)}}) \in E^{in}_{v'_{y(s)}};$$

Split $\forall e_{m(s)} = (v_{i(s)}, v_{y(s)}; L_{e_{m(s)}}) \in E^{in(NT)}_{v_{y(s)}}$ into two edges as:

$$e'_{m(s)} = (v_{i(s)}, v'_{y(s)}; L_{e_{m(s)}}) \in E^{in(NT)}_{v'_{y(s)}} \text{ and}$$

$$e'_{m(s*)} = (v_{i(s)}, v'_{y(s*)}; L_{e_{m(s)}}) \in E^{depend}_{v'_{w(s*)}} \text{ (defined below)};$$

end

Figure 2. Action inconsistency elimination algorithm.

simplicity, only loops with single entry/exit nodes are considered in this paper. Note that syntactically endless loops are not considered in this paper.

If by traversing an edge $e_k = (v_x, v_y)$, of a graph with loops, an action inconsistency is detected with one of the edges in $E^{reachable}_{v_y}$, then one of the following cases is true:

- **Case 1**: $v_x \notin V^{Loop_{v_i}}$.
- **Case 2**: $v_x, v_y \in V^{Loop_{v_i}}$ and $y \neq i$ (i.e., $\text{tail}(e_k)$ is not the loop entry/exit node).
- **Case 3**: $v_x, v_y \in V^{Loop_{v_i}}$ and $y = i$ (i.e., $\text{tail}(e_k)$ is the loop entry/exit node).

The method of graph splitting for Cases 1 and 2 is similar to that of the loop-free graphs except that if $Loop_{v_i}$ is split, a copy of each edge $e_r \in E^{out}_{Loop_{v_i}}$

is included in all new subgraphs. For Case 3, the following steps are taken to eliminate action inconsistencies:

- The loop is advanced one iteration by duplicating all nodes of $V_{Loop_{v_i}}^{reachable}$ and the edges of $E_{v_y}^{reachable}$ as: $V_{v_{y(s)}}^{reachable}$, $V_{v_{y(s*)}}^{reachable}$, $E_{v_{y(s)}}^{reachable}$, and $E_{v_{y(s*)}}^{reachable}$, respectively. Since a loop must have only one entry node, $\forall\, e_r = (v_x, v_y) \in E_{Loop_{v_i}}^{in}: v_x \notin V^{Loop_{v_i}}$ are not duplicated.
- The AUM pairs of $Loop_{v_i}$ are not updated.
- The tail node of $e_{k(s)}$ is changed to $Loop_{v'_{i(s*)}}$.
- To determine the exit condition(s) of the loop, the feasibility of the conditions of each edge in $E_{Loop_{v'_{i(s)}}}^{out}$ is investigated. If the loop exit criterion is satisfied, the conditions of the edges in $\left(E_{Loop_{v'_{i(s)}}}^{out} - E_{Loop_{v'_{i(s)}}}^{exit} \right)$ become infeasible. Otherwise, the edges in $E_{Loop_{v'_{i(s)}}}^{exit}$ are infeasible. An edge whose condition is found to be infeasible is removed from the graph.

As in the case of loop-free graphs, these steps are repeated after each graph split.

Let us introduce the following definitions for any node $v_i \in V$:

- $AUM(v_i, J)[v_i \overset{SE}{\rightsquigarrow} v_x]$: the resulting $AUM(v_x, J)$ after applying symbolic execution on $AUM(v_i, J)$ in the paths between v_i and $v_x \in V$.
- $E_{v_i}^{conf(e_x \in E_{v_i}^{in}, e_r)} \subseteq E_{v_i}^{in(T)}$: set of traversed incoming edges of v_i, where each edge conflicts with e_r after obtaining $AUM(v_i, J)[v_i \overset{SE}{\rightsquigarrow} head(e_r)]$.
- $E_{e_k \rightsquigarrow e_r}^{conf(e_x \in E_{e_k \rightsquigarrow e_r}^{out}, e_r)}$: set of edges in the paths from $tail(e_k)$ to $head(e_r)$, where each edge conflicts with e_r after obtaining $AUM(tail(e_k), J)[tail(e_k) \overset{SE}{\rightsquigarrow} head(e_r)]$.
- $E_{v_i \rightsquigarrow head(e_r)}^{inf(v_i, head(e_r))}$ set of edges, in the paths between v_i and $head(e_r)$, whose conditions are infeasible after obtaining $AUM(v_i, J)[v_i \overset{SE}{\rightsquigarrow} head(e_r)]$.

After an action is detected and eliminated, the resulting graph $G'(V', E')$ is characterized as:

$$V' = (V - V_{v_{y(s)}}^{reachable} - \{v_{y(s)}\}) \cup V_I' \cup V_{II}', \text{ where}$$

$$V_I' = (\{v_{y(s)}'\} \cup V_{v_{y(s)}'}^{reachable}) \text{ and } V_{II}' = (\{v_{y(s*)}'\} \cup V_{v_{y(s*)}'}^{reachable})$$

$$E' = (E - E_{v_{y(s)}}^{reachable} - E_{v_{y(s)}}^{in}) \cup E_I' \cup E_{II}', \text{ where}$$

$$E'_I = (E^{reachable}_{v'_{y(s)}} - E^{inf(v'_{y(s)},tail(e_r))}_{v'_y \rightsquigarrow tail(e_r)}) \cup E^{in(NT)}_{v'_{y(s)}} \cup$$

$$(E^{in(T)}_{v'_{y(s)}} - E^{conf(e_x \in E^{in}_{v'_{y(s)}}, e_r)}_{v'_{y'(s)}}) \text{ and}$$

$$E'_{II} = (E^{reachable}_{v'_{y(s*)}} - E^{conf(e_x \in E^{out}_{e_k \rightsquigarrow e_r}, e_r)}) \cup E^{in}_{v'_{y(s*)}} \cup (\{e_k\} \cup E^{depend}_{v'_{w(s*)}})$$

Due to the removal of edges with infeasible conditions from $G'(V', E')$, unreachable subgraphs may result. A simple DF graph traversal can be used to eliminate unreachable subgraphs from $G'(V', E')$. Furthermore, if the condition of an edge e_i cannot be satisfied due to the actions of another edge e_j, it is removed from the graph during the condition inconsistency analysis as described next.

3.2.2   Condition Inconsistencies.

In this section, the detection and elimination of condition inconsistencies (if any) is considered, which is the next step after the action inconsistencies are eliminated from the EFSM model.

A test sequence generated from an EFSM should avoid including two or more edges with conflicting conditions. Since the conditions of the edges of a test sequence constitute a system of constraints, algorithms available for solving linear programming problems can be used in deciding whether a certain path predicate is feasible [1].

The edge conditions in a path from the starting node v_0 to a node v_i can be represented in matrices. A triplet of matrices are defined as C (mxp), $\tilde{O}P$ $(px1)$, and $\tilde{D}$ $(px1)$, where m is the number of variables, p is the number of conditions in the path from v_0 to v_i, C is the coefficient matrix, $\tilde{O}P$ is the operator vector containing the relations of $=, <, >, \neq \cdots$, etc., and $\tilde{D}$ is the scalar vector containing the scalar values of the conditions in the path.

The AUM pairs discussed in Section 3.2.1 are applied to the edge conditions of the EFSM graph as follows. A single condition of an edge $e_r = (v_i, v_j)$ is in the form of $\tilde{C} * \tilde{V}(\tilde{O}P)\tilde{D}$. The condition of e_r will be modified based on the symbolic values of the variables var_0 through v_{m-1}, which are represented by the AUM(v_i, J). The current values of the variables including all the modifications represented by an AUM pair of v_i are in the form of: $\tilde{V} = A_{v_i,k} * \tilde{V} + \tilde{B}_{v_i,k}$. Substituting $\tilde{V}$ values in an edge condition will result in $\tilde{C}(A_{v_i,k} * \tilde{V} + \tilde{B}_{v_i,k})(\tilde{O}P)\tilde{D}$, which simplifies as $\tilde{E} * \tilde{V}(\tilde{O}P)f$, where $\tilde{E} = \tilde{C} * A_{v_i,k}$ is an m-element vector and f is a scalar. An edge $e_r = (v_i, v_j)$ whose condition is infeasible based on the AUM pairs of v_i is deleted from the graph. The values assumed by the variables used in the condition of e_r can be determined from:

$$C * \tilde{V} = C * (A_{v_i,k} * \tilde{V} + \tilde{B}_{v_i,k})$$

where C is the coefficient matrix for the condition of e_r and $0 \leq k < J$, where J is the number of AUM pairs associated with v_i.

The accumulated different conditions of the paths leading to v_i can be represented by a set of *Accumulated Condition Matrix* (ACM) triplets: $\text{ACM}(v_i, J)$ = $(C_{v_i,0}, \ \tilde{OP}_{v_i,0}, \ \tilde{D}_{v_i,0}, \ C_{v_i,1}, \tilde{OP}_{v_i,1}, \ \tilde{D}_{v_i,1}, \cdots, \ C_{v_i,J-1}, \tilde{OP}_{v_i,J-1},$ $\tilde{D}_{v_i,J-1})$, where $C_{v_i,k}, \tilde{OP}_{v_i,k}, \tilde{D}_{v_i,k}$, and J are the kth coefficient matrix, k^{th} operator matrix, k^{th} scalar value matrix $(0 \leq k < J)$, and the number of the ACM triplets associated with v_i, respectively.

3.2.2.1   Detection of Condition Inconsistencies.

The condition inconsistency detection is performed by traversing the graph in a depth-first (DF) manner. Condition inconsistencies can be detected and eliminated by focusing on the ACM triplets of a node v_y, $\text{ACM}(v_y, J)$, one triplet at a time. Let $\text{ACM}(e_r)$ be the triplet representing the condition of $e_r \in E_{v_y}^{reachable}$. Furthermore, let $VAR_{e_i}^{con-used} \in VAR$ be the set of variables used in the conditions of $e_i \in E$. Once a node v_y is visited by traversing an edge $e_k = (v_x, v_y)$, each edge $e_r = (v_w, v_z)$, where $v_w \in V_{v_y}^{reachable}$, such that

$$VAR_{e_r}^{con-used} \cap VAR_{e_k}^{con-used} \neq \emptyset$$

is identified. The consistency between the conditions of e_k and e_r is then checked by concatenating $\text{ACM}(e_r)$ to the $\text{ACM}(v_y, p)$, where $0 \leq p < J$.

Let us introduce the following definition:

$$\text{Feas}(\text{ACM}(v_y, p), e_r) = \begin{cases} 1, & \text{if } \text{ACM}(v_y, p) \# \text{ACM}(e_r) \text{ has a solution} \\ 0, & \text{otherwise} \end{cases}$$

where $\#$ denotes concatenation of the two ACM triplets.

If the new constraints formed by $\text{ACM}(v_y, p) \# \text{ACM}(e_r)$ has a solution, e_r is said to be *consistent* with the edges whose conditions constitute $\text{ACM}(v_y, p)$. Otherwise e_r is *inconsistent* with the edges whose conditions constitute $\text{ACM}(v_y, p)$. The algorithm in Figure 3 stops when either a condition inconsistency is found or the DF graph traversal is completed.

3.2.2.2   Elimination of Condition Inconsistencies.

A condition inconsistency between two edges $e_k = (v_x, v_y)$ and $e_r = (v_w, v_z)$ is eliminated by splitting the nodes $(\{v_y\} \cup \{v_i \in V_{tail(v_y)}^{reachable} : v_i \rightsquigarrow v_w\})$, with their incoming and outgoing edges, into two subgraphs such that each subgraph contains either e_k or e_r, but not both. The outgoing edges of the nodes $\{v_i \in V_{tail(v_y)}^{reachable} : v_i \rightsquigarrow v_w\}$ will be referred to as $E_{v_y \rightsquigarrow v_w}^{out}$. The condition inconsistency elimination algorithm is given in Figure 4.

Since the consistency of the edges of $E_{v_y}^{in(NT)}$ with $E_{v_w}^{out}$ cannot be decided during the graph splitting, copies of these edges are included temporarily in the

Condition Inconsistency Detection

begin

    input: $G(V, E)$, **output:** E^{conf}

    goal: to detect a condition inconsistency

    $E^{conf} = \emptyset \ \ v_x = v_0, \ \ V = V - \{v_x\};$

    DONE_DF = False;

    while $((NOT \ DONE\_DF) \wedge (V \neq \emptyset))\{$

        while $((E_{v_x}^{out} \neq \emptyset) \wedge (NOT \ DONE\_DF))\{$

            traverse $e_k = (v_x, v_y) \ \in E_{v_x}^{out};$

            $E_{v_x}^{out} = E_{v_x}^{out} - \{e_k\};$

            if $((\exists \ e_r = (v_w, v_z) \in E_{v_y}^{reachable}$ where

                $(VAR_{e_k}^{con-used} \cap VAR_{e_r}^{con-used} \neq \emptyset) \wedge \ \ (\text{Feas}(ACM(v_y, p)e_r) = 0)))\{$

                $E^{conf} = \{e_k\} \cup \{e_r\};$

                DONE_DF = True;

            }

        }

        select a new node $v_x \ \in \ V$ as determined by the DF;

        $V = V - \{v_x\};$

    }

    return $E^{conf};$

end

Figure 3. Condition inconsistency detection algorithm.

same subgraph with $e_k = (v_x, v_y)$. The influence of $E_{v_w}^{depend}$ on the conditions of the outgoing edges of $v_w \in V_{tail(e_k)}^{reachable}$ must be analyzed later when each edge in $E_{v_w}^{depend}$ is traversed, where $E_{v_w}^{depend}$ is the set of edges which are temporarily included in the subgraph containing e_k. An edge $e'_{i(s*)} \in E_{v'_{w'(s*)}}^{depend}$ whose condition is found to be inconsistent with the condition of an outgoing edge of $v'_{w(s)}$ is removed from the graph.

After a condition inconsistency is detected and eliminated, the resulting graph $G'(V', E')$ is defined as:

$$V' = (V \ - \ \{v_i \mid v_i \in V_{v_{y(s)}}^{reachable} \wedge v_i \rightsquigarrow v_w\} - \{v_{y(s)}\}) \ \cup V'_I \ \cup V'_{II}$$

where $V'_I = (\{v'_{y(s)}\} \cup V_{v'_{y(s)}}^{reachable})$ and $V'_{II} = (\{v'_{y(s*)}\} \cup V_{v'_{y(s*)}}^{reachable})$

$$E' = (E \ - \ E_{v_y \rightsquigarrow v_w}^{out} - E_{v_{y(s)}}^{in}) \ \cup E'_I \ \cup E'_{II}, \text{ where}$$

$$E'_I = E_{v'_{y(s)}}^{in(NT)} \cup E_{v'_{y(s)}}^{in(T)} \ \cup \ E_{v'_{y(s)}}^{reachable} \text{ and}$$

Condition Inconsistency Elimination

input: $e_k = (v_x, v_y)$ and $e_r = (v_w, v_z)$, where a condition inconsistency exists between e_k and e_r

goal: to eliminate the inconsistency between e_k and e_r by creating two subgraphs such that each subgraph contains either e_k or e_r, but not both

begin

Split $v_{y(s)}$ into two nodes as: $v'_{y(s)}$ and $v'_{y(s*)}$;

Split $\forall\, v_{i(s)} \in V^{reachable}_{v_{y(s)}} : v_i \rightsquigarrow v_w$ into two nodes as:

$$v'_{i(s)} \in V^{reachable}_{v'_{y(s)}} \text{ and } v'_{i(s*)} \in V^{reachable}_{v'_{y(s*)}};$$

Split $\forall\, e_{m(s)} = (v_{i(s)}, v_{j(s)}; L_{e_{m(s)}}) \in E^{reachable}_{v_{y(s)}}: tail(e_{r(s)}) \rightsquigarrow v_w$ into two edges as:

$$e'_{m(s)} = (v'_{i(s)}, v'_{j(s)}; L_{e_{m(s)}}) \in E^{reachable}_{v'_{y(s)}} \text{ and}$$

$$e'_{m(s*)} = (v'_{i(s*)}, v'_{j(s*)}; L_{e_{m(s)}}) \in E^{reachable}_{v'_{y(s*)}};$$

Split $\forall\, e_{m(s)} = (v_{i(s)}, v_{y(s)}; L_{e_{m(s)}}) \in (E^{in(T)}_{v_{y(s)}} - \{e_k\})$ into two edges as:

$$e'_{m(s)} = (v_{i(s)}, v'_{y(s)}; L_{e_{m(s)}}) \in E^{in(T)}_{v'_{y(s)}} \text{ and}$$

$$e'_{m(s*)} = (v_{i(s)}, v'_{y(s*)}; L_{e_{m(s)}}) \in E^{in}_{v'_{w(s*)}} \text{ where } v_{w(s)} \text{ is the node whose outgoing}$$

edges are inconsistent with $e_{k(s)}$;

Split $\forall\, e_{m(s)} = (v_{i(s)}, v_{y(s)}; L_{e_{m(s)}}) \in E^{in(NT)}_{v_{y(s)}}$ into two edges as:

$$e'_{m(s)} = (v_{i(s)}, v'_{y(s)}, L_{e_{m(s)}}) \in E^{in(NT)}_{v'_{y(s)}} \text{ and}$$

$$e'_{m(s*)} = (v_{i(s)}, v'_{y(s*)}, L_{e_{m(s)}}) \in E^{depend}_{v'_{w(s*)}};$$

end

Figure 4.   Condition inconsistency elimination algorithm.

$$E'_{II} = E^{depend}_{v'_{w(s*)}} \cup E^{in(T)}_{v'_{y(s*)}} \cup \{e_k\} \cup (E^{reachable}_{v'_{y(s*)}} -$$

$$\{e_i \in E^{reachable}_{v'_{y(s*)}} : \text{Feas}(\text{ACM}(v_y, p)\#\text{ACM}(e_i)) = 0\})$$

As for the action inconsistency detection and removal case, the algorithms restart after each graph split.

3.3   The Complexity of the Algorithms

The complexity of the action inconsistency detection and elimination is contributed by a two-phase MBF graph traversal, constructing the number of AUM pairs for each node, and executing the linear programming for each edge for each AUM pair.

The complexity for the two-phase MBF graph traversal is $(O(E^2)$. For each node v_i, the number of AUM pairs is $\sum_1^{|V|-1} |E^{v_j \to v_i}| \times |\text{AUM}(v_j, J)|$ (where $|E^{v_j \to v_i}|$ is the number of edges from v_j to v_i such that $\exists\, e_k = (v_j, v_i)$. Linear programming takes $\min(m^2, S^2)$ steps where m is the number of variables and S is the number of constraints [1].

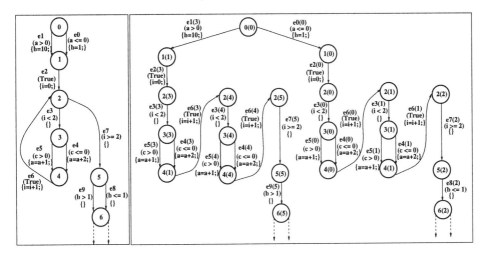

Figure 5. a) An EFSM graph *b)* The resulting EFSM graph after the action inconsistencies are eliminated from the EFSM graph shown in Figure 5.a.

The complexity for the condition inconsistency detection and elimination is bounded by the number of times the graph is split. Therefore, for the general case, the complexity of algorithms for handling the action inconsistencies is exponential with respect to the number of simple paths (i.e., the number AUM pairs). Similarly, the condition inconsistency elimination can be exponential with respect to the number of graph splits. However, based on our experience with several protocols (even with nested and/or concatenated loops), the complexity of both algorithms and, hence, the size of the consistent graph are bounded by the number of different values each condition variable assumes.

4. EXAMPLE

In the EFSM graphs in all the figures, the condition and actions of an edge are inclosed in parentheses "$(\cdots)$" and curly braces "$\{\cdots\}$", respectively.

In Figure 5 the AUM(v_1, J) associated with v_1 is based on the AUM$(v_0, 1)$ and the actions of the edges between v_0 and v_1 (i.e., e_0 and e_1). The first AUM pair of v_1 $(A_{v_1,0}, \tilde{B}_{v_1,0})$

$$A_{v_1,0} = \begin{bmatrix} 1 & 0 & 0 & 0 & 0 \\ 0 & 0 & 0 & 0 & 0 \\ 0 & 0 & 1 & 0 & 0 \\ 0 & 0 & 0 & 1 & 0 \\ 0 & 0 & 0 & 0 & 1 \end{bmatrix} \qquad \tilde{B}_{v_1,0} = \begin{bmatrix} 0 \\ 1 \\ 0 \\ 0 \\ 0 \end{bmatrix}$$

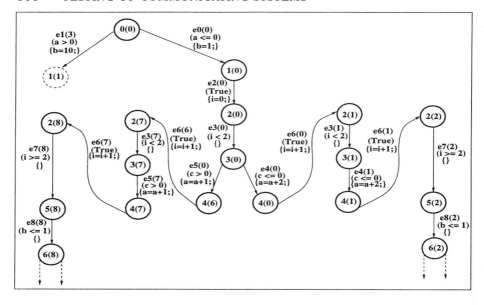

Figure 6.   The EFSM graph after the graph of Figure 5.b is split due to the condition of $e_{4(0)}$ (the subgraph starting from node $v_{1(0)}$ is shown).

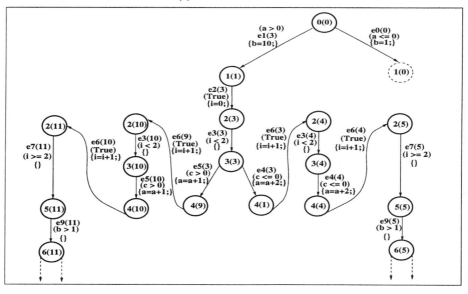

Figure 7.   The EFSM graph after the graph of Figure 6 is split due to the condition of $e_{4(3)}$ (the subgraph starting from the node $v_{1(0)}$ is identical to the subgraph starting from $v_{1(0)}$ of Figure 6).

is formed when e_0 is traversed. The values assumed by a, b, c, i, and *True* after e_0 is traversed can be determined from:

$$\tilde{V} = A_{v_1,0} * \tilde{V} + \tilde{B} \tag{2}$$

where $\tilde{V}(m\text{x}1)$ is the vector for the variables. Equation (2) yields $a=a$, $b=1$, $c=c$, $i=i$, and *True* = *True*. Similarly, due to the action associated with e_1, a new AUM pair $(A_{v_1,1}, \tilde{B}_{v_1,1})$, where $b=10$, is formed for v_1 when e_1 is traversed. P1-MBF continues until the loop is advanced two iterations as shown in Figure 5b. The P2-MBF graph traversal is invoked upon visiting v_1 of Figure 5b. P2-MBF proceeds until $v_{5(2)}$ is visited. An action inconsistency is detected between the edges leading to $v_{1(0)}$ and the outgoing edges of $v_{5(2)}$. As a result, the graph of Figure 5b is split as shown in Figure 5.b. There are no action inconsistencies in the EFSM graph of Figure 5.b. The DF graph traversal of Figure 5.b proceeds with the edges of $e_{2(0)}$, $e_{3(0)}$, $e_{4(0)}, \cdots$. The ACM triplet resulting from the traversal of the path $e_{0(0)} \cdot e_{2(0)} \cdot e_{3(0)} \cdot e_{4(0)}$ is:

$$\begin{bmatrix} 1 & 0 & 0 & 0 & 0 \\ 0 & 0 & 0 & 0 & 1 \\ 0 & 0 & 0 & 1 & 0 \\ 0 & 0 & 1 & 0 & 0 \end{bmatrix} \begin{bmatrix} \leq \\ = \\ < \\ \leq \end{bmatrix} \begin{bmatrix} 0 \\ 1 \\ 2 \\ 0 \end{bmatrix}$$

It is found that the edges $e_{4(1)}$ and $e_{5(1)}$ use the variable c in their conditions, which is also used in the condition of $e_{4(0)}$. Since the ACM($v_{4(0)}$, 1)# ACM($e_{5(1)}$) system of constraints does not have a solution, there is a condition inconsistency between $e_{4(0)}$ and $e_{5(1)}$. To eliminate this condition inconsistency, the graph of Figure 5.b is split such that $e_{4(0)}$ and $e_{5(1)}$ are placed into two separate subgraphs as shown in Figure 6.

Similarly, when $e_{4(3)}$ of the graph of Figure 6 is traversed, the system of constraints formed by ACM($v_{4(3)}$, 1)#ACM($e_{5(4)}$) has no feasible solution. As a result, the EFSM graph of Figure 6 is split. The graph of Figure 7 contains no inconsistencies.□

5.   CONCLUSIONS

The method presented in this paper enables the generation of realizable test sequences from a class of EFSMs. Due to the variable interdependencies among the actions and conditions, certain paths of an EFSM model may not be feasible. Algorithms to eliminate these inconsistencies from a class of EFSMs to produce consistent EFSMs are presented. The elimination of inconsistencies is achieved by using a graph splitting technique based on symbolic execution and linear programming. To avoid unnecessary state explosion, new nodes and edges are created only when needed during the inconsistency elimination. Although the well-known state explosion cannot be avoided for all EFSMs, the growth of the state space is limited. For the cases where the state explosion is unavoidable, the size of the new graph is constantly monitored as the algorithms eliminate the inconsistencies.

Since consistent EFSMs behave like FSMs, the FSM-based test generation techniques can be used to generate feasible test sequences from the consistent EFSMs.

Currently this methodology is being applied to generate tests for the MIL-STD 188-220 protocol [7].

References

[1] I. Adler and N. Megiddo, "A Simplex Algorithm Whose Average Number of Steps is Bound Between Two Quadratic Functions of the Smaller Dimension," *Journal of the ACM*, vol. 32, No. 4, Oct. 1985, pp. 871-895.

[2] E. Brinksma, "A Theory for the Derivation of Tests," In *Proc. IFIP Protocol Specif. Test. Verif. (PSTV)*, Amsterdam: North Holland, 1988.

[3] S. Chanson and J. Zhu, "A Unified Approach to Protocol Test Sequence Generation," Proc. of *IEEE INFOCOM* 1993, pp. 1d.1.1-1d.1.9.

[4] K. T. Cheng and A. S. Krishnakumar, "Automated Generation of Functional Vectors Using the Extended Finite State Machine Model", *ACM Trans. on Design Automation*, vol. 1, No. 1, Jan. 1996, pp. 57-79.

[5] P. D. Coward, "Symbolic Execution Systems - a Review," *Software Eng. Jour.*, Nov. 1988, pp. 229-239.

[6] A. En-Nauaary, R. Dssaouli, F. Khendek, and A. Elqortobi, "Timed Test Cases Generation Based on State Characterisation Technique," *Proc. IEEE Real-Time Syst. Symp. (RTSS)*, pp. 220-229, Madrid, Spain, Dec. 1998.

[7] M. Fecko, U. Uyar, A. Duale, P. Amer, "Test Generation in the Presence of Conflicting Timers," The IFIP 13th International Conference on Testing of Communicating Systems, Ottawa, Canada, 2000.

[8] M. Fecko, U. Uyar, P. Amer, A. Sethi, T. Dzik, R. Menell, and M. McMahon, "A Success Story of Formal Description Techniques: Estelle Specification and Test Generation for MIL-STD 188-220," In R. Lai, ed, *FDT in Practice*, vol. 23 of Comput. Communic. (special issue), Spring 2000 (in press).

[9] S. Fujiwara, G. Bochmann, F. Khendek, M. Amolou, and A. Ghedamsi, "Test Selection Based on Finite State Machine Models," *IEEE Trans. on Software Engr.*, 17(6):591-603, June 1991.

[10] T. Higashino, A. Nakata, K. Taniguchi, and A. R. Cavalli, "Generating Test Cases for a Timed I/O Automaton Model," In *Proc. IFIP Int'l Workshop on Test Commun. Syst. (IWTCS)*, pp. 197-214, Budapest, Hungary, Sept. 1999.

[11] D. Hogrefe, "On the Development of a Standard for Conformance Testing Based on Formal Specifications," *Comput. Stand. Interf.*, 14(3):185-190, 1992. Test Cases from SDL Specifications," SDL '89: The Language at Work, Proc. Fourth SDL Forum, pp. 267-279, 1989.

[12] X. Li, T. Higashino, M. Higuchi, and K. Taniguchi, "Automatic Generation of Extended UIO Sequences for Communication Protocols in an EFSM Model," In Proc. of 7th Int'l Workshop on Protocol Test Systems, Tokyo, Japan, Nov. 1997, pp. 225-240.

[13] G. Luo, G. Bochmann, and A. F. Petrenko, "Test Selection Based on Communicating Non-deterministic Finite State Machines Using a Generalized Wp-Method," *IEEE Trans. on Software Engr.*, 20(2):149-162, 1994.

[14] R. Miller and S. Paul, "Generating Conformance Test Sequences for Combined Control Flow and Data Flow of Communication Protocols," In Proc. of 12th International Symposium of Protocol Specification, Testing, and Verification, 1992, pp. 12-27.

[15] K. Salah, H. Ural, and A. Williams, "Test Generation Based on Control and Data Dependencies within Systems Specified in SDL," (to appear in Computer Communications).

[16] B. Sarikaya, G. Bochmann, and E. Cerny, "A Test Design Methodology for Protocol Testing," *IEEE Trans. on Software Eng.*, Vol. SE-13, No. 5, May 1987, pp. 518-531.

[17] J. Tretmans, "Conformance Testing with Labelled Transitions Systems and Test Generation," *Comput. Networks ISDN Syst.*, 29(1):49-79, 1996.

[18] H. Ural, "Formal Methods for Test Sequence Generation," *IEEE Trans. on Commun.*, 39(4):514-523, 1992.

[19] H. Ural, "A Test Derivation Method for Protocol Conformance Testing," Proc. Sixth Int' Conf. Protocol Specification, Testing and Verification, 1989, pp. 347-358.

[20] H. Ural and B. Yang, "A Test Sequence Selection Method for Protocol Testing," *IEEE Trans. on Communications*, vol., 39, No., 4, Apr. 1991, pp. 514-523.

[21] M. U. Uyar and A. Y. Duale, "Resolving Inconsistencies in VHDL Specifications," *In Proc. IEEE Milit. Commun. Conf. (MILCOM)*, Atlantic City, NJ., Oct. 1999 , No. 5.1.3.

[22] M. U. Uyar and A. Y. Duale, "Modeling VHDL Specifications as Consistent EFSMs," *In Proc. IEEE Milit. Commun. Conf. (MILCOM)*, Monterey, CA., Oct. 1997, pp. 740-744.

PART III

INTEROPERABILITY TESTING OF INTERNET PROTOCOLS

7

EXPERIMENTS ON IPV6 TESTING

Tibor Csöndes, Sarolta Dibuz, Péter Krémer
Ericsson Research
Conformance Center, Ericsson Ltd.
H-1037 Budapest, Laborc u. 1., Hungary
E-mail: { Tibor.Csondes, Sarolta.Dibuz, Peter.Kremer } @eth.ericsson.se

Abstract   IPv6 lays the groundwork for the next generation of networking, and IPv6 implementations will play vital importance in network-dependent business in the future. This vital role enhances the need for quality and interoperability assurance of IPv6 implementations for which automated testing with test tool – and vendor independent test suites is a promising solution.

The paper presents experiences in testing IPv6 implementations automatically with test cases written in TTCN. The paper also shows via examples how the conformance type of testing can be used to analyze errors occurring during interoperability testing.

Keywords: Internet, IPv6, TTCN, Conformance test, Interoperability test

1.   INTRODUCTION

The convergence of the telecom and Internet technologies brings up several questions, one of them is bridging the gap between the different testing approaches of the telecom and datacom world. Reliability and robustness of telecommunication products are assured through test campaigns during which the products are conformance tested according to a strictly standardized process and in most of the cases with standardized test suites.

A test notation was also standardized for writing test tool independent test suites that can be applied by all vendors. Comparable and repeatable test results are gained as the same test suites are used for different implementations and the conditions of testing are described in documents. TTCN (Tree and Tabular Combined Notation) [10], [1] was developed for conformance testing and Abstract Test Suites (ATS) have been standardized for several telecommunication protocols in TTCN mainly by ETSI and ITU-T. Most of the telecom vendor companies use TTCN in their internal test procedures because of its usefulness in protocol testing. Tools exist supporting the writing and execution of TTCN test cases as well as generating them from formal specifications written in SDL (Specification Description Language) or MSC (Message Sequence Chart).

Telecom service providers are also executing conformance testing on the products they buy to check if it fulfills the requirements described in the protocol standards. This is a main step assuring the interoperability of the heterogeneous products built in their network. In the datacom world this kind of strict testing is not applied though interoperability of products of different vendors is becoming to be as important as for telecom networks with the growing number of datacom product vendors and the growing reliability requirements. The interoperability events use ad hoc test scenarios, and no test results are really available for the potential customers to support their decision which product to buy.

This paper presents an application of the conformance testing methodology on IPv6 testing. In section two we present the main features of IPv6. Section 3 summarizes ideas why conformance testing would be useful for the Internet protocols. In section 4 we describe the test configuration we have used, in section 5 the test cases and in section 6 the test results of conformance testing of some IPv6 implementations. In section 7 we compare the test results with interoperability problems of the examined IPv6 implementations. Section 8 is the conclusion.

2.   IPV6 OVERVIEW

IP version 6 (IPv6) [3] is a new version of the Internet Protocol designed as the successor to IP version 4 (IPv4). The changes from IPv4 to IPv6 fall primarily into the following categories:

- *Expanded Addressing Capabilities:* IPv6 increases the IP address size from 32 bits to 128 bits (increase the address space by a factor of 2^{96}), to support more levels of addressing hierarchy, a much greater number of addressable nodes, and simpler auto-configuration of addresses. The scalability of multicast routing is improved by adding a "scope" field to multicast addresses. A new type of address called an "anycast address" is defined, used to send a packet to any one of a group of nodes.

- *Header Format Simplification:* Some IPv4 header fields have been made optional or have been dropped, to reduce the common-case processing cost of packet handling and to limit the bandwidth cost of the IPv6 header. This simplifies and speeds up router processing of IPv6 packets compared to IPv4 datagrams. The IPv6 packet header is fixed-length whereas the IPv4 header is variable-length. Again, the IPv6 design simplifies processing.

- *Improved Support for Extensions and Options:* Changes in the way IP header options are encoded allows for more efficient forwarding, less stringent limits on the length of options, and greater flexibility for introducing new options in the future.

- *Flow Labeling Capability:* A new capability is added to enable the labeling of packets belonging to particular traffic "flows" for which the sender requests special handling, such as non-default quality of service or "real-time" service. IPv4 provides minimal assistance in this area.

- *Authentication and Privacy Capabilities:* Extensions to support authentication, data integrity, and (optional) data confidentiality are specified for IPv6. IPv4 provides no security capabilities other than an optional security label field.

2.1   IPv6 Header Format

An IPv6 protocol data unit has the general form shown in Fig. 1.

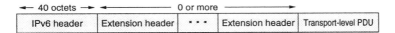

Figure 1.   IPv6 PDU form

The only header that is required is referred to simply as the IPv6 header (Fig. 2). This is of fixed size with a length of 40 octets, compared to 20 octets for the mandatory portion of the IPv4 header.

Ver	Class		Flow Control	
Payload Length			Next Head	Hop Limit
Source Address				
Destination Address				

Figure 2.   IPv6 Header

The fixed 40 octets length of the IPv6 header consists of the following fields (Fig. 2):

- *Version (4 bits):* IP version number; the value is 6.

- *Traffic class (8 bits):* The Class field lets the originating host or the forwarding router identify the class or priority of the packet.

- *Flow label (20 bits):* The Flow Control field lets the source host label a sequence of packets (i.e., a flow) that requires special handling by intermediate routers when the packets travel from source to destination.

- *Payload length (16 bits):* Length of the remainder of the IPv6 packet following the header, in octets. In other words, this is the total length of all the extension headers plus the transport-level PDU.

- *Next header (8 bits):* Identifies the type of header immediately following the IPv6 header.

- *Hop limit (8 bits):* The remaining number of allowable hops for this packet. The hop limit is set to some desired maximum value by the source, and decremented by 1 by each node that forwards the packet. The packet is discarded if hop limit is decremented to zero. This is a simplification over the processing implied for the time-to-live field of IPv4. The consensus was that the extra effort in accounting for time intervals in IPv4 added no significant value to the protocol. In fact, IPv4 routers, as a general rule, treat the time-to-live field as a hop limit field.

- *Source address (128 bits):* The address of the originator of the packet.

- *Destination address (128 bits):* The address of the intended recipient of the packet. This may not in fact be the intended ultimate destination if a *Routing header* is present.

2.2   IPv6 Extension Headers

The following extension headers have been defined (the numbers show the suggested order):

1. *Hop-by-hop options header:* Defines special options that require hop-by-hop processing
2. *Destination options header[1]:* Contains optional information to be examined by the destination node
3. *Routing header:* Provides extended routing, similar to IPv4 source routing
4. *Fragment header:* Contains fragmentation and reassembly information
5. *Authentication header:* Provides packet integrity and authentication
6. *Encapsulating security payload header:* Provides privacy
7. *Destination options header[2]:* For options to be processed only by the final destination of the packet

In Fig. 3 we can find an example of an IPv6 packet's extension headers.

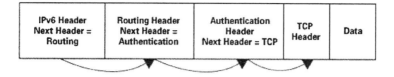

Figure 3.   IPv6 packet's extension headers

2.3   Internet Control Message Protocol Version 6 (ICMPv6)

ICMPv6 [2] is the control protocol of IPv6. It contains error and informational messages. One group of the messages describe the different error types (the numbers indicate the message types):

1 Destination Unreachable
2 Packet Too Big
3 Time Exceeded or Hop Limit Exceeded
4 Parameter Problem

The informational messages are mainly used for diagnostic purposes (the numbers indicate the message types):

128 Echo Request
129 Echo Reply

A **Destination Unreachable** message should be generated by a router, or by the IPv6 layer in the originating node, in response to a packet that cannot be delivered to its destination address for reasons other than congestion.

A **Packet Too Big** must be sent by a router in response to a packet that it cannot forward because the packet is larger than the MTU (Maximum Transmission Unit) of the outgoing link. The information in this message is used as part of the Path MTU Discovery process [8].

If a router receives a packet with a `Hop Limit` of zero, or a router decrements a packet's `Hop Limit` to zero, it must discard the packet and send an ICMPv6 **Time Exceeded** message with `Code` 0 to the source of the packet. This indicates either a routing loop or an initial too small `Hop Limit` value.

If an IPv6 node processing a packet finds a problem with a field in the IPv6 header or extension headers such that it cannot complete processing the packet, it must discard the packet and should send an ICMPv6 **Parameter Problem** message to the packet's source, indicating the type and location of the problem.

Every node must implement an ICMPv6 Echo responder function that receives **Echo Requests** and sends corresponding **Echo Replies**. A node should also implement an application-layer interface for sending **Echo Requests** and receiving **Echo Replies**, for diagnostic purposes.

3.   ADVANTAGES OF CONFORMANCE TESTING IPV6

There has been trials for applying conformance testing for Internet protocols [7], [9], [4], [5] and these papers show that it is possible to use the methodology with minor modifications for non-OSI protocols like those Internet protocols described in the above mentioned references. On interoperability events the cause of several errors could be found more easily if conformance test was executed on every product beforehand. This would make interoperability testing more efficient.

This is true for testing IPv6 implementations as well, and as these network elements will play a key role in next generation networks their conformance to the protocol requirements is also a key issue. To achieve a smooth IPv4 to IPv6 migration it is essential that the IPv6 nodes support those features that enable them to communicate with IPv4 ones. This can be assured with executing conformance tests on the appropriate features.

There are initiatives already existing for conformance testing of IPv6. The TAHI project [12] develops conformance and interoperability tests for IPv6. Unfortunately, most of the documentations are in Japanese. Only partial information is available in English. The test configuration consists of one testing node, which has to have multiple connection to IUT (two different Ethernet segments and one serial line). There are some other minor restrictions on IUT, such as its global address, root password, shell and prompt. These limitations make the tests platform specific and the execution more difficult.

Another interesting approach is made by [13], which described test purposes for IPv6 in a similar way as ETSI produces the Test Suite Structure and Test Purpose standards for protocols. These test purposes can be implemented in any language a test tool can execute. Using a standardized test notation like TTCN gives the advantage to be able to execute the test with several test tools.

4.   TEST METHODS AND CONFIGURATION

We have used the remote abstract test method that do not contain Upper Tester. We have taken the advantage of open source operating systems like Linux. We have developed a kernel module, which helped us to create and capture any kind of Ethernet packets (i.e. an IPv6 packet with non-existing Ethernet addresses). The main difference to packet filters is that instead of making copies of the original packets, this module redefines a packet's route in the kernel. Doing so we can work with the "original" packets and do not have to cope with packet losses, filter errors or other ambiguities.

For the execution of our tests we have used System Certification System (SCS), which is a TTCN test executor developed in Ericsson. SCS consists of a set of tools: TTCN Translator translates MP files into EXTEL format, which is then used by Test Component Executor (TCE), which interprets and executes test cases selected from TTCN Manager. The most interesting feature of SCS is the Test Port concept. All Abstract Service Primitives (ASPs) of a given Point of Control and Observation (PCO) have to be mapped onto the Service Primitives (SPs) of the underlying service provider in order to exchange data with IUT. Since different protocols can be based on different services, which use different types of connections. This mapping needs serious programming skills. In order to simplify the task, SCS provides the so called Test Ports. A Test Port is a piece of software that holds routines, which are called after an ASP is received from or before ASPs are sent to TCE. Thus, we get the full control over the SPs.

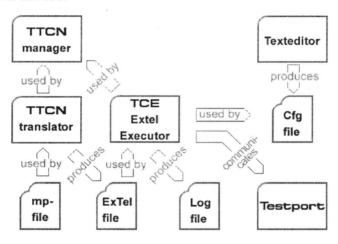

Figure 4. The structure of SCS

We have developed a new test port for IPv6 tests, which can provide two ASPs for communication with the socket interface: "recv" can only be received, and "send" that can only be sent. IPv6 PDUs are carried by these ASPs through the Test Port. The checksums are calculated in the Test Port, so we do not have to waste time on defining them in the constraints.

5. TEST CASES

The main objective of this work is to show the applicability of TTCN in the field of testing Internet Protocols, especially IPv6. The test purposes for the base specification of IPv6 are written in [13]. It contains 43 test purposes (for the base specification) altogether, which we have arranged into the following five groups:

1. Extension header processing (9 test cases)
2. Option processing (11 test cases)

3. Routing header processing (7 test cases)
4. Reassembly (12 test cases)
5. Fragment header processing (4 test cases)

We have written TTCN test cases for these test purposes. Besides the correct behaviour they contain test cases for invalid events (inopportune test cases are not developed yet). We have developed the necessary tools to be able to execute the tests and evaluated three different IPv6 implementations. Moreover, the next sections will show minor and major problems we have found.

For demonstrating the structure of the test cases let us see an example (Fig. 5). This test case (unrecognized next header) verifies that a node discards a packet containing an unknown value in the `Next Header` field and transmits an ICMPv6 **Parameter Problem** message to the source of the packet.

Test Case Dynamic Behaviour					
Test Case Name : unrecognized_next_header					
Group : Base_Specification/Host/					
Purpose :					
Configuration :					
Default : DEFAULT_BASE (TR1_lila_addr, TR1_addr, '1'B)					
Comments :					
Nr	Label	Behaviour Description	Constraints Ref	Verdict	Comments
1		+RESET			
2		+GEN_PREAMBLE (TR1_lila_addr, TR1_addr, TR1_addr_solnode, '1'B)			
3		IPv6 ! send (paylen:=0) START T1s	s_head (TR1_lila_addr, NUT_lila_addr, ipv6_header_s (138, TR1_addr, NUT_addr))		
4		IPv6 ? recv CANCEL T1s	r_head_data (NUT_lila_addr, TR1_lila_addr, ipv6_header_r (NH_ICMPV6, NUT_addr, TR1_addr), parameter_problem_r (1, 6))	(P)	
5		+GEN_POSTAMBLE (TR1_lila_addr, TR1_addr)			
6		?TIMEOUT T1s		(F)	
7		+GEN_POSTAMBLE (TR1_lila_addr, TR1_addr)			
Detailed Comments :					

Figure 5.　Example of a Test Case behaviour

The structure of the test case is very simple, we were following the traditional OSI concept: Preamble, Test Body, Postamble, and - of course - the Default. First, we reset the IPv6 test port and then start the GEN_PREAMBLE test step, which is generally used in every test case. It contains the following:

- sending a **Router Advertisement** message,

- sending an **Echo Request** message (containing 1400 bytes of data),

- upon receiving a **Group Membership Report** or a **Group Membership Reduction** message we discard it,

- upon receiving a **Neighbor Solicitation** message we send the appropriate **Neighbor Advertisement** message (with valid IPv6 and link layer addresses),

- waiting for an **Echo Reply** message, which contains the very same values in the fields of `Identifier`, `Sequence Number` and `Echo Data` as we sent out.

In this preamble we inform IUT that a new router (a TTCN test component) is attached to its segment. To be able to address each other the exchange of the link layer (Ethernet) addresses is necessary, this is done by **Neighbor Solicitation** and **Neighbor Advertisement** messages. The arrival of the **Echo Reply** assures us that we can communicate with IUT and test is ready to begin.

In the main body, after the preamble, we send an IPv6 Header with the unknown value of 138 in the `Next Header` field. Then, we wait for a **Parameter Problem** message whose ICMPv6 `Code` field equals to 1 (unrecognized `Next Header` type encountered) and ICMPv6 `Pointer` field equals to 6 (offset of the `Next Header`).

At the end we close the test case with GEN_POSTAMBLE, which sends a **Router Advertisement** message to the so-called "All-Nodes" multicast address. In this message, a value of zero in the `Router Lifetime` field indicates that this router cannot be used any more as a default router. This only message cleans up the whole mess that was caused by the test case.

The DEFAULT_BASE step is also generally used in almost every test case like GEN_PREAMBLE. In this default step we are waiting for a **Neighbor Solicitation** message and after receiving it we send back an appropriate **Neighbor Advertisement**. This procedure is important in default step because an IPv6 implementation can send **Neighbor Solicitation** messages at any time – in order to ensure its neighbors' reachability – and we have to reply immediately.

6. CONFORMANCE TEST RESULTS

In this section we give detailed description on the results of the test execution on three different platforms (FreeBSD, Linux, Solaris). There is not too much to say about the passed test cases, so we will mainly expound the test cases that failed. The summary of the results is shown in Table 1.

Table 1.   Results of Test Cases

Test Groups	FreeBSD PASS	FreeBSD FAIL	Linux PASS	Linux FAIL	Solaris PASS	Solaris FAIL
Extension header processing	9		8	1	8	1
Option processing	11		11		11	
Routing header processing	6	1	5	2	3	4
Reassembly	12		6	6	11	1
Fragment header processing	4		3	1	4	
Total	42	1	33	10	37	6

6.1   FreeBSD

FreeBSD 2.2.8 was used together with `KAME kit RELEASE 19991213`. This was the most reliable and most usable implementation. The installation and configuration was also the easiest. Only one test case gave fail verdict during the execution. The purpose of the test case is to wait for a **Hop Limit Exceeded** message after sending a packet with `Hop Limit = 1` and its *Routing header* contains two more segments. The implementation does not forward the packet (which is the correct behaviour) but omits to send back the appropriate error message. Note that none of the implementations have passed this test case.

6.2   Linux

The Linux kernel version 2.2.12 was used together with net-tools-1.52 (it contains the necessary tools for network configuration).

Unfortunately, Linux has major deficiencies in the process of the *Routing header*. Let us see two examples: a node must discard a packet whose *Routing header* contains a multicast address at any position. Contrary to this, Linux forwarded the packet to a multicast address. None of the examined implementations send back a **Hop Limit Exceeded** message when they receive a packet with the value of 1 in the `Hop Limit` field. Linux processed the *Routing header* because it sent back **Parameter Problem** messages with correct values but not always correctly (see the examples above).

There are problems with the reassembly process as well. The implementation recognized the situations where the `Fragment ID` field, the source address or the destination address[1] of the fragments were different.

Linux was also able to handle out-of-order packets but failed one test case according to the inappropriate use of the `m` bit (this flag indicates the existence

---

[1] A node should have more than one address and must differentiate one from the other. If it receives one half of an **Echo Request** to one if its addresses and the other half of the message to another of its addresses, an **Echo Reply** should not be sent.

of more fragments) in the *Fragment header*. A fragmented packet is acceptable if it consists of only one fragment, which means that both the `m` bit and the `Fragment Offset` field has a value of 0. A proper implementation must send an **Echo Reply** to a single-fragmented **Echo Request** but Linux did not behave so.

The most serious problems were generated by the "Time Exceeded" timer. Normally, it is set to 60 seconds, which means that all segments must be received within 1 minute. On the reception of the first segment of a fragmented packet every node starts a 60 second-long timer. When this timer expires and the node did not received all the segments a **Time Exceeded** message must be sent. During the execution we measured 33, 34 and 36 seconds instead of 60 seconds.

This implementation did not keep an eye on the size of the reassembled packet. A `Payload Length` of 278 and a `Fragment Offset` of 65528 would exceed 65535 (which is the maximum size of an IP packet) but Linux did not send a **Parameter Problem** message in this case. The implementation did not send this message and neither did it when a fragment contained an unknown `Destination Option`. However, Linux responded with a **Parameter Problem** message to the same invalid option when it was in a single packet.

6.3   Solaris

Solaris 7.0 (SunOS 5.7) was extended with patch 107788-01, which contains the prototype of IPv6 core functionality. The prototype was stable and worked very well except the handling of *Routing headers* and packet forwarding.

The implementation is not able to recognize invalid or erroneous values in the following fields: `Routing Type`, `Segments Left`, `Header Extension Length`. Solaris passed only those *Routing header* related test cases which did not require the process of the *Routing header* at all. *Routing header* appears in nine test cases and only three of them gave `PASS` verdict (the rest failed).

Two of the passed test cases contain zero in the `Segments Left` field, which means that there are no useful information in the *Routing header* any more (the packet already reached the given destinations). A correct implementation must skip the *Routing header* and proceed to process the next header. The only difference between the two test cases is that in one case the `Routing Type` contains a valid value, in the other case it contains an invalid value. The correct behaviour is the same, an error-free implementation must not send a **Parameter Problem** message in the latter case. There is a multicast address in the *Routing header* in the third test case. The RFC says that a node must silently discard such a packet.

It can be seen very easily that among the above mentioned conditions a correct implementation and a faulty one (which skips every *Routing header* irrespective of its content) behave the very same. Moreover, this implementation has not forwarded the packets which were correctly source routed, either. Based on these signs we came to the conclusion that the IPv6 prototype of Solaris totally ignores the *Routing header*.

7.   EFFECTS OF THE ERRORS ON INTEROPERABILITY CAPABILITIES

In this section we show the experienced and the assumed effects of the found errors on interoperability capabilities of each implementation. Since a thorough interoperability test requires much larger network than conformance test, we had to assume the effect of a certain error. In the following we will conclude from the errors found during the conformance testing on possible errors revealed in interoperability testing. We think that some of these bugs would stay undetected using the traditional[2] interoperability testing.

7.1   FreeBSD

The default value of the `Hop Limit` field is set by the sender's default routers, which means that the increase of the `Hop Limit` by the node itself is not possible. Besides this, an implementation must pass the incoming **Hop Limit Exceeded** messages to the upper-layer process. It can be seen that the degree of the problem depends on the upper-layer process heavily. In case it maintains a timer (or more timers) for the purpose of destination reachability, then the lack of such a message is unable to cause severe problems. But in the other case – when the upper-layer count on this message – the situation could be more dangerous. An interoperability test can detect this fault only in the case of using certain upper-level applications. The analysis of the traces is also possible if the huge amount of traces of *tcpdump* enables.

7.2   Linux

There are two problematic fields in Linux's IPv6 code. The first one is the already presented reassembly timer, which is set to 60 seconds in normal case. Imagine the following situation: there is a slow or congested link between the two communicating endpoints and the last fragment of a packet arrives 40 seconds after the first. Linux – as the receiver – sends back a **Time Exceeded** message and drops the arrived fragments after 33-36 seconds. If the delay is permanent this packet will never be delivered! The exploration of this bug

---

[2]In this case there are no test scripts and no packet generators. Implementors interconnect their implementations and try them out. The main focus is on interworking, thus they do not examine the correct behaviour either the implementation's answer to faulty attitude or erroneous packets. For example the error qwith the reassembled packet's size in Linux would have been stay undetected.

presumes slow network connection and an application or a method, which is able to split the packets into more fragments.

The most exciting feature of Linux is that it can act as a traffic generator. Sometime Linux floods the network with hundreds of unnecessary packets but only those times when the IPv6 module is loaded into the kernel. We tried other versions of the kernel (2.2.14 and 2.3.47) and re-installed the OS but nothing helped. Very rarely Linux got into an infinite loop, it sent a valid message (for example an **Echo Request**), received the correct answer and started it over again. In other cases it sent out the packet for 3-5 seconds and then stopped. This phenomenon has several impacts: the network is temporarily down and telnet or other types of connections freeze during that time; if the repeated packet is a correct one, then it may totally confuse all computers (including itself as well) on that link; since this symptom occurs rather frequently all connections slow down. Most of the time the user can only see that the network is slow but nothing else. It is quite easy to notice this "feature" since everything slows down and the same packet appears many times in the traces. Certainly, this is the biggest and the most important problem that need to be fixed very soon.

7.3 Solaris

However, Solaris gave more `FAIL` verdict than FreeBSD, all of them were related to the using of the *Routing header* and thus the process of source routing. Note, that source routing is a restriction: a packet must go through given routers or nodes (they are supplied by the *Routing header*. Besides, source routing is not used generally, an implementation can easily communicate with any other node without using it. If the ISP's (Internet Service Provider) policy or the network's topology is not the reason for source routing, the lack of this feature is acceptable. Since this method is rarely used and it is hard to create such packets without a packet generator, the detection of this error in an interoperability test could be very difficult.

8. CONCLUSION

We strongly believe that more thorough testing of datacom products on the protocol requirements defined in the RFCs is becoming more and more important with the growth of the datacom market and with the appearance of IP-based telecommunication networks. Operators used to the traditional telecom market will most probably support the idea of using test suites to trace the capabilities and requirements of products in interoperability events. Using common test suites has also the advantage to produce comparable and repeatable results.

Test suites for the Internet protocols contain usually less test cases so test suites are not as compound as standardized test suites of telecom protocols, like

ISDN. For the success of this test approach in the Internet world the de facto type of test standardization should be worked out to be able to follow the quick changes of the protocol standards with the tests.

TTCN is further developing. TTCNv3 will be standardized in ETSI until the end of 2000. It will be more easy to learn and understand the new version of TTCN and it will be suitable for not only conformance testing but also for other testing types. This shows further possibilities to utilize test suites written for protocol conformance test in function testing and in end-to-end testing of systems. These new features of TTCN will further strengthen its application in testing next generation networks.

References

[1] B. Baumgarten, A. Giessler: OSI conformance testing methodology and TTCN, North Holland, 1994.

[2] A. Conta, S. Deering: Internet Control Message Protocol (ICMPv6) for the Internet Protocol Version 6 (IPv6) Specification, RFC 2463, IETF Network Working Group, December 1998.

[3] S. Deering, R. Hinden: Internet Protocol Version 6 (IPv6) Specification, RFC 2460, IETF Network Working Group, December 1998.

[4] R. Gecse: Conformance testing methodology of Internet protocols, Testing of Communicating Systems, Tomsk, Russia, September 1998.

[5] R. Gecse, P. Krémer: Automated test of TCP congestion control algorithms, Testing of Communicating Systems, Budapest, Hungary, September 1999.

[6] Robert M. Hinden: IP Next Generation Overview, Communications of the ACM, Vol. 39, No. 6, June 1996.

[7] T. Kato, T. Ogishi, A. Idoue and K. Suzuki: Design of protocol monitor emulating behaviours of TCP/IP protocols, Testing of Communicating Systems, Cheju Island, Korea, September 1997.

[8] J. McCann, S. Deering, J. Mogul: Path MTU Discovery for IP version 6, RFC 1981, IETF Network Working Group, August 1996.

[9] T. Ogishi, A. Idoue, T. Kato and K. Suzuki: Intelligent protocol analyzer for WWW server accesses with exception handling function, Testing of Communicating Systems, Tomsk, Russia, September 1998.

[10] OSI - Open System Interconnection, Conformance testing methodology and framework, ISO/IEC 9646, 1997.

[11] William Stalling: IPv6: the New Internet Protocol, http://www.cs-ipv6.lancs.ac.uk/ipv6/documents/papers/stallings/

[12] TAHI project, http://www.tahi.org/

[13] University of New Hampshire, InterOperability Lab: IP Consortium Test Suite, Internet Protocol Version 6, Technical Document, January 2000.

8

INTEROPERABILITY TEST SUITE GENERATION FOR THE TCP DATA PART USING EXPERIMENTAL DESIGN TECHNIQUES

Jiwon Ryu[†], Myungchul Kim[†], Sungwon Kang[††], and Soonuk Seol[†]
† Information and Communications University
{jwryu, mckim, suseol}@icu.ac.kr
†† Korea Telecom Research & Development Group
kangsw@sava.kotel.co.kr

Abstract   Test derivation methods suitable for interoperability testing of communication protocols were proposed in the literature and applied to the TCP and the ATM protocols. The test cases that were generated by them deal with only the control part of the protocols. However, in real protocol testing, the test cases must manage the data part of them as well. For complete testing, in principle all possible values of the data part must be tested although it is impractical to do so. In this paper, a method is presented for generating the interoperability test suite for both the data part and the control part with the example of TCP connection establishment. In this process, experimental design techniques from industrial engineering are used to reduce the size of test suite while keeping a well-defined level of test coverage. Experimental design techniques have been used for protocol conformance testing but not for interoperability testing so far. We generate the test suite for the TCP data part by this method and show a possibility that we can test interoperability of protocols with the reduced number of test cases with a well-defined level of test coverage.

Keywords: protocol testing, interoperability testing, test suite generation, control part, data part, TCP, experimental design techniques, orthogonal arrays, factor, level

1.   INTRODUCTION

With the wide spread of the telecommunication equipment and services, it is considered essential to guarantee interoperability of communication protocol implementations residing in them. As an attempt to guaranteeing interoperability, conformance testing methodology has been published as international standards in ITU-T X.290 Series [6] and ISO/IEC 9646 [5], and used world-wide. Protocol conformance testing is used for confirming whether or not the behavior of Implementation Under Test (IUT) conforms to its standards and specifications, and promotes the probability of interoperation among IUTs. However, even though equipment and services successfully passed conformance testing, they do not often interoperate owing to the variety of mandatory parameters, optional parameters and control scopes. For this reason, interoperability testing is needed to check the interaction behavior among IUTs implemented according to their specifications. So far, no accepted common methodology for interoperability testing has existed.

Paper [8] proposed an algorithm for generating the interoperability test suite for communication protocols, and [13] and [14] respectively applied it to the Internet TCP protocol and the ATM/B-ISDN signaling protocol. The derived test cases are based on Finite State Machine (FSM) and the test generation methods consider only the control part of the protocols. So far, there has been no work on generating the interoperability test suite considering the data part of protocols.

Because most protocols have a variety of data variables, for real protocol testing the test suite generation has to deal with the data part as well as the control part of them. The size of test suite generated with the test derivation method considering the data part of protocols becomes much larger than that considering only the control part of them. In order to reduce the size of test suite while maintaining a well-defined level of test coverage, we make use of experimental design techniques from industrial engineering. They are used for planning experiments so that one can extract the maximum possible information from as few experiments as possible. Adopting these techniques leads to faster detection of non-interoperation of protocols while maintaining a well-defined level of test coverage. Experimental design techniques have been used for protocol conformance testing but not for interoperability testing so far. In this paper, we present an interoperability test suite derivation method considering the data part of protocols. It is illustrated with the example of TCP connection establishment. We then apply experimental design techniques to the test suite and compare the size of the test suite generated by the techniques with the size of test suite generated without the techniques.

This paper is organized as follows: in Section 2, as background we describe the interoperability test generation method for the control part of the TCP in paper [13], experimental design techniques, and the specifications of TCP connection establishment. In Section 3, with the experimental design techniques under certain assumptions, we generate the interoperability test suite for the data part of TCP connection establishment. Finally in Section 4, we discuss the conclusion and the future work.

2.   RELATED WORK

In this section, we survey the interoperability test suite generation method for the TCP control part, explain experimental design techniques to be used to reduce the size of the interoperability test suite for the TCP data part, and describe the specification of TCP connection establishment.

2.1   Interoperability Test Suite Generation for the TCP Control Part

Paper [13] implemented the algorithm for testing interoperability for the class of communication protocols proposed in paper [8] and applied it to the TCP. The implemented program generates the interoperability test suite after an FSM is given. Figure 1 represents the connection establishment phase of the TCP FSM shown in paper [13]. Input is given by the application program running on TCP and output is presented by selecting one or more bits among URG, ACK, PSH, RST, SYN, and FIN in the 6-bit control field of TCP packet. TCP controls the connection establishment by selecting these bits.

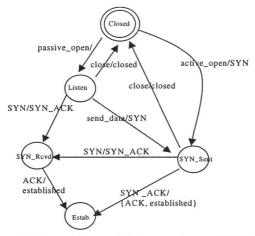

Figure 1. Connection establishment phase of TCP FSM.

As the result of executing the program of which input is TCP FSM in Figure 1, twelve interoperability test cases shown in Appendix 1 are derived. The following test case is for a three-way handshake and equivalent to item <1> of Appendix 1:

(Closed,Listen)   --   active_open_a/[<,i_SYN,,i_SYN_ACK>,
<established,i_ACK,established,>] → (Estab,Estab)

(Closed,Listen) is the starting stable state[1], 'active_open_a' before '/' is an input symbol to IUT A, and '[<,i_SYN,,i_SYN_ACK>, <established,ACK,established,>]' after '/' is a set of output symbols. Corresponding to the input symbol with postfix 'a' (or 'b') represents a message transmitted by Tester A (or Tester B) via interface A (or interface B) respectively. 'i_' is used to indicate internal messages as opposed to external messages. As shown in Figure 2, outputs are represented as a vector <u1, u2, u3, u4>, where u1, u2, u3, and u4 respectively represent messages transmitted by IUT A via interface A, IUT A via interface C, IUT B via interface B, and IUT B via interface C respectively. '→' is the transition relation and (Estab,Estab) is the arrival stable state. The test cases like this cannot be executed in real testing because they do not have values for the data part.

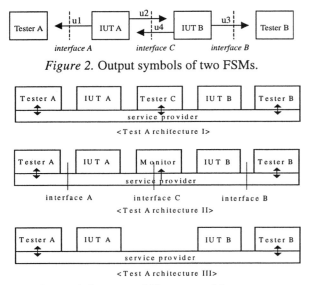

Figure 2. Output symbols of two FSMs.

Figure 3. Interoperability test architectures.

Paper [14] presents three test architectures for the interoperability testing as shown in Figure 3. In Test Architecture I, Tester A sends a message to IUT A and then IUT A can send messages to IUT B and/or Tester A. If IUT B receives the message, it can also send messages to IUT A and/or Tester B. Tester C located between IUT A and IUT B can observe/modify messages

---

[1] A stable state is defined in paper [1] as a system state that is reachable from the initial state adhering to the single stimulus principle, and from which no change can occur without another stimulus.

between IUT A and IUT B. In Test Architecture II, however, Monitor located between IUT A and IUT B can only observe messages between IUT A and IUT B. In Test Architecture III, it is impossible to read these messages owing to the absence of Tester C or Monitor. Architecture III was considered in generating the test suite of the control part in paper [13]. In order to generate the test suite including data part, we have to observe values of internal messages between IUT A and IUT B. Therefore Architecture II is considered in this paper.

2.2   Experimental Design Techniques

Experimental design is a statistical technique for planning experiments and for choosing and analyzing data so that one can extract the maximum information from as few experiments as possible [9].

After deciding on the purpose of the experiment, we must choose factors and levels of the factors. The factors are defined as the various parameters of interest and the levels are defined as the values taken for each parameter. For example, if the temperature is chosen as a factor, 150°C or 200°C can be a level of the factor. The number of the different levels of the factor is defined as the space of the level.

When too many test cases are derived, we need to reduce the number of test cases while achieving the purpose of the experiment. In order to make the number of the test cases as small as possible, an assumption is needed such that the interactions of three or more factors virtually do not exist in the experiment. The assumption is that the risk of an interaction among three or more fields is balanced against the ability to complete system testing within a reasonable budget. An analysis of field data at Bellcore indicated that most field faults are caused by interactions of one or two fields [3, 4]. In the protocol world, it is also felt that most problems are caused by the interactions of a few state variables [3]. Thus the assumption for the experimental design techniques is satisfied in protocol testing area. This paper investigates this approach, using the method of orthogonal arrays to determine the test suite that cover all two-way interactions. Orthogonal arrays are test sets such that, for any pair of factors, all combinations of levels occur. The test suite using orthogonal array has more well-defined level of test coverage than other test suites with the same size [9, 10].

At Bellcore, the Automatic Efficient Test Generator (AETG) [2, 3, 4] was developed based on ideas of statistical experimental design theory to reduce the number of tests. The AETG is a system that generates test suite from user defined test requirements. AETG was used in Bellcore for screen testing and protocol conformance testing such as ISDN protocol conformance testing: call rejection and channel negotiation [1].

Paper [15] presented a guide to the theory and practical application of the method of orthogonal Latin squares to generate system test configurations that achieve pairwise parameter coverage.

2.2.1   Orthogonal Arrays

Orthogonal array design is a method requiring as few experiments as possible in an experiment with many factors [10]. Orthogonal array designs are test sets such that, for any pair of factors, all combinations of levels occur and every pair occurs the same number of times. So orthogonal array designs produce a test set of a manageable size that still covers the interactions that cause most of the field faults.

Orthogonal arrays are available with a variety of levels from 2 to 5. Depending on the levels, a method for making orthogonal arrays is different. For the majority of purposes, orthogonal arrays consisting of two or three levels should be sufficient. In this paper, two-level orthogonal arrays are referred. 2^m is the number of test cases and 2^m-1 is not only the number of columns but also the maximum factor to enable to arrange. The variable, m, is an integer of 2 or above. Thus two-level orthogonal arrays for 2^m-1 factors as the maximum enable to test out with 2^m test cases while covering interactions that causes most field faults.

To elaborate on these designs, consider a situation where three factors have two levels per a factor, say 1 and 2. In this case, the exhaustive test set has eight test cases, namely, (1,1,1), (1,1,2), (1,2,1), (2,1,1), (1,2,2), (2,1,2), (2,2,1), and (2,2,2). These test cases cover the interactions of three factors. A corresponding orthogonal array has four test cases, namely, (1,1,1), (1,2,2), (2,1,2), and (2,2,1). These test cases cover pairwise interactions.

Table 1. Orthogonal array for 3 factors, 2 levels.

Experimental	Column number (factor)		
Number	1	2	3
1	1	1	1
2	1	2	2
3	2	1	2
4	2	2	1

Table 2. Number of test cases and breadth of coverage.

Experimental example	Number of test cases needed		Breadth of coverage (%)	
	Exhaustive testing	Orthogonal array	Exhaustive testing	Orthogonal array
3 factors, 2 levels testing	8	4	100	100

Table 1 represents the orthogonal array to enable to test three factors as the maximum with four test cases, as substituted 2 for m. In orthogonal arrays, the arrangement of factors and levels are randomly chosen. In this example, compared to the exhaustive testing, there is a 50% reduction in the number of test cases. It is possible to test with only four test cases because in the real world most problems are caused by the interactions of two factors as illustrated in section 2.2. As shown in Table 1, two columns chosen randomly include pairwise interactions: (1,1), (1,2), (2,1), and (2,2), so orthogonal array designs include all pairwise combinations of the test factors. Table 2 summarizes the number of test cases needed and the breadth of coverage for the exhaustive testing and orthogonal array designs. The breadth of coverage is defined here as the percentage of all pairwise combinations of the test factors.

2.3   TCP Specification for Connection Establishment

TCP consists of three phases, i.e., connection establishment, connection release, and data transmission. In this paper, since we generate the test suite for TCP connection establishment phase, we only give the specification of connection establishment phase. Figure 4 shows the format of the TCP header [12].

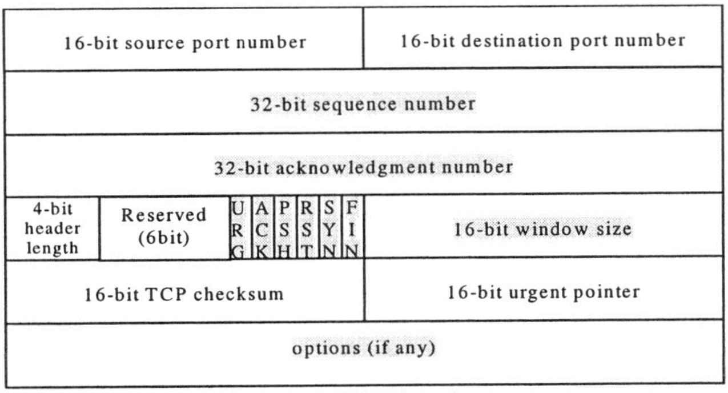

16-bit source port number							16-bit destination port number
32-bit sequence number							
32-bit acknowledgment number							
4-bit header length	Reserved (6bit)	URG	ACK	PSH	RST	SYN	FIN	16-bit window size
16-bit TCP checksum							16-bit urgent pointer
options (if any)							

Figure 4. TCP header.

Since TCP is a connection-oriented protocol, a connection must be established between two ends before either end can send data to the other. TCP establishes connections with a procedure known as a three-way handshake. The TCP packets for this procedure include sequence number (seq), acknowledgement number (ack), window size (win), and maximum segment size (mss) as well as the control field. These fields are presented by the shadow in Figure 4 and are the factors related to test suite generation for the TCP data part.

Since the sequence number and the acknowledgement number are 32-bit fields, their space ranges from 0 to $2^{32}-1$. Since the space is finite, they cycle from $2^{32}-1$ to 0 again. When each end sends its SYN to establish a connection, it chooses an initial sequence number (ISN) for the connection. The ISN should change over time so that each connection has a different ISN. The ISN should be viewed as a 32-bit counter that increments by one every 4 microseconds because the sequence numbers on the clock are increased about every 4 microseconds [11].

TCP's flow control begins by each end advertising a 16-bit window size. Since the window size is significant only when combined with an acknowledgement number, this field is meaningful only when the acknowledgement field is valid. When the ACK bit in the control field is set, the requesting end sends a SYN segment with the window size. Some applications change their buffer sizes to increase performance, but the window size need not change its default because any data is not exchanged for the connection establishment phase.

The maximum segment size for option field is a 16-bit field and TCP uses this option only during connection setup. The sender advertises the maximum segment size and does not want to receive TCP segments larger than this value. This is normally to avoid fragmentation. For Ethernet this implies the maximum segment size of up to 1460 bytes.

Figure 5 shows the tcpdump [7] output for the segments for TCP connection establishment.

```
1 svr4.1037 > bsdi.discard: S 1415531521 : 1415531521(0) win 4096 <mss 1024>
2 bsdi.discard > svr4.1037: S 1823083521 : 1823083521(0) ack 1415531522 win 4096
  <mss 1024>
3 svr4.1037 > bsdi.discard: . ack 1823083521 win 4096
```

Figure 5. tcpdump output for TCP connection establishment.

These three TCP segments contain only TCP headers. No data is exchanged. For TCP segments, each output line begins with 'source > destination: flags' where flags represent the control bits. '1415531521:1415531521(0)' means the sequence number of the packet is 1415531521 and the number of data bytes in the segment is 0. In line 2 the field 'ack 1415531522' shows the acknowledgement number. This is printed only if the ACK flag in the header is on. The field 'win 4096' in every line of output shows the window size being advertised by the sender. The final field '<mss 1024>' in the lines 1 and 2 shows the maximum segment size option specified by the sender.

3.   INTEROPERABILITY TEST SUITE GENERATION FOR THE TCP DATA PART

In this section, we describe a method to derive the test suite for the TCP data part from the test suite previously generated in paper [13]. Figure 6 shows stages for deriving the interoperability test suite for the TCP data part. In paper [13], the test suite for the control part was generated as the result of giving TCP FSM to the implemented program. The generated test suite for the control part has the effect to exclude impossible behavior sequences. In this work, the test suite for the data part is based on that for control part. In Section 3.1, we lay down some assumptions for test suite generation for the data part. In Section 3.2, based on the assumptions of Section 3.1 we generate test suite for the data part. In Section 3.3, by using orthogonal arrays, we again generate test suite from the test suite derived in Section 3.2. In Section 3.4, we calculate and compare the sizes of the test suites generated at various stages.

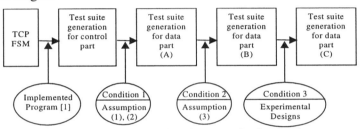

Figure 6. Stages for generating the test suite for the data part.

3.1   Assumptions

We need assumptions for the test purpose to generate test suite for the data part based on the test suite for the control part.

(1) We consider six factors: the sequence numbers, the window sizes, and the maximum segment sizes of two TCPs.

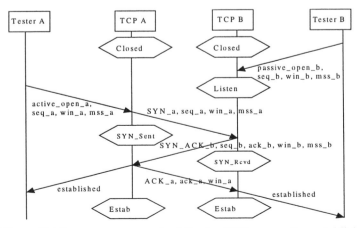

Figure 7. Message sequence chart for the TCP connection establishment.

Figure 7 shows the general procedure for TCP connection establishment in which each segment has several factors. We need not assign an initial value of acknowledgement number because TCP acknowledges the other TCP's SYN by ACKing the other's ISN plus one. Since the acknowledgement number and the next sequence number are fixed by the initial sequence number and the next window size maintains the default, each tester sends initial values of three factors to TCP.

(2) As shown in Figure 7, Tester A (Tester B) assigns initial values of factors to TCP A (TCP B) respectively. When 'active_open', 'passive_open', or 'send_data' message is given by the application program running on TCP, values of the three factors in the TCP packets are initialized. When the packets between TCPs are exchanged, these values are transmitted to the other TCP.

(3) All factors have two levels and levels of the factors are shown in Table 3.

Because the level space is usually very large, it is impossible to sample all level values. So in order to provide better sampling coverage, levels need to be chosen properly. The method to choose levels follows the three principles [9]: (i) for the level space, it is better to choose the minimum and maximum values and to partition it into homogeneous spaces for middle values, (ii) levels are better to include values used in current systems and expected to be the optimal solution, and (iii) the suitable size of level space is from 2 to 5 because over 5 level space makes the domain of factors be hard to be managed. In this work, for the sequence number, we partition the level space into homogeneous spaces and then choose values including the minimum and maximum. For the window size and the maximum segment size, we choose values used in current systems. These values of each factor can be assigned to the tester by the implementer through the Protocol Implementation Conformance Statement / Protocol Implementation eXtra Information for Testing (PICS/PIXIT).

Table 3. Levels for factors.

Factor Level	seq of TCP A	win of TCP A	mss of TCP A	seq of TCP B	win of TCP A	mss of TCP A
1	2823083521	2048	1460	1415531521	4096	1024
2	0	8192	256	4294967295	16384	256

3.2   Test Suite Generation

To generate the test suite for the data part, we consider the assumption (1) and (2) (Condition 1). Then the derived test suite has two types. For example, the item <1> of Appendix 1 which is an example of test case for the message interaction in Figure 7 is the first type and as follows:

(Closed,Listen) -- (active_open_a,seq,win,mss)/[<,(i_SYN,seq,win,mss),,
(i_SYN_ACK,seq,ack,win,mss)>, <established,(i_ACK,ack,win),
established,>] → (Estab,Estab)

Judging from the fact that (Closed,Listen) is the starting stable state, TCP B has already received 'passive_open' message and initial values of its factors from Tester B. TCP A receives 'active_open' message and initial values of it's factors from Tester A. Thus this example of test case is a test case with six factors because both TCPs use three factors. Like this, in generating the test suite for the data part, eight test cases with six factors are represented by the items <1> to <8> of Appendix 1.

The item <9> of Appendix 1, for example, is the second type. It is an example of test case for the message interaction in Figure 8 and as follows:

(Closed,Closed) -- (active_open_a,seq,win,mss)/[<,(i_SYN,seq,win,
mss),,>] → (SYN_Sent,Closed)

When TCP B is not ready to receive packets from TCP A because of not receiving 'passive_open' message from Tester B, TCP B is in state Closed. Thus this test case have only three factors. In generating the test suite for the data part, four test cases with three factors like this is represented by the items <9> to <12> in Appendix 1.

Let us calculate the size of test suite after choosing all possible values of the factors. The 32-bit sequence number field has 2^{32} levels and the 16-bit window size and the 16-bit maximum segment size fields have 2^{16} levels. The eight test cases with six factors have 2^{128} (= $2^{32} \times 2^{32} \times 2^{16} \times 2^{16} \times 2^{16} \times 2^{16}$) test cases for the data part respectively and the four test cases with three factors have 2^{64} (= $2^{32} \times 2^{16} \times 2^{16}$) test cases respectively. Thus the total number of test cases for the data part is $2^{131} + 2^{66}$ (= $8 \times 2^{32} \times 2^{32} \times 2^{16} \times 2^{16} \times 2^{16} \times 2^{16} + 4 \times 2^{32} \times 2^{16} \times 2^{16}$).

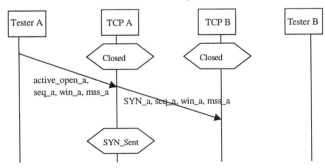

Figure 8. Message sequence chart in case of using only factors of one TCP.

Let us calculate the size of test suite after considering the assumption (3) (Condition 2) as well as the assumptions (1) and (2) (Condition 1). The eight test cases with six factors have 64 (= 2^6) test cases for the data part respectively and the four test cases with three factors have 8 (= 2^3) test cases respectively. Thus the total number of test cases is 544 (= $8\times2^6+4\times2^3$).

3.3   Test Suite Generation using Experimental Design

The size of test suite derived in Section 3.2 can be reduced by using orthogonal arrays (Condition 3). Table 4 represents 2-level orthogonal array substituting 2 for m and we randomly arrange six factors to the columns. We chose the column numbers 1 to 6 for the arrangement of the factors. The arrangement of levels of each factor, 1 and 2, is randomly decided and follows Table 3 in this paper. Table 4 means that it is possible to test interoperability for the TCP data part with only 8 test cases without 64 (= 2^6) with all combinations of test factors.

Table 4. Orthogonal array for 6 factors, 2 levels.

Experiment al number	Column number (factor)						
	1	2	3	4	5	6	7
1	1	1	1	1	1	1	1
2	1	1	1	2	2	2	2
3	1	2	2	1	1	2	2
4	1	2	2	2	2	1	1
5	2	1	2	1	2	1	2
6	2	1	2	2	1	2	1
7	2	2	1	1	2	2	1
8	2	2	1	2	1	1	2
Arrangement	seq of TCP A	win of TCP A	mss of TCP A	seq of TCP B	win of TCP B	mss of TCP B	

For the eight test cases with six factors, we arrange their factor levels by using the orthogonal array in Table 4 and for the four test cases with three factors by using the orthogonal array in Table 1. In case of using Table 1, the sequence number, the window size, and the maximum segment size are arranged in the column numbers 1, 2, and 3 respectively. Therefore the total number of test cases is 80 (= $8\times8+4\times4$). Appendix 2 shows eight and four test cases respectively from the items <1> and <9> of Appendix 1. This test case for the item <1> of Appendix 2 has levels assigned by the experimental number 1 of Table 4.

(Closed,Listen) -- (active_open_a seq=2823083521,win=2048, mss=1460)/[<,(i_SYN,seq=2823083521,win=2048, mss=1460),, (i_SYN_ACK,seq=1415531521,ack=2823083522,win=4096,mss=1024)

>, <established,(i_ACK_a,ack=1415531522,win=2048),established,>]
→ (Estab,Estab)

TCP A sends SYN segment with its sequence number 2823083521, window size 2048, and maximum segment size 1460 to TCP B. TCP B responds with its own SYN segment containing TCP B's sequence number 1415531521, window size 4096, and maximum segment size 1024. TCP B also acknowledges TCP A's SYN by ACKing TCP A's ISN plus one. TCP A acknowledges this SYN from TCP B by TCP B's ISN plus one. So this is the test case to establish connection by the three-way handshake.

3.4   Assessment

Table 5 shows size of test suite for the data part when we add each of three conditions in order. By Condition 1, the size of test suite becomes $2^{131}+2^{66}$ and by the addition of Condition 2, it becomes 544. Also by Condition 3 using orthogonal arrays, the size of it is reduced to 80 while a well-defined level of test coverage is maintained in terms of the observation illustrated in Section 2.2. Thus compared to (B), there is an 85% reduction in the number of test cases and compared to (A), there is a reduction of more than 99.999%. Because most field faults are caused by the interactions of two factors, 80 test cases covering all pairwise combinations have nearly the same test coverage, compared to the 544 test cases.

Table 5. Comparison of test suite size at each stage.

Indication of Figure 6	Conditions for generating the test suite for data part	Size of test suite	Note
(A)	Condition 1. Considering the assumptions (1) and (2)	$2^{131}+2^{66}$	(C) : 85% reduction compared to (B) and reduction of more than 99.999% compared to (A)
(B)	Condition 2. Considering the assumption (3)	544	
(C)	Condition 3. Using orthogonal arrays of Experimental Designs	80	

4.   CONCLUSION AND FUTURE WORK

In this paper, we presented the test generation method suitable for testing interoperability of the data part for TCP connection establishment. For this work, we laid down three assumptions for the data part and considered experimental designs in order to reduce the size of the test suite while maintaining a well-defined level of test coverage. The 80 test cases were finally generated for testing the data part. Thus compared to the case of choosing all possible values as the level space, there is a reduction of more than 99.999% in the number of test cases and compared to the case of reducing the level space, there is an 85% reduction. This method leads to a faster detection of non-interoperation, which would help to get a higher quality of products in a shorter development interval.

As further work, this method will be applied to the TCP connection release and data transmission phases. We need to develop an algorithm for generating the test suite for the data part and to implement the algorithm. Also we will demonstrate the feasibility of the algorithm by comparing its application result with the test suite derived manually in this paper and the generality of the algorithm by applying it to the other protocols.

REFERENCES

[1]   K. Burroughs, A. Jain, and R. L. Erichson, "Improved Quality of Protocol Testing Through Techniques of Experimental Design," Supercomm/ICC '94, 1994.

[2]   D. M. Cohen, S. R Dalal, M. L. Fredman, and G. C. Patton, "The AETG System: An Approach to Testing Based on Combinatorial Design," IEEE Transactions on Software Engineering, Volume 23, Number 7, pp.437-444, July 1997.

[3]   D. M. Cohen, S. R. Dalal, A. Kajla, and G. C. Patton, "The Automatic Efficient Test Generator (AETG) System," Proc. 5th Int'l Symp. Software Reliability Eng., IEEE CS Press, pp. 303-309, 1994.

[4]   D. M. Cohen, S. R. Dalal, J. Parelius, and G. C. Patton, "The combinatorial design approach to automatic test generation," IEEE Software Volume: 13, Issue: 5, pp. 83-88, September 1996.

[5]   ISO/IEC/9646, OSI Conformance Testing Methodology and Framework Parts 1-7, 1994.

[6]   ITU-T X.290 Series, Conformance Testing Methodology and Framework, 1994.

[7]   V. Jacobson, C. Leres, and S. McCanne, The Tcpdump Manual Page, Lawrence Berkeley National Laboratory, Berkeley, CA., June 1997.

[8]   S. Kang and M. Kim, "Interoperability Test Suite Derivation for Symmetric Communication Protocols," IFIP Joint International Conference on Formal Description Techniques for Distributed Systems and Communication Protocols (FORTE X) and Protocol Specification Testing and Verification (PSTV XVII), pp. 57-72, November 1997.

[9]   D. C. Montgomery, Design and Analysis of Experiments, 4th Ed., John Wiley & Sons, Inc., 1997.

[10]   G. S. Peace, Taguchi Methods: A Hands-On Approach, Addison-Wesley, 1993.

[11]   J. B. Postel, "Transmission Control Protocol," RFC 793, September 1981.

[12]   W. Richard Stevens, TCP/IP Illustrated, Volume1: The Protocols, Addison-Wesley, 1995.

[13] S. Seol, M. Kim, S. Kang, and Y. Park, "Interoperability Test Suite Derivation for the TCP," IFIP TC6 WG6.1 Joint International Conference on Formal Description Techniques for Distributed Systems and Communication Protocols (FORTE XII) and Protocol Specification, Testing and Verification (PSTV XIX), October 5-8, Beijing, China, 1999.

[14] J. Shin and S. Kang, "Interoperability Test Suite Derivation for the ATM/B-ISDN Signaling Protocol," Testing of Communicating Systems, Vol. 11, Kluwer Academic Publishers, pp. 313-330, 1998.

[15] A. W. Williams and R. L. Probert, "A Practical Strategy for Testing pair-wise Coverage of Network Interfaces," Proceedings of the 7th International Conference on Software Reliability Engineering (ISSRE '96), White Plains NY USA, pp. 246-254, October 1996.

Appendix 1. Test suite related to the control part of TCP connection establishment.

<1> (Closed,Listen)   --   active_open_a/[<,i_SYN,,i_SYN_ACK>,
     <established,i_ACK,established,>] → (Estab,Estab)

<2> (Listen,Closed)   --   active_open_b/[<,i_SYN_ACK,,i_SYN>,
     <established,,established,i_ACK >] → (Estab,Estab)

<3> (Closed,SYN_Sent)   --   active_open_a/[<,i_SYN,,i_SYN_ACK>,
     <established,i_ACK,established,>] → (Estab,Estab)

<4> (SYN_Sent,Closed)   --   active_open_b/[<,i_SYN_ACK,,i_SYN>,
     <established,,established,i_ACK>] → (Estab,Estab)

<5> (Listen,Listen)   --   send_data_a/[<,i_SYN,,i_SYN_ACK>,
     <established,i_ACK,established,>] → (Estab,Estab)

<6> (Listen,Listen)   --   send_data_b/[<,i_SYN_ACK,,i_SYN>,
     <established,,established,i_ACK >] → (Estab,Estab)

<7> (Listen,SYN_Sent)   --   send_data_a/[<,i_SYN,,i_SYN_ACK>,
     <established,i_ACK,established,>] → (Estab,Estab)

<8> (SYN_Sent,Listen)   --   send_data_b/[<,i_SYN_ACK,,i_SYN>,
     <established,,established,i_ACK >] → (Estab,Estab)

<9> (Closed,Closed) -- active_open_a/[<,i_SYN,,>] → (SYN_Sent,Closed)

<10> (Closed,Closed) -- active_open_b/[<,,,i_SYN>] → (Closed,SYN_Sent)

<11> (Listen,Closed) -- send_data_a/[<,i_SYN,,>] → (SYN_Sent,Closed)

<12> (Closed,Listen) -- send_data_b/[<,,,i_SYN>] → (Closed,SYN_Sent)

Appendix 2. Test suite for the TCP data part from the items <1> and <9> in
Appendix 1.

<1> (Closed,Listen) -- (active_open_a,seq=2823083521, win=2048,mss=1460)/
     [<,(i_SYN,seq=2823083521,win=2048,mss=1460),,(i_SYN_ACK,
     seq=1415531521,ack=2823083522,win=4096,mss=1024)>, <established,

(i_ACK,ack=1415531522,win=2048),established,>] → (Estab,Estab)

<2> (Closed,Listen) -- (active_open_a,seq=2823083521, win=2048,mss=1460)/
[<,(i_SYN,seq=2823083521,win=2048,mss=1460),,(i_SYN_ACK,
seq=4294967295,ack=2823083522,win=16384,mss=256)>, <established,
(i_ACK,ack=0,win=2048),established,>] → (Estab,Estab)

<3> (Closed,Listen) -- (active_open_a,seq=2823083521, win=8192,mss=256)/
[<,(i_SYN,seq=2823083521,win=8192,mss=256),,(i_SYN_ACK,
seq=1415531521,ack=2823083522,win=4096,mss=256)>,<established,
(i_ACK,ack=1415531522,win=8192),established,>] → (Estab,Estab)

<4> (Closed,Listen) -- (active_open_a,seq=2823083521, win=8192,mss=256)/
[<,(i_SYN,seq=2823083521,win=8192,mss=256),,(i_SYN_ACK,
seq=4294967295,ack=2823083522,win=16384,mss=1024)>, <established,
(i_ACK,ack=0,win=8192),established,>] → (Estab,Estab)

<5> (Closed,Listen) -- (active_open_a,seq=0,win=2048,mss=256)/[<,(i_SYN,
seq=0,win=2048,mss=256),,(i_SYN_ACK,seq=1415531521,ack=1,
win=16384,mss=1024)>, <established,(i_ACK,ack=1415531522,win=2048),
established,>] → (Estab,Estab)

<6> (Closed,Listen) -- (active_open_a,seq=0,win=2048,mss=256)/[<,(i_SYN,
seq=0,win=2048,mss=256),,(i_SYN_ACK,seq=4294967295,ack=1,win=4096,
mss=256)>, established,(i_ACK,ack=0,win=2048),established,>] →
(Estab,Estab)

<7> (Closed,Listen) -- (active_open_a,seq=0,win=8192,mss=1460)/[<,(i_SYN,
seq=0,win=8192,mss=1460),,(i_SYN_ACK,seq=1415531521,ack=1,
win=16384,mss=256)>, <established,(i_ACK,ack=1415531522,win=8192),
established,>] → (Estab,Estab)

<8> (Closed,Listen) -- (active_open_a,seq=0,win=8192,mss=1460)/[<,(i_SYN,
seq=0,win=8192,mss=1460),,(i_SYN_ACK,seq=4294967295,ack=1,
win=4096,mss=1024)>, <established,(i_ACK,ack=0,win=8192),
established,>] → (Estab,Estab)

<9> (Closed,Closed) -- (active_open_a,seq=2823083521,win=2048,mss=1460)/
[<,(i_SYN,seq=2823083521,win=2048,mss=1460),,>] →
(SYN_Sent,Closed)

<10> (Closed,Closed) -- (active_open_aseq=2823083521,win=8192,mss=256)/
[<,(i_SYN,seq=2823083521,win=8192,mss=256),,>] → (SYN_Sent,Closed)

<11> (Closed,Closed) -- (active_open_a,seq=0,win=2048,mss=256)/
[<,(i_SYN,seq=0,win=2048,mss=256),,>] → (SYN_Sent,Closed)

<12> (Closed,Closed) -- (active_open_a,seq=0,win=8192,mss=1460)/
[<,(i_SYN,seq=0,win=8192,mss=1460),,>] → (SYN_Sent,Closed)

9

INTEROPERABILITY TESTING SYSTEM OF TCP/IP BASED COMMUNICATION SYSTEMS IN OPERATIONAL ENVIRONMENT

Toshihiko Kato, Tomohiko Ogishi, Hiroyuki Shinbo, Yutaka Miyake, Akira Idoue and Kenji Suzuki
KDD R&D Laboratories, Inc.

Abstract   Recently, TCP/IP protocols are widely used. Here, although the protocols are realized in operating systems and users of communication systems pay no attention to their detail, some system errors are actually reported. Since those errors occur only in specific communication situations, the interoperability testing by watching communication by a testing system is appropriate to detect them. This interoperability testing has the following requirements. Firstly, since the system that includes system errors is not unidentified, the testing system needs to check all communication systems attached to a network. Secondly, for the similar reason, all protocols in the TCP/IP protocol stack need to be checked. Thirdly, the testing system needs to discriminate operational failures, such as server down and mis-configuration of parameters in clients, from system errors, when it detects any problems in the communication. Based on these considerations, we have designed an interoperability testing system of TCP/IP based communication systems applied in an operational environment. This paper describes the detailed design of our testing system and the testing algorithm for DHCP (Dynamic Host Configuration Protocol), for which some system errors are reported.

Keywords: interoperability testing, TCP/IP, operational environment, multiple protocols

1.   INTRODUCTION

Recently, the Internet becomes widely spread and many computer systems communicate with each other using TCP/IP protocols [5]. Here, the communication software is considered as a fundamental function such as an operating system, and most of users do not pay any attentions to its detail. Actually, commercial TCP/IP software is stable and, in most cases, users can access to WWW servers and send e-mails without any problems.

However, it is pointed out so far that there are some system errors of TCP/IP based communication systems. Examples are that some version of Solaris TCP software has problems in the initial value of retransmission timeout [4] and that some version of DHCP (Dynamic Host Configuration Protocol) [2] software of Macintosh has some bugs in the client operation [1]. Those errors are difficult to detect, because they occur only in exceptional situations such that the transmission delay increased dynamically due to network congestion, and that the combination of versions of DHCP software is one of special cases in a DHCP client and a DHCP server.

So far, the framework of protocol testing [3] identifies two testing approaches: the conformance testing and the interoperability testing. In order to detect the system errors in the TCP/IP communication systems described above, the interoperability testing approach is appropriate because they can be detected only while those systems are actually communicating with various systems and because the testing needs to be performed for relatively a long time.

In order to realize the interoperability testing in the operational environment, the following requirements need to be considered:

- Since any TCP/IP based communication systems attached to the network have the possibility to include system errors, all of which need to be the targets of testing.
- In the TCP/IP based communication systems, multiple protocols cooperate together to realize communication. They include the protocols related to specific applications, such as FTP and HTTP, and the protocols that play a supplementary role, such as DHCP and DNS (Domain Name System). There are possibilities that any of them have system errors, and therefore all of them need to be investigated.
- Users of TCP/IP based communication systems often suffer from failures due to the operational conditions, such as the server downs and the mis-configuration of parameters in clients. That is, the reason of a detected problem may be sometimes system errors and otherwise operational failures. Therefore, a testing system for an interoperability testing needs to check both possibilities described above.

Based on these considerations, we have designed an interoperability testing system, which detects system errors in the operational environment. Our testing system has the following features:

- We adopt an approach of network monitoring [4]. That is, the testing system attached to a network captures all packets exchanged among communication systems in a network, and it looks for any problems.
- Our testing system detects both system errors and operational failures, and identifies their reasons.
- Our testing system analyzes the protocol procedures of communicating systems to detect system errors. It also estimates the network configuration, operational status of terminals and servers, and parameter settings, in order to detect operational failures [5].

We have designed an interoperability testing system for TCP/IP based communication systems. This paper describes the detailed design of our testing system. The next section and section 3 describe the design principles and the overall design of the testing system, respectively. Section 4 describes the testing algorithm for DHCP and the result of preliminary testing. We make a conclusion in section 5.

2.   DESIGN PRINCIPLES

(1)   As shown in Figure 1, our testing system is attached to a LAN segment (subnet) and captures packets transmitted by communication systems such as terminals, servers and routers. If a network is divided into multiple subnets, one testing system is attached to an individual subnet.

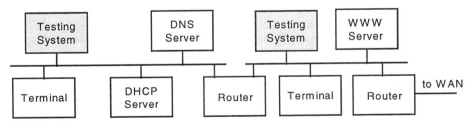

Figure 1. Network Configuration

(2)   By using captured packets, our testing system performs the following:
- It identifies systems attached to the subnet automatically.
- It estimates the network configurations, such as the IP addresses of DHCP/DNS servers and routers (*configuration estimation*).
- For an individual captured packet, it emulates the protocol behavior of systems which sent or received the packet, and checks whether the behavior is conforming to the protocols (*protocol emulation*). If a

problem is detected, it estimates that the problem comes from system errors or operational failures.

(3)   As described above, the testing is required to operate for a long time. Therefore, our testing system needs to continue capturing packets transmitted by any system on the subnet and analyzing them to find system errors and operational failures.   However, in the protocol emulation the correspondence between requests and responses needs to be examined and this requires our testing system to wait for responses when it detects requests.   Therefore, we adopt the following approach:

- Our testing system keeps capturing packets in an online manner and saves them in a buffer, called the *packet buffer*.
- After a specific time of period (e.g. 120 seconds) has passed from the capturing of a packet, it considers that the related packets are also in the buffer and performs the protocol emulation for the packet.

(4)   The first version of our testing system supports the interoperability testing the TCP/IP protocols including ARP, IP, ICMP, TCP, UDP, DHCP, DNS and SMTP.   The protocol emulation is performed from lower layer to higher layer as shown in Figure2.

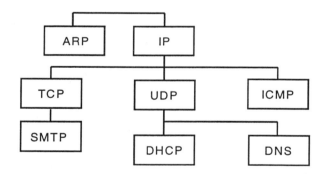

Figure 2. TCP/IP Protocols Handled in Testing System

3.   OVERALL DESIGN

3.1   Software Structure

Figure 3 shows the software structure of our testing system.

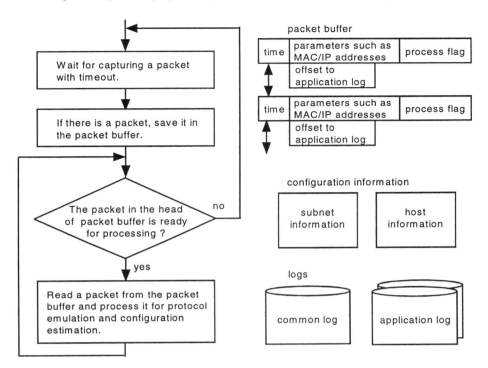

Figure 3. Software Structure

When the software captures a packet, it saves the information on it in the packet buffer. The packet buffer includes the time when this packet is captured and the parameters included in the packet. After that, the software checks if the packet saved in the head of the packet buffer is ready for processing, that is, a specific time elapsed after it was captured. If so, the software reads out the packet information from the buffer and performs the configuration estimation and the protocol emulation.

The result of configuration estimation is saved in the configuration information. This information consists of *subnet information* and *host information*. The subnet information includes the list of terminals and servers (hosts) attached to this subnet and the list of servers (DHCP, DNS and SMTP servers) which the hosts in this subnet have access to. The host information includes the information of an individual terminal or server, such as the MAC address and IP address.

The protocol emulation checks the behavior of the host sending the packet and detects operational failures and system errors. When some problems are detected, the results are displayed on the screen and stored in the logs.

Basically, individual packets are handled one by one in the protocol emulation. However, for client-server type application protocols such as DHCP and DNS, the software pre-reads the corresponding response when a request is detected, and checks the mapping between the request and response. In order to realize this pre-reading, the log is prepared for individual application protocols and the packet buffer includes process flag indicating that the specific application protocol is processed before hand related to this packet.

3.2   Information for Configuration Estimation

As described above, the configuration information includes subnet information and host information. The subnet information includes the following information:
- Host list: each element includes its IP address and the pointer to host information.
- Router list: each element includes its IP address and working flag.
- DNS server list: each element includes its IP address, working flag and domain name table.
- DHCP server list: each element includes its IP address, working flag and available IP addresses.
- SMTP server list: each element includes its IP address, working flag and user list.

The host information includes following information.
- MAC address
- IP address
- State of IP address, which is used for inspecting duplicated IP address
- Working flag
- ARP cache table: each element includes IP address, hardware address and updated time
- List of routers the host uses: each element includes the router IP address, destination IP address and default router flag
- Use/nonuse of DHCP
- DHCP server address

When our testing system processes a captured packet, the information described above is updated. The configuration information helps our testing system to detect operational failures and system errors.

3.3   Overview of Protocol Emulation and Configuration Estimation Procedures

This subsection gives overviews of protocol emulation and configuration estimation for ARP, IP and TCP as examples.

3.3.1   Procedures for ARP

(1)   At first, our testing system deals with exceptional cases such as operational failures. For example, sender IP address is out of the subnet, which comes to be a configuration error in subnet mask, or a response to gratuitous ARP is detected, which comes to be a duplicated use of IP address.

(2)   Our testing system identifies a host attached to the subnet from a pair of sender hardware address and sender IP address of ARP packets. If the identified host is not registered in the host list, it is registered. The testing system also detects an operational failure by duplicated use of IP address.

(3)   Our testing system checks the behavior of ARP for corresponding ARP Request and ARP Reply and estimates its current ARP cache table of the host. If there are no responses to ARP Request, then the testing system recognizes that the target host is down, when it is registered in the subnet information. Otherwise, the testing system considers that the IP address of the host may be mis-configured at the sending host.

3.3.2   Procedures for IP

(1)   Our testing system identifies a host attached in the subnet together with its hardware address and IP address from the source pair or the destination pair.

(2)   When IP address represents a host out of the subnet, our testing system considers that the hardware address represents that of a next-hop router. If the router is not registered in the subnet information, it is registered. The testing system also updates the list of routers of the host.

(3)   Then our testing system processes the user data of this packet according to the protocol ID. ICMP, UDP and TCP emulation is performed according to its value. If the protocol ID is not that of above, it produces an emulation result and performs the emulation of the next packet in the packet buffer.

3.3.3   Procedures for TCP

(1)   As for the processing of TCP, we depend on the Intelligent TCP analyzer, which we developed previously [4]. It emulates the TCP behaviors in the communication systems by using captured TCP packets. It maintains the state and internal variables such as "SND.NXT" and estimates TCP's internal procedures such as slow start and congestion avoidance algorithm. As a result, it detects the following system errors:

- No initial slow start
- No slow start after retransmission timeout
- Uninitialized congestion window
- Too small initial retransmission timeout
- Miscalculation of retransmission timeout

(2)   If a captured TCP packet is "SYN" and "SYN+ACK" and the destination port of "SYN" is a well-known port of an application server, our testing system considers that the host focused on is working as a server. If the host is not registered in the list of the server, it is registered in the subnet information. Similarly, our testing system maintains the subnet information in response to the abrupt release of TCP and no response to "SYN".

(3)   Then our testing system processes the user data of this packet according to the port number. Currently we are designing SMTP handling over TCP.

4.   DETAIL DESIGN FOR DHCP

In this section, we describe more detailed design of our system by taking DHCP as an example.

4.1   Overview of DHCP

DHCP provides the function of allocating an IP address dynamically to a host. DHCP has the following procedures in which a host requests an IP address to a DHCP server:

(1) Allocating a new IP address:

A client that does not have its IP address broadcasts a DISCOVER message by setting all 0 address for its source IP address and all 1 address for its destination IP address. (Multiple) DHCP servers that can allocate an IP address respond to it with an OFFER message (see Figure 4).

The client which receives several OFFER messages chooses one server. It then broadcasts a REQUEST message by specifying the selected server ID and IP address provided by the server. In the end, the selected DHCP sever responds to it with an ACK message specifying the allocated IP address and its lease time. These messages are transmitted by UDP and the related messages are identified by the parameter, transaction ID.

It should be noted that OFFER and ACK messages will be broadcasted or unicasted according to the BROADCAST flag included in the DISCOVER and REQUEST messages respectively.

(2) Reusing a previously allocated IP address:

If a client remembers and wishes to reuse a previously allocated IP address, it sends a REQUEST message containing the IP address. If a DHCP server can allocate the requested IP address, it responds with an ACK message. There are three cases here. The first one is that a client requests the IP address during the lease time to the DHCP server which allocated it, which is called "RENEWING". In this case, the REQUEST message is unicasted. The second one is that a client broadcasts a REQUEST message to require the reallocation of the IP address during its lease time, which is called "REBINDING". The third one is that a client broadcasts a REQUEST message to expect the reallocation of the IP address after its lease time finished, which is called "INIT-REBOOT". These three cases are differentiated by the parameters included in the REQUEST message.

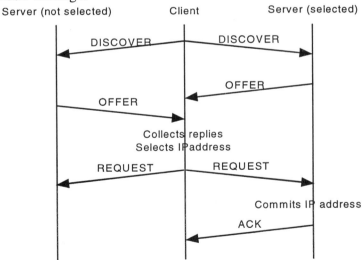

Figure 4. Sequence of DHCP When Allocating a New IP Address

4.2   Algorithm for DHCP Testing

DHCP is a complex protocol due to the facts that it combines the unicast and broadcast communication and that the multiple procedures need to be realized by changing the combinations of parameters in the same message. Actually, some system errors related to DHCP are reported as described in section 1. In order to perform the interoperability testing of DHCP, we have designed the algorithm shown in Figures 5 and 6. This algorithm is for testing the case that a new IP address is allocated. It is summarized in the following way:

(1) When a packet being processed is a DISCOVER message, our testing system collects the related DHCP messages from the packet buffer. The related messages are those that have the same transaction ID as that of this DISCOVER message and whose process flag is not set. Those messages are copied in another buffer called the *DHCP message buffer* and their process flags in the packet buffer is set.

(2) The handling of DISCOVER message is to update the host and subnet information and check its format.

(3) Then the related OFFER messages are collected, and their format check and the update of the configuration information are performed. As described above, the BROADCAST flag in the DISCOVER message and the transmission of the related OFFER message that indicates broadcast or uni-cast need to be matched. So this matching is also examined. If there are no OFFER messages, a system error or an operational failure is reported. Here, it needs to be mentioned that the logs generated when our testing system handles pre-read packets are stored in the application log, and that the offset to the logs are stored in the packet buffer to be handled when the testing system processes the IP level later.

(4) Then, the related REQUEST messages are collected. If there are no REQUEST messages, a system error or an operational failure is reported. If there is one, the format check and the update of subnet information are performed.

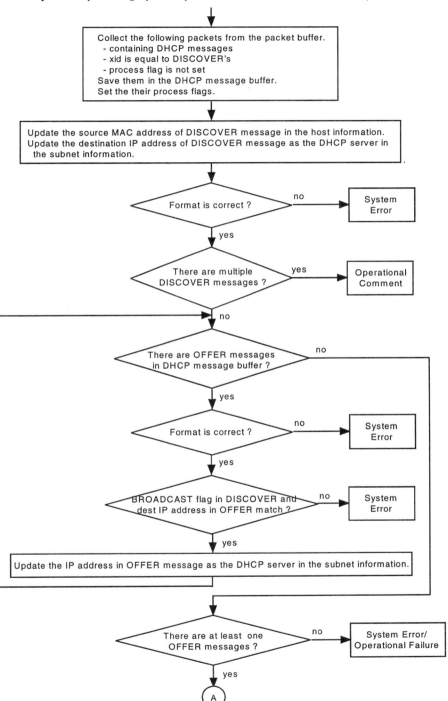

Figure 5. Algorithm for DHCP (1)

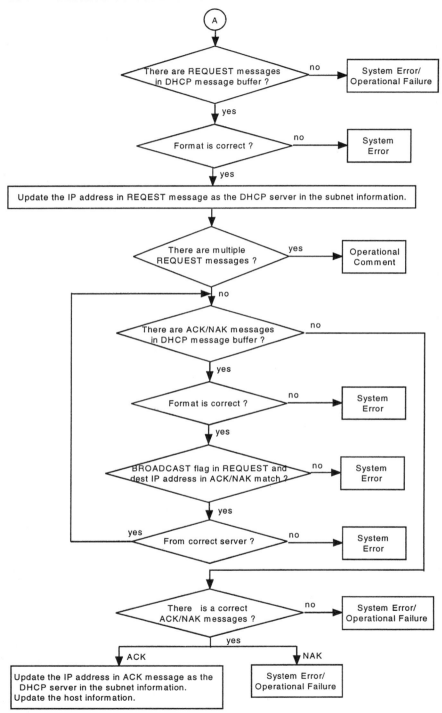

Figure 6. Algorithm for DHCP (2)

(5)   Then, the related ACK or NAK messages are collected and checked. The procedure is similar to that for OFFER messages. In addition to it, whether the ACK or NAK is sent by the DHCP server requested in the REQUEST message or not is checked. In the end, the host and subnet information is updated in the case of ACK message, and a system error or an operational failure is reported in the case of NAK message.

For the procedure for reusing a previously allocated IP address, the testing algorithm is designed similarly. In this case, the format check is important for discriminating RENEWING, REBINDING and INIT-REBOOT cases.

4.3   Preliminary Experiment

We have performed a preliminary experiment for the interoperability testing of DHCP based on the algorithm described in the previous subsection. As a result, we have found several system errors related to DHCP. The first example is shown in Figure 7. This is summarized in the as follows:

- When a Windows 98 personal computer is used as a DHCP client, it expects a broadcasted OFFER message even if it sends a DISCOVER message with its BROADCAST flag unset.
- When a Windows NT personal computer is used as a DHCP server, it broadcasts an OFFER message even if the BROADCAST flag in the corresponding DISCOVER message is unset.
- As a result, a Windows based client and server can communicate well.
- If a DHCP server works correctly such as the case of FreeBSD, that is, it unicasts an OFFER message, a Windows based client does not respond to it and retransmits a DISCOVER message. After the third OFFER message is received, it will respond to the OFFER message.

Figure 8 shows another example. This is summarized as follows.

- When a Macintosh personal computer is a DHCP client and an NTT router is a DHCP server, a Macintosh sends a DISCOVER message with a large value of IP address lease time. For that, an NTT router responds by an OFFER message with a smaller value of lease time.
- However, a Macintosh sends a REQUEST message with the same lease time as that of the DISCOVER message.
- As a result, the NTT router sends a NAK message.
- After that, a Macintosh again sends the same DISCOVER message and both of them continue the same procedure.

Our interoperability testing system can detect those system errors that will occur in an specific combination of communication systems.

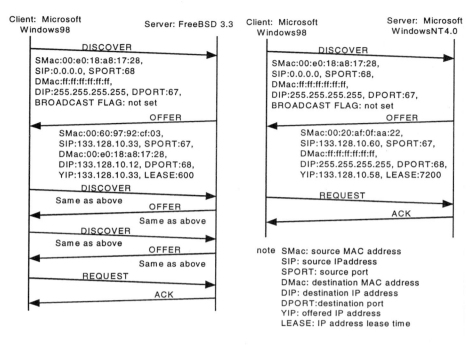

Figure 7. Example of DHCP Error (1)

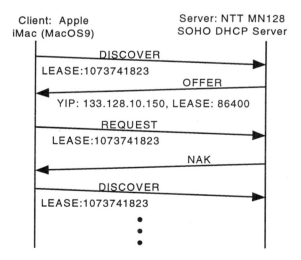

Figure 8. Example of DHCP Error (2)

5.   CONCLUSIONS

In this paper, we described the design of the interoperability testing system for TCP/IP based communication systems. Actually, most of TCP/IP based communication systems have only few system errors. However, some systems have bugs, which are difficult to find because they can be detected only in specific situations. In order to detect those errors, our testing system has the features to check any system attached to a network by considering the operational failures, such as server downs and parameter mis-configurations.

This paper gives a detailed design of our testing system, focusing on the DHCP testing by which it is possible that some systems have bugs. The results of the preliminary testing based on the design show that our testing system can detect system errors related to DHCP, which occur in a specific combination of clients and servers.

References

[1]   Apple Care Tech Info Library, "Mac OS: DHCP Issues With Open Tansmport 2.5.1 and 2.5.2," http://til.info.apple.com/techinfo.nsf/ artnum/n25049, Nov. 1999.

[2]   R. Droms, "Dynamic Host Configuration Protocol," RFC 2131, March 1997.

[3]   ITU-T, "OSI Conformance Testing Methodology and Framework for Protocol Recommendations for CCITT Applications," X.290, 1988.

[4]   T. Kato, T. Ogishi, A. Idoue and K. Suzuki, "Design of Protocol Monitor Emulating Behaviors of TCP/IP Protocols," Proc. of 10th IWTCS, Sept. 1997.

[5]   T. Ogishi, A. Idoue, T. Kato and K. Suzuki, "A Protocol Analizer for Operational Testing of TCP/IP based Communication Systems," in Proc. of ICCC '99, Sept. 1999.

[6]   V. Paxon, "Automated Packet Trace Analysis of TCP Implementations," in Proc. SIGCOMM '97, Aug. 1997.

[7]   W. Stevens, "TCP/IP Illustrated, Vol. 1: The Protocols," Addison Wesley, 1994.

PART IV

TTCN EVOLUTION AND APPLICATION

10

ON THE DESIGN OF THE NEW TESTING LANGUAGE TTCN-3

Jens Grabowski[a], Anthony Wiles[b], Colin Willcock[c], Dieter Hogrefe[a]

[a]*Institute for Telematics, University of Lübeck, Ratzeburger Allee 160, D-23538 Lübeck, Germany*
e-mail: {jens,hogrefe}@itm.mu-luebeck.de

[b]*European Telecommunication Standards Institute, 650, Route des Lucioles, F-06921 Sophia Antipolis, France*
e-mail: Anthony.Wiles@etsi.fr

[c]*Nokia Research Center, Bochum,Meesmannstr. 103, D-44807 Bochum, Germany*
e-mail: Colin.Willcock@nokia.com

Abstract   This paper gives an overview of the main concepts and features of the new testing language TTCN version 3 (TTCN-3). TTCN-3 is a complete new testing language built from a textual core notation on which a number of different presentation formats are possible. This makes TTCN-3 quite universal and application independent. One of the standardised presentation formats is based on the tree and tabular format from previous TTCN versions and another standardised presentation format is based on MSCs. TTCN-3 is a modular language and has a similar look and feel to a typical programming language. However, in addition to the typical programming constructs it contains all the important features necessary to specify test suites.

Keywords: Conformance Testing, ETSI, ITU-T, Programming Languages, TTCN, Test Specification, Telecommunication Systems

1.   INTRODUCTION

TTCN-3 is designed in such a way that a broader user community is addressed. The syntax looks similar to typical implementation languages and should therefore be easy to understand and apply for someone familiar with software development.

The core language is freed from the peculiarities specific to OSI and conformance testing. This makes TTCN-3 flexible, applicable to the specification of all types of reactive system tests over a variety of communication interfaces. Typical areas of application are protocol testing (including mobile and Internet protocols), service testing (including supplementary services), module testing and testing of CORBA based platforms.

The TTCN-3 standard is separated into three parts. The first part [5] defines the core language. The second part [6] defines the tabular presentation format which is similar in appearance and functionality to TTCN-2 [8]. The third part [7] describes an MSC [9] based presentation format.

The core language serves three purposes. Firstly the core language can be used as a generalised text based test language in its own right. Secondly it is used as a standardised interchange format of TTCN test suites between TTCN tools and thirdly the core language provides the semantic basis for the all the various presentation formats.

The core language is defined by a complete syntax and operational semantics. It contains minimal static semantics which do not restrict the use of the language due to some underlying application domain or methodology.

The next section of this paper provides an overview of presentation formats and then the rest of the paper concentrates on the design of the TTCN-3 core language.

2.   PRESENTATION FORMATS AND ATTRIBUTES

Presentation formats provide alternate ways to specify and visualise TTCN-3 test suites, as shown in Figure 1. The tabular format and the MSC format are the first in an anticipated set of different presentation formats. Further presentation formats may be standardised formats or be proprietary formats defined by TTCN-3 users themselves. Use and implementation of all presentation formats is based on the core language.

A presentation format is defined by specifying the graphical representation required for the test suite and the mapping necessary to convert between this graphical form and the TTCN-3 core language.

To enable this mapping, language elements within the core langauge may have attributes associated with them. There are three kinds of attributes:

1.   **encode**: allows references to specific encoding rules;

2. **display** : allows the specification of display attributes related to specific presentation formats;

3. **extension**: allows the specification of user-defined attributes.

Attributes are associated with TTCN-3 language elements by means of **with** statements. In the core language the syntax for the argument of the **with** statement (i.e. the actual attributes) is simply defined as a free text string. Special attribute strings related to the display attributes for the tabular (conformance) presentation format can be found in [6]. Special attribute strings related to the display attributes for the MSC presentation format can be found in [7].

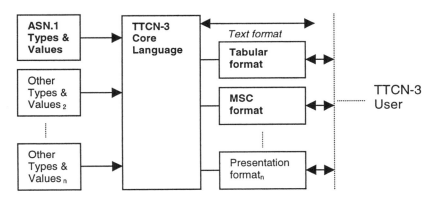

Figure 1. User's view of the core language and the various presentation formats

3.   TYPES AND VALUES

TTCN-3 supports a number of predefined basic, structured and special types. Basic types include ones normally associated with a programming language, such as **integer**, **boolean** and string types, as well as TTCN-3 specific ones such as **objectidentifier**, **verdicttype** and **duration**. Structured types such as **record** types, **set** types and **enumerated** types can be constructed from these base types. Special types associated with configurations such as **port** and **component** may be used to define the architecture of the test system as described in Section 5.

TTCN-3 is strongly typed and there are a number of predefined functions to handle type conversion.

A special kind of data value called a template provides parameterisation and matching mechanisms for specifying test data to be sent or received over the test ports. The operations on these ports provide both asynchronous and synchronous communication capabilities (Section 7).

TTCN-3 is fully harmonised with ASN.1 [11,12,13,14] which may optionally be used with TTCN-3 modules as an alternative data type and value syntax. The approach used to combine ASN.1 and TTCN-3 could be applied to support the use of other type and value systems with TTCN-3.

4.   MODULES

The principle building-blocks of the TTCN-3 core language are modules. A module is a self-contained and complete specification, i.e. it can be parsed and compiled as a separate entity. A module consists of an (optional) definitions part, and an (optional) module control part.

The module definitions part specifies the top-level definitions of the module. These definitions may be used elsewhere in the module, including the control part.

The module control part describes the execution order (possibly repetitious) of the actual test cases. A test case shall be defined in the module definitions part and then called in the control part. For example:

```
module MyTestSuite {
   // Definitions part
   testcase MyTestcase1()…
   testcase MyTestcase2()…
      :
   // Control part
   control {
      var boolean MyVariable; //Localcontrol variable
         :
      MyTestCase1(); // sequential execution of test cases
      MyTestCase2();
         :
   }
}
```

TTCN-3 does not support the declaration of global variables, therefore declarations of dynamic language elements such as **var** or **timer** is only allowed locally within functions and test cases.

It is possible to re-use definitions specified in other modules using the **import** statement. TTCN-3 has no explicit export construct thus, by default, all module definitions in the module definitions part may be imported. An **import** statement can only be used in the module definitions part.

5.   TEST CONFIGURATIONS

TTCN-3 allows the specification of (dynamic) concurrent test configurations. A test configuration consists of a set of interconnected *test components* with well-defined *communication ports* and an explicit *test system interface* which defines the borders of the test system.

Within every test configuration, there is one *Main Test Component* (MTC). All other test components are *called Parallel Test Components* (PTCs). The MTC is created automatically at the start of each test case execution and the behavior defined in the body of the test case (Section 6) is executed on this component. During execution of a test case PTCs can be created and stopped dynamically by the explicit use of **create** and **stop** operations. The conceptual view of a typical TTCN-3 testing configuration is shown in Figure 2.

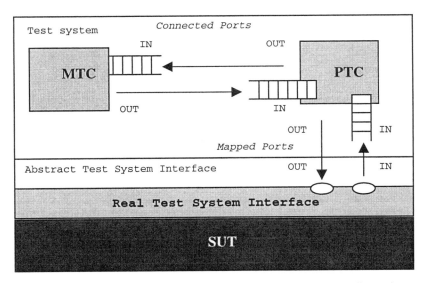

Figure 2. Conceptual view of a typical TTCN-3 testing configuration

5.1   Communication Ports

Ports facilitate communication between test components and between test components and the test system interface. There are no restrictions on the number of connections a component may have, but a component shall not be connected to itself. One-to-many connections are allowed, but TTCN-3 only supports one-to-one communication, i.e. during test execution the communication partner has to be specified uniquely. Each port is modeled as an infinite FIFO queue which stores the incoming messages or procedure calls until they are processed by the component owning that port.

TTCN-3 ports are either message-based or procedure-based. Message-based ports are used for asynchronous communication by means of message exchange. Procedure-based ports are used for synchronous communication by means of remote procedure calls. Ports are directional and each port may have an **in** list (for the *in* direction), an **out** list (for the *out* direction) or an **inout** list (for both directions) of allowed messages or procedures. For example:

```
// Message-based port allowing MsgType1 and MsgType2 to be
// received, MsgType3 to be sent and integer values to be
// send and received.
type port MyMessagePortType message {
    in      MsgType1, MsgType2;
    out     MsgType3;
    inout   integer
}
```

5.2   Component Types and the Test System Interface

A test case consists of one or more test components. The test case behavior is executed on these components. The **component** type defines which ports are associated with a component. For example:

```
// Component type with three ports
type component MyPTCType {
    MyProcedurePortType     PCO1;
    MyMessagePortType       PCO2;
    MyAllMesssagesPortType  PCO3
}
```

The port names in a component type definition are used in the component behavior definition to address the different ports. Port names are local to a component, i.e. another component may have a port with the same (local) name.

A **component** type definition is also used to define the test system interface, because conceptually component type definitions and test system interface definitions have the same form, i.e. both are collections of ports defining possible connection points.

5.3   Configuration Operations

Configuration operations are concerned with setting up and controlling test components. During the execution of a test case, the actual test configuration of components, the connections among them and the connections between the components and the test system interface are created dynamically by performing configuration operations. Configuration operations are **create**, **connect**, **map**, **start**, **stop**, **mtc**, **system**, **self** and **done**.

The create operation

The MTC is the only test component which is created automatically when a test case starts. All other test components are created explicitly during test execution by **create** operations. Since all components and ports are destroyed at the end of a test case, each test case must completely create its required configuration of components and connections.

As shown in Figure 3, the **create** operation returns a unique reference to the newly created instance. The reference can be used for connecting instances and for communication purposes, i.e. for addressing individual components.

```
// usage of create
var component MyNewComponent := MyComponentType.create;
:
// usage of connect and mtc
connect(MyNewComponent.Port1, mtc.Port3);
:
// usage of map, self and system
map(self.Port2, system.PCO1);
:
// usage of start operation
MyNewComponent.start(MyCompBehaviour(…));
:
// usage of done
if (MyNewComponent.done) {
    :    // Do something
}
 :
// usage of stop
if (date = 1.1.2000) stop;
```

Figure 3. Usage of configuration operations

Components can be created at any time during a test run providing full flexibility with regard to dynamic configurations, i.e. any component can create any other component. Component references are local to the scope of their creation. In order to reference a component outside its scope of creation, the component reference can be passed as a parameter to a function or can be sent in a message.

The mtc, system and self operations

The operations **mtc** and **system** return the references (or addresses) of the MTC and the system interface. The **self** operation allows a test component to retrieve its own reference, i.e. **self** returns the reference of the component in which **self** is called. The operations **mtc**, **system** and **self** can be used for addressing purposes in communication operations or, as shown in Figure 3 , in configuration operations.

The connect and map operations

The ports of a test component can be connected to ports of other components or to the ports of the test system interface. The connection between two test components is done by means of the **connect** operation. When linking a test component to a test system interface, the **map** operation shall be used. As illustrated in Figure 2, the **connect** operation directly connects one port to another with the *in* side of the one port connected to the out side of the other, and vice versa. The **map** operation on the other hand can be seen as a pure name translation defining how communications streams should be referenced. In Figure 3 examples for the usage of **connect** and **map** operations are shown.

The start operation

Once a component has been created and connected the execution of its behavior has to be started. This is done by using the **start** operation. The reason for the distinction between **create** and **start** is to allow connection operations to be done before actually running the test component. The **start** operation binds the behavior to a component by referring to a function (Section 6). An example for the usage of the start operation can be found in Figure 3.

The stop and done operations

By using the **stop** operation, a test component is able to stop itself. A stopped component disappears from the configuration. The **done** operation allows a test component to ascertain whether another test component has completed, i.e. is stopped. Examples for the usage of **done** and **stop** operations can be found in Figure 3.

6.   TEST CASES AND FUNCTIONS

Behavior in TTCN-3 is related to the definition of test cases, functions and named alternatives. Named alternatives are a special form of macros and will be explained in Section 8.

6.1   Test Cases

The test cases define the behaviour which has to be executed in order to judge whether an implementation under test passes the test or not. Test cases are defined in the module definitions part and called in the module control part. Each test case returns a test verdict of either **none, pass, fail, inconclusive** or **error**. This means a single test case can be considered to be a special kind of function returning a test verdict.

An example of a test case definition is shown in Figure 4. The test case is called *MyTestCase* and has the **inout** parameter *MyPar* of type **integer**. The **runs on** clause following the parameter defines the type of the MTC. The **system** clause specifies the type of the test system interface. The definition body defines the behavior of the MTC and will be started automatically when the test case is called. The MTC type is required to make the port names of the MTC visible inside the behavior definition. The type of the system interface is mandatory, if during the test run several test components are created and stopped dynamically. If the MTC performs the whole test on its own, the type of the test system interface is identical to the MTC type and can be omitted.

```
testcase MyTestCase(inout integer MyPar)
runs on MyMtcType1        // defines the type of the MTC
system  MyTestSystemType  // defines test system interface
{
  :        // The behaviour defined here executes on the MTC
}
```

Figure 4. Example for a test case definition

6.2   Functions

In TTCN-3, functions are used to express test behaviour or to structure computation in a module, for example, to calculate a single value or to initialize a set of variables. A function may be parameterized and may return a value. As shown in the function definition of *MyFunction* in Figure 5, the return value is defined by the **return** keyword followed by a type identifier.

If no **return** is specified then the function result is void. An explicit keyword for void does not exist in TTCN-3.

If a function defines test behavior, the type of the test component on which the behavior is executed has to be specified by means of a **runs on** clause. This type reference makes the port names of the component type visible inside the behavior definition of the function. This is shown in the definition of function *MyBehaviour* in Figure 5.

```
// Definition of MyFunction which has no parameters
function MyFunction () return integer {
    return 7   // returns 7 when the function terminates
}

function MyBehaviour (inout integer MyPar)
runs on MyPTCType
{   :                          // MyFunction3 make use of
    var integer MyVar := 5 * MyPar; // the port operation send
    PCO1.send(MyVar);          // and therefore requires a
    :                          // runs on clause to resolve
    :                          // the port identifiers
}
```

Figure 5. Examples for function definitions

7.   COMMUNICATION OPERATIONS

TTCN-3 supports message-based (asynchronous) and procedure-based (synchronous) communication. As illustrated in Figure 6 asynchronous communication is non-blocking on the **send** operation, where processing in the MTC continues immediately after the **send** operation. The SUT is blocked on the **receive** operation until it receives the send message.

Synchronous communication in TTCN-3 is related to remote procedure calls. As illustrated in Figure 7, the synchronous communication mechanism is blocking on the **call** operation, where the **call** operation blocks processing in the MTC until either a reply or an exception is received from the SUT. Similar to the asynchronous **receive** operation, the **getcall** blocks the SUT until the call is received.

7.1   Asynchronous Communication

For asynchronous communication, TTCN-3 provides the **send** and **receive** operations. The **send** operation is used to place a value on an outgoing message-based port. The value may be specified by referencing a template, a variable or a constant, or can be defined in-line in form of an expression (which of course can be an explicit value). When defining the value in-line, the optional type field can be used to avoid any ambiguity of the type of the value being sent.

Figure 6. Illustration of the asynchronous **send** and **receive** operations

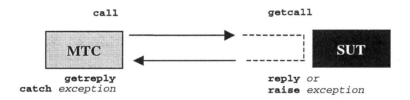

Figure 7. Illustration of a complete synchronous call

The **receive** operation is used to receive a value from an incoming message port queue. If the top message in the port satisfies all matching criteria associated with the **receive** operation, it is removed from the queue. The matching criteria may be related to the value of the message or the sender of the message. If the match is not successful, the top message is not removed, i.e. an alternative **receive** operation is required to remove the message from the port queue. For example:

```
MyCL.send(integer:5);
// Sends integer value 5 is sent via port MyCL

MyCL.receive(MyTemplate(5, MyVar));
// Reception of a value which fulfils the conditions defined
// by template MyTemplate with actual parameters 5 and MyVar

MyCL.receive(MyType:*) from MyPartner -> value MyVar;
// Receives an arbitrary value of MyType from a MyPartner.
// The received value is assigned to MyVar.
```

7.2   Synchronous Communication

As is shown in Figure 7, for synchronous communication, the calling side and the called side have to be distinguished. In order to test both, TTCN-3 provides communication operations for both sides.

The communication operations for the calling side are the **call** operation to call a remote procedure, the **getreply** operation to handle replies (or answers) to calls and the **catch** operation to handle exceptions which in case of exceptional situations may be received instead of a reply. In addition, TTCN-3 provides special **timeout** exception to cope with situations where the called party neither replies nor raises an exception. For example:

```
// Calls remote procedure MyProc via MyCl. A timeout
//   exception will be raised after 30 ms
MyCL.call(MyProc(5,MyVar), 30ms) to MyPartner {
   [] MyCl.getreply(MyProc:{MyVar1, MyVar2}) ->
          value MyResult param(MyPar1, MyPar2);
      // Handles a reply to the call. The return value is
      // assigned to MyResult. The out/inout parameters
      // are assigned to MyPar1 and MyPar2.

   [] MyCL.catch(MyProc, MyExceptionOne) { // Exception
         stop                      // Stop of component
      }
   [] MyCL.catch(MyProc, MyExceptionTwo); // Second exception

   [] MyCL.catch(timeout) {   // Timeout exception
         verdict.set(fail);
         stop;
      }
}
```

For the called side, TTCN-3 provides the **getcall** operation to accept calls, the **reply** operation to reply to calls and the **raise** operation to raise exceptions. For example:

```
MyCL.getcall(MyProc:{5, MyVar}) -> sender MySenderVar;
// Accepts a call of MyProc. The calling party is retrieved
// and stored in MySenderVar.

MyCL.reply(MyProcTemp(20,MyVar2) value 20) to MySender;
// Replies to the accepted call above.

MyCL.raise(MyProc, MyVar + YourVar - 2) to MySenderVar;
// Raises an exception for an accepted call with a value
// which is the result of the arithmetic expression.
```

7.3   The Check Operation

The **check** operation is a generic operation that permits to read the top element of message-based or procedure-based incoming port. The **check** operation has to handle values at message based ports and to distinguish between calls to be accepted, exceptions to be caught and responses from previous calls at procedure-based ports. This is done by using the operations **receive**, **getcall**, **getreply** and **catch** together with their matching and assignment parts to define the condition which has to be checked and to extract the value or values of its parameters if required. Examples:

```
MyAsyncPort.check(receive(integer:5) from MyPartner);
// Check for an integer value of 5 from MyPartner in port
//MyAsyncPort.

MyAsyncPort.check(receive(integer:*) -> value MyVar);
// Checks for any integer value at port MyAsyncPort.
```

7.4   Controlling Communication Ports

TTCN-3 provides the **clear**, **start** and **stop** operations to control communication ports. The **clear** operation removes the contents of an incoming port queue. The **start** operation starts listening at and gives access to a port. The **stop** operation stops listening and disallows **send**, **call**, **reply** and **raise** operations at the port.

8.   SPECIAL BEHAVIOR STATEMENTS IN TTCN-3

The configuration operations, the communication operations, and the verdict operations have already been explained in the previous sections. The basic program statements such as **if-else**, **while**-loop, **for**-loop or **goto**, and the timer operations **set cancel** and **readtimer** are well known from other programming and specification languages and need no special explanation. Only the handling of alternatives and the handling of defaults are special to TTCN-3 and need some explanation.

8.1   Alternative Behaviour

The alternative behavior statement (or **alt** statement for short) describes branching of control flow due to the reception of communication and timer events, i.e. the **alt** statement is related to the use of the TTCN-3 operations **receive, getcall, getreply, catch, check** and **timeout**.

An example of an **alt** statement is shown in Figure 8. The different branches of the **alt** statement start with square brackets which may include nothing, a boolean expression or the keywords **expand** or **else**. The brackets can be seen as a sort of boolean guard for the following receiving event. Empty brackets denote the value **true**. An **alt** statement is evaluated from top to bottom. A branch is selected when the boolean guard evaluates to **true** and the following **receive, trigger, getcall, getreply, catch, check** or **timeout** operation can be executed. A selected branch is executed in the expected manner.

The keyword **expand** denotes a macro expansion and is described in the next section. The keyword **else** is an unconditional exit of an **alt** statement. The else branch does not have to start with a receiving operation and is always taken if none of the previous branches can be selected.

```
alt {
    []        L1.receive(MyMessage1) {
                  : // Do something
              }
    [x>1]     L2.receive(MyMessage2);    // boolean guard

    [x<=1]    L2.receive(MyMessage3);    // boolean guard

    [expand]  MyNamedAlt;                // macro expansion

    [else]  stop                         // else branch
}
```

Figure 8. Example of an **alt** statement

8.2   Named Alternatives

An **alt** statement which is used in several places can be defined in a named alternative denoted by the keyword pair **named alt**. A **named alt** is a macro definition and causes a textual replacement when it is referenced. It can be referenced at any place in a behavior definition where it is valid to include a normal **alt** construct. Furthermore, it can be used to add alternative branches in an **alt** statement as shown in Figure 8. For example:

```
named alt MyNamedAlt {
    [] PCO2.receive(DL_EST_IN);
    [] PCO2.receive(DL_EST_CO);
}
```

8.3 Default Handling

In TTCN-3 defaults are used to handle communication events which may occur, but which do not contribute to the test objective. For example, when testing the call forwarding feature of an ISDN system, charging information may be received at any time. This information is not relevant for the testing objective and thus, can be ignored in the test evaluation. During the test, execution messages or calls containing such information may be received and have to be handled. This can be done by means of defaults.

The default concept of TTCN-3 is related to the macro expansion concept of named alternatives, i.e. an activated default expands automatically all named alternatives referenced in an **activate** statement. It is also possible to deactivate defaults by using the **deactivate** statement.

9. CONCLUSIONS AND OUTLOOK

We have presented here a simple and general core testing language called TTCN-3. The language is currently in the standardisation process at ETSI and ITU-T with the plan to be published in the year 2000 as an EN by ETSI (under the work program of Technical Committee MTS) and in the year 2001 as ITU-T standard Z.140. The next steps will then be the publication of the presentation formats. There are concrete plans for the tree and tabular presentation format and the MSC presentation format.

A number of tool makers have already shown interest in implementing the language. Some of the tools are embedded in an environment together with SDL [10]. Therefore it is necessary to consider the interworking between an SDL specification and a TTCN-3 test suite. This will also enable mechanisms for automated test case generation.

There is already research on the way for including real-time [2,3] and performance [4] aspects into TTCN. With this new version it seems feasible to base the performance and real-time extensions on the core language. New research projects are just in the process of being started in this direction

Acknowledgements

There are many individuals who contributed to TTCN-3 in several ways. Listing names at this point brings the risk, that someone will be forgotten. We therefore constrain ourselves, with one exception, to listing the main contributing organizations: Danet, Ericsson, Expert Telecoms, France Telecom, Fraunhofer Gesellschaft (FhG), GMD Fokus, Motorola, NMG Telecoms, Nokia, Nortel, Tektronix, Telelogic, University of Lübeck (Institute for Telematics). However, out of all the all the individuals who contributed, we would like to highlight the engagement of Os Monkewich from Nortel.

TTCN-3 is currently being developed under the work program of ETSI TC MTS (Methods for Testing and Specification). This proposed standard has not yet been published. For the official version of the TTCN-3 standard please contact the ETSI publications office at publications@etsi.fr.

References

[1] Jens Grabowski, Dieter Hogrefe: An Introduction to TTCN-3. Invited Presentation of the '12th International Workshop on Testing Communicating Systems' (IWTCS'99), Budapest, September 1999.

[2] Thomas Walter, Jens Grabowski.: A Framework for the Specification of Test Cases for Real Time Distributed Systems. In: Information and Software Technology, vol. 41, Elsevier, July 1999.

[3] [Thomas Walter, Jens Grabowski: Real-time TTCN for testing real-time and multimedia systems. In: Testing of Communicating Systems (Editors: M. Kim, S. Kang, K. Hong), volume 10, Chapman & Hall, 1997.

[4] Ina Schieferdecker, et al: PerfTTCN, a TTCN Language Extension for Performance Testing. . In: Testing of Communicating Systems (Editors: M. Kim, S. Kang, K. Hong), volume 10, Chapman & Hall, 1997.

[5] EN00063-1 (provisional),TTCN-3 Core Language

[6] EN00063-2 (provisional), TTCN-3 Tabular Presentation Format

[7] EN00063-3 (provisional), TTCN-3 MSC Presentation Format

[8] ISO/IEC 9646-3 (1998): Information technology - Open systems interconnection – Conformance testing methodology and framework - Part 3: The Tree and Tabular combined Notation (TTCN).

[9] ITU-T Recommendation Z.120 (2000): Message sequence Chart (MSC).

[10] ITU-T Recommendation Z.100 (2000): Specification and Description Language (SDL).

[11] ITU-T Recommendation X.680 (1997): I ISO/IEC 8824-1:1998,Abstract Syntax Notation One (ASN.1): Specification of basic notation.

[12] ITU-T Recommendation X.681 (1997) I ISO/IEC 8824-2:1998, Information technology – Abstract Syntax Notation One (ASN.1): Information object specification.

[13] ITU-T Recommendation X.682 (1997) I ISO/IEC 8824-3:1998, Information technology – Abstract Syntax Notation One (ASN.1): Constraint specification.

[14] ITU-T Recommendation X.683 (1997) I ISO/IEC 8824-4:1998, Information technology – Abstract Syntax Notation One (ASN.1): Parameterisation of ASN.1 specifications.

NOTE: The EN-00063 numbers are only provisional ETSI Work Item numbers (the actual EN numbers will not be the same)

11

HTTP PERFORMANCE EVALUATION WITH TTCN

Roland Gecse, Péter Krémer, János Zoltán Szabó
Ericsson Research
Conformance Center, Ericsson Ltd.
H-1037 Budapest, Laborc u. 1., Hungary
E-mail: { Roland.Gecse, Peter.Kremer, Szabo.Janos } @eth.ericsson.se

Abstract   This paper presents an approach, an implementation and an example for using TTCN in performance tests. The idea of using TTCN in performance test originates from PerfTTCN. The concept of PerfTTCN is excellent. The realization, however, does not seem to get acceptance as – to our knowledge – no vendor supports its non-standard tables. This is the point where the implementation described herein differs. The extra performance related tables are replaced by services abstracted in ASPs. The functionality behind them is then imported from or implemented as part of the MoT. This approach results in greater flexibility and wider range of applicability. An example is also shown using HTTP/1.0 and HTTP/1.1. Performance measurement results of these versions of the protocol are also presented. Finally we draw a conclusion.

Keywords: Performance test, TTCN, HTTP, Internet

1.   INTRODUCTION

The performance aspects in today's telecommunication networks became as important as the functional correctness. In order to get repeatable and comparable results we need formalized and standardized methods not only for conformance tests but for performance measurements as well. A proposal to this is the PerfTTCN [1], an extension of TTCN (Tree and Tabular Combined Notation) [4], [5].

The basic idea behind conformance testing is to verify a protocol implementation – Implementation Under Test (IUT) – whether it fulfills the requirements described in its specification or not. TTCN is a standardized test description language used for black box testing of different protocols that are based on the OSI reference model, and it is mostly used for conformance testing of telecom protocols and applications. The usage of TTCN seems to be an evident candidate since performance testing presumes the functional correctness (i.e. conformance) of the implementation and the conformance testing methodology is also applicable for IP-based protocols as it was shown in [17], [12].

PerfTTCN adds some extra functions to the base TTCN system. It defines a performance test configuration, which consists of one or more IUTs, Foreground Test Components (FTCs) communicating with IUTs and Background Test Components (BTCs). BTCs communicate with each other according to given traffic models in order to emulate the real load on the network, and they can monitor continuously the state of the network, too. PerfTTCN supports measurement and analysis as well: FTCs can measure for example the response times of IUT.

We chose the Hypertext Transfer Protocol (HTTP) for the evaluation of PerfTTCN, mainly because of its simplicity and widespread use. We know that there are plenty of well-established research studies related to the performance aspects of different HTTP versions (1.0 and 1.1) and we did not want to produce a "yapa" (yet another paper about ...). Our aim was to demonstrate the capabilites and to prove the usefulness of PerfTTCN. We think that the most effective way is to show that this approach gives comparable results.

This paper is organized as follows. The next section presents former work on HTTP performance evaluation outlining some problems and solutions. Section 3 describes PerfTTCN, our modifications on the theory and the implementation. After, we illustrate the usage of PerfTTCN with an example of HTTP (both 1.0 and 1.1) and present some test cases with explanation. The results of the tests can be found in section 4, where we compare the output of different configurations (HTTP/1.0 serial, HTTP/1.0 parallel, HTTP/1.1 persistent, HTTP/1.1 pipelined). Finally, we draw a conclusion and describe the future directions of our work.

2.   RELATED WORK

The main purpose of this paper is to show how our PerfTTCN tool can be used for performance analysis. Our intention is to show consistent measurements taken with our tool rather than arguing on existing research results or introducing some new phenomenon. We chose HTTP as a base for our demonstration because there are a number of well-founded research results available.

In this section we discuss related work in the field of HTTP performance analysis. The first paper on HTTP performance problems [10] was published in 1994. It pinpointed certain design errors in HTTP protocol features to interfere with TCP. Single retrieval per request combined with small document size yield short lifespan connections, which cause performance and server scalability problems. Diminished protocol performance can be expressed in terms of packet overhead caused by continuous TCP connection establishment and termination that requires 7 packets. Moreover, the short living connections cannot utilize the available network bandwith efficiently because they are still in slow-start phase when finishing transmission. On the server side, this results in numerous connections being in TIME_WAIT state, which makes servers run out of resources soon. Unfortunately, the author did not suggest any workaround to the presented problems.

Another significant paper aiming on decreasing HTTP latency is [9]. Basically it presented the same problems as [10] but also suggests methods for avoiding the problem, such as long-lived connections and the introduction of two new methods. The GETALL method was intended to retrieve a document with all embedded objects using only one request. The second one is the GETLIST method that was the basic idea behind the pipelining feature of HTTP/1.1. The authors also compared their experimental implementation with existing HTTP applications and measured gains both in number of transmitted packets and latency.

[8] describes further investigations on proposed persistent HTTP extensions with wide focusing simulation studies. It also goes into detail in interaction between HTTP and TCP with respect to experimental protocol extensions. Transaction TCP tries to get rid of connection handling overhead and problems related to that too many server connections are in TIME_WAIT state. This paper has been a serious contribution that led to changes in HTTP dynamics included in HTTP/1.1.

Later, [14] showed that persistent HTTP itself does not solve the latency problem. This is especially true on low bandwidth links, where it can hardly accomplish gain even in optimistic case.

[7] pointed out additional performance bottlenecks in TCP-HTTP interactions, which might, in the worst case, lead up to 20 times slower download time using persistent connections. Delayed acknowledgements as instructed by [15] combined with Nagle may cause serious delays that affects persistent HTTP traffic when small amount of data is transferred. The authors raised the problem of restarting idle TCP connections in combination with HTTP.

There have been continous investigations whether HTTP/1.1 can outperform HTTP/1.0 or not. [6] compared parallel 1.0, (persistent) 1.1 and 1.1 pipelined implementations. The authors took measurements on a sample web site using two different httpds. They also observed the impact of changing web content (e.g. using PNG graphics and CSS1 style sheets instead of GIF) as well as considered speed gains using HTTP/1.1's caching mechanism. Content compression techniques and different QoS links were also investigated. Their measurements proved that there are serious gains in overhead in terms of packets transmitted. However, in latency – which is the most interesting factor from the user's perspective – there are no dramatic improvements even if pipelining is implemented.

Some papers address HTTP-TCP interaction from TCPs point of view. [16] deals with the HTTP latency problem. It proposed a new method for recovering from idle TCP connections. TCP congestion control did not regulate clearly what should happen with cwnd after recovering from a long inactive period. Some implementations use Jacobson's conservative approach of resetting the restart window (RW) to 1 segment size. Others simply use the previous value of cwnd and restart the connection with a burst. The latter technique would be very advantageous for HTTP latency. Nevertheless, this behaviour jeopardizes network stability. Authors of this paper gave a new proposal which is a good compromise between the two extremities mentioned before.

The latest version of TCP congestion control [18] gives new guidelines for TCP implementors. Among other serious issues, such as slow start and congestion avoidance now it also deals with restarting idle connections. However, the modifications suggested by the previous paper have not been included. [18] requires to set RW = IW ($\leq 2 * SMSS$ bytes or 2 segments). And that means, that recovering from idle persistent connections will remain a bottleneck in HTTP. This also justifies the findings of [6] that pipelining is needed in HTTP/1.1 implementations in order to outperform HTTP/1.0 with many simultaneous connections.

3.   PERFTTCN

As PerfTTCN is only a proposal there will not be any commercial software supporting it in the near future. Therefore, our aim was to design and realize a performance testing environment, which has the similar functionality as the original PerfTTCN. Since there are many commercial implementations, which can execute test suites written in TTCN, it was obvious that we can add these extra functions as extensions to such a software.

The first implementation of PerfTTCN was presented together with the proposal in [1]. The researchers of GMD FOKUS used the GCI compiler of ITEX, which generated an executable C program code from the TTCN test suite. Then the generated code was completed with functions, which realized PerfTTCN's extensions.

We did not want to use the non-standard extension tables of PerfTTCN because it is very difficult – if possible at all – to make commercial tools interpret them. Thus, our performance test suites consist of only the standard TTCN tables though they contain all information about the background traffic and measurements as well.

These test suites have two extra PCO types (for the background traffic generation and measurements) and several ASP definitions. Our developed modules – which handle these ASPs – can be considered as a part of the test execution environment. An advantage of this architecture is that existing conformance test suites can be easily converted into performance test suites by adding the needed definitions and ASP events.

3.1   Extensions to the Original Proposal

In the implementation of GMD FOKUS the extensions cannot be handled independently from the test suite. Thus the modification of the generated source code is always needed and it cannot be done automatically. One needs to know the structure of ASPs and PDUs in order to perform measurements. Furthermore, the configuration and parameters of the background traffic is stored separately from the test suite. In our solution all parameters are defined in the test suite and the PerfTTCN's extensions can be applied to any protocol without any changes.

Our extension module can evaluate the performed measurements run-time, during the execution of our tests. This means the measured values are accessible from the test suite and the performance test verdicts can be determined immediately at the end of the test case.

In our implementation (Figure 1), the parameters of the background traffic are set in the dynamic part during the execution of the test. Since we do not need to hard-wire these values, adaptive test cases can be developed, which can change the characteristics of the generated load based on the measured values.

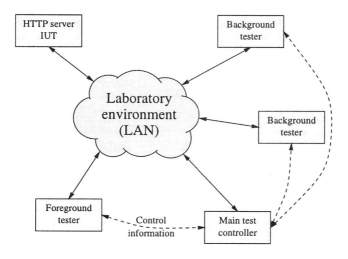

Figure 1.   Test configuration of PerfTTCN

The original implementation limited the background traffic models to Markov Modulated Poisson Processes. But there are many protocols – especially in the application layer – whose traffic cannot be modelled with MMPPs. So we used traffic generators which do not have this limitation. With the current version of our extension module we can describe and generate arbitrary IP-based traffic and interfaces to any traffic generator systems can be easily developed.

3.2   The Implementation

For the execution of our tests we use Ericsson's System Certification System – SCS. It has an interpreter-based TTCN test executor, which provides a generic framework for testing various protocols. SCS has to be adapted to the tested protocol using a so called test port. A test port is a library used by the test executor which maps the TTCN ASPs onto SPs of the underlying service provider in order to exchange data with IUT.

Each test port consists of a C++ class, which implements the ASPs of a given PCO type. The ASPs to be sent or received are mapped to functions, which must be written by the user. The ASN.1 and TTCN data types are converted into C structures.

Since we did not want to change the format of the TTCN test suite by adding new tables, we have defined a new interface for PerfTTCN through ASPs. We have developed two test ports that implement the extensions of PerfTTCN instead of communicating with IUT.

The main idea of this conception is that these test ports provide a service for the test executor which can be accessed through ASPs. In order to use their services first we have to define some objects (like measurements, traffic models, etc.) inside the test port. The definitions are done by sending special declaration ASPs usually in the preamble of the test case. Each object must have a unique name that identifies it in further operations.

3.3   Measurements

One of the developed test ports deals with measurements and analysis (Figure 2(a)). These functions are used by FTCs, which communicate directly with IUT according to the TTCN dynamic behaviour specification. This test port can carry out performance measurements during our test (e.g. the elapsed time between the sending of a request and the IUT's response can be measured) and it can perform statistical calculations automatically. We need these functions to accomplish performance tests because the base TTCN language lacks in such measurements. There are timers and timeout events, which can only detect error situations. The derived parameters, such as throughput rates, can be simply calculated from the measured times.

This test port supports three types of objects: measurements, characteristics and performance constraints.

A measurement object means a simple measurement of elapsed time between two test events based on the system time of the test executor equipment. Such a measurement behaves like a stop-watch, it is to be started and stopped by sending special ASPs to the test port at the corresponding test events. A measurement has a unit – like TTCN timers – which should be set in the declaration. The test port supports parallel measurements as well, so the different objects are treated independently. The last value of a measurement can be queried at any time.

The characteristic stands for a statistical property – like mean or peak value – of one given measurement. The declaration of a characteristic should consist of the name of the corresponding measurement and the type of the property. We can set a minimal sample size or a minimal sampling time interval as well. The actual value of a characteristic is automatically calculated by the test port according to the repeated measurements. This value – providing that the characeristic has a valid one – can be accessed at any time.

A performance constraint is a logical expression which contains the condition that the IUT has to fulfil to pass the performance test case. The expression should be given in a GeneralString argument of the ASP. It may refer to characteristics or other performance constraints and may use the usual arithmetical, logical and comparison operators. A performance constraint – like a verdict in TTCN – can have one of the three possible values: pass, fail or inconclusive. The inconclusive value means that one or more of the referenced characteristics still do not have a value because of the insufficient number of samples. The value can be queried at any time, but it is useful at the end of the test case's postamble to set the final verdict.

The test port makes a log file which contains the most important events including the measured values to enable further data processing and evaluation.

3.4   Background Traffic Generation

The TTCN-based implementation of BTCs is quite difficult if possible at all. Our approach was to build ready-made traffic generators into the TTCN test executor system. Since the protocol layer used by the traffic generators should match the tested layer in most cases, we need to use different traffic generators for testing different layers. Thus, our purpose was to develop a uniform control interface to different traffic generators. There are traffic generators available, which can be controlled remotely from a central point, therefore we have integrated such a system first.

The traffic generator that we built into our PerfTTCN environment is called NetSpec [13]. It was developed at the University of Kansas and its source code

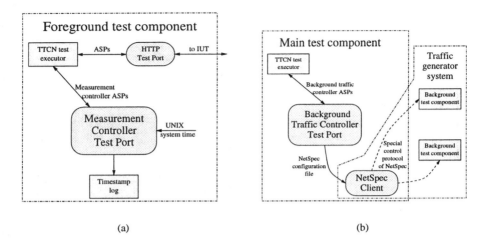

Figure 2.   Measurement controller test port (a), background traffic controller test port (b)

is freely available. It is portable to several UNIX platforms (Solaris, FreeBSD, Linux, etc.) and its components that are running on different platforms can also co-operate. The base system can generate traffic flows over TCP or UDP and there are extension modules for ATM, CORBA, etc. NetSpec has a distributed client-server architecture, which uses a special – TCP-based – control protocol. The only thing that is needed for traffic generation is to run a main server daemon on each computer that we want to participate in our configuration. The traffic models and configuration is described in a text file using a simple block-oriented language. The downloading of traffic parameters and the synchronization are done by the server daemons automatically. NetSpec has a limitation that it can generate only one-way data flows, so symmetric TCP flows cannot be modelled with it.

Our second test port is responsible for the control of background traffic (Figure 2(b)). The main purpose of this module is to make the process of performance testing fully automated. Since the traffic flows are described and controlled from a TTCN test suite, there is no need for user interaction during the test.

The parameters of background traffic can be defined in the test port with three types of objects: traffic models, components and loads. These objects are independent of the used traffic generator. The NetSpec-specific parameters are described in ASN.1 CHOICE constructs, which can be extended to support the features of other load generators.

In our approach a traffic model covers all stochastic behaviours of a single one-way flow. The model describes only the data source and is independent of the type of the underlying service. The destination node is assumed to be an ideal sink. With this abstraction we can generate both connectionless datagram (UDP) and connection-oriented stream (TCP) traffic based on the same model.

NetSpec supports user defined and application-layer traffic models. The application-layer models emulate the generated traffic of the most commonly used Internet applications like WWW, FTP or telnet. These models have few parameters (e.g. average request interarrivals or document sizes) because they assume the distribution type of the traffic parameters fixed. A user defined traffic model enables to specify the base stochastic characteristics of the traffic. The request interarrival time, the number of blocks per request and the block size can be defined using mathematical or empiric distributions.

NetSpec has another type of model called full. This is the simplest one of all. In this case the source node tries to send as much data as possible to utilize all the available bandwidth. This is useful to cause artifical overload situations on a part of the network.

In the test port first we have to define our models. The parameters of a model are described in nested ASN.1 SEQUENCE and CHOICE constructs. For example, several types of distributions with their parameters are represented as the alternatives of a CHOICE construct.

A traffic component is a one-way flow based on a given model with definite source and destination nodes and transport protocol. Thus, the parameters of the component-definition ASP contain the used protocol (TCP or UDP), the two endpoints and port numbers of the corresponding flow. NetSpec enables to set some protocol specific features, like the window size of TCP. These can be specified in the TTCN test suite as well.

During a performance test the background load is generated by numerous flows. These loads have to be defined in the test port, too. Such a load consists of a set of the components. It is possible that a load includes several instances of the same component. In this case the test port can assign unique port numbers to the multiple instances of the same component.

After the definitions but still in the preamble, the TTCN main test component should start the generated load by sending an ASP. At this point our test port generates a NetSpec configuration file according to the model and component definitions and passes it to the NetSpec client program.

Since NetSpec cannot stop its traffic generators at any time, we have to specify a duration for each load. Of course, this duration must be longer than the estimated execution time of the test case.

4.   TEST RESULTS

We performed our initial measurements in laboratory conditions on a separate 10 Mbps Ethernet segment with four machines (a Sun Sparc and three PCs with Linux) as shown in Figure 3(a). All stations were connected to a hub, which did not have any external connection. Heintel (Sun workstation) played the roles of FTCs and the main test controller.

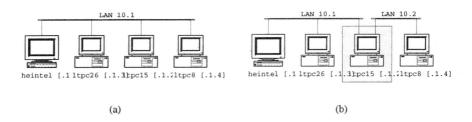

(a)   (b)

Figure 3.   Configuration of the test network with one segment (a) and with two segments (b)

Traffic was generated only between ltpc15 and ltpc26. These two PCs were the BTCs. The Apache Web Server (version 1.3.4) running on ltpc8 acted as IUT.

The background load was generated using NetSpec's "full" traffic model with a blocksize of 8192 bytes (this is the default value in NetSpec). We always generated the same number of TCP flows in both directions. We ensured that all measurements have been executed after the generated background load became stable. Thus the given number of TCP connections occupied fairly all available Ethernet bandwidth.

We increased the number of simultaneous TCP connections up to 100. Additional increase would not be worth since in our 10 Mbps Ethernet the effective bandwidth is 5 kbyte/s for each TCP connection.

$$\frac{10Mbps * 0.8(eff\_rate)}{2 * 100} = 40kbit/s = 5kbyte/s$$

It means that a TCP flow can send only 3 packets in one second because the MTU is 1536. This amount is too small so the congestion window could not grow above 9. Situations like this does not allow a TCP to perform well. Our test network collapsed when we tried to increase the number of TCP connections to 200.

We set up a test web "site" consisting of 13 files: an HTML page and a dozen embedded GIF images. The main HTML page has the size of 29 kbytes, about the half of the images are shorter than 3 kbytes, while the others are between 3 and 20 kbytes. The total size of about 80 kbytes can be considered as a typical web document.

We developed test cases (TCs) for downloading these files from IUT using different features of the two versions of HTTP. The only FTC of the TC named "HTTP/1.0 serial" downloads the documents consecutively using the version 1.0. It opens different TCP connections for each page. This method is the worst of all and is used for reference. The TC "HTTP/1.0 parallel" also uses version 1.0 but it uses parallel TCP connections to decrease download time. The main difference between these two TCs is that the latter uses four FTCs and four parallel TCP connections while the former is satisfied with only one of each. Furthermore, in the parallel TC the main test component assigns the documents to the FTCs dynamically in order to achieve the best latency among stochastic conditions.

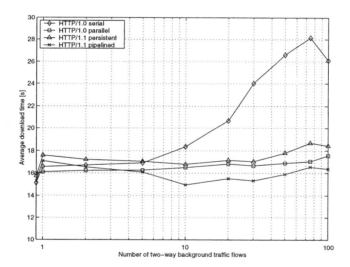

Figure 4.   Measurement of HTTP download times against network load using the conformance test port of HTTP (on the single segment configuration)

The TCs "HTTP/1.1 persistent" and "HTTP/1.1 pipelined" use the last version of HTTP. Both of them use persistent connections, thus they download all the documents using the same TCP connection. The latter TC uses the pipelining feature to further enhance download speed.

The initial results are shown in Figure 4. The presented results show the mean values of 10 successive measurements as calculated by our measurement controller test port.

At first sight we found that our results satisfied our expectations. We achieved similar results as the majority of cited papers. The best performing "HTTP/1.1

pipeline" always outperformed HTTP/1.0 using four parallel connections. The latency of "HTTP/1.0 serial" grew with the increasing number of background traffic components.

It is interesting, that this is not the case for the rest of our measurements. Another, more significant problem is the download time. It takes more than 15 seconds to download the 80 kB test documents, which is pretty slow on a dedicated Ethernet line. The bottleneck in our configuration turned out to be the HTTP test port in our TTCN executor. This test port was originally developed for conformance test for [11]. It uses very detailed ASN.1 descriptions of HTTP data structures including grammar extensions for building all kinds of invalid PDUs. Additional slowing factor is that we also validate the received data to check if there is mismatch between expected/received data (HTML and GIF content). Unfortunately, the execution platform was not fast enough to cope with these problems.

4.1   Performance Test Port

Finally, we decided to replace our HTTP test port with another one, specially designed for performance measurement called performance test port. We disabled the payload examination and replaced the detailed ASN.1 PDU description part with a simple string based structure. All other parameters, the executed TCs and the configuration were the same. The results, however, showed significant improvement as it can be seen in Figure 5. The download time on the unloaded LAN decreased under 700 ms for all cases. HTTP/1.1 with pipelining was the winner this time, too. Then HTTP/1.0 using multiple connections follows before the persistent HTTP/1.1. Again the original HTTP/1.0 performed worst. It is interesting to note that after a short ramp up phase all curves become flat. That is, over a certain number of competing parallel connections the download times do not change significantly.

A serious drawback of the previous configurations is that both of them measure CSMA/CD performance. Since the traffic generators run on other machines than IUT and FTCs the generated load wipes out the chance of the access to the shared medium only. But our initial goal was to point out differences between implementation details in realistic situations. Thus, we set up a second configuration having two subnets, such as Figure 3(b), where we also introduced an intermediate router powered by Linux 2.2.12.

In addition to performing regression tests using our new configuration, this time we also performed the measurements using real WWW clients in order to compare our results. Netscape Navigator 4.5 runs 4 parallel "Keep-Alive" HTTP/1.0 connections (which is similar to a "HTTP/1.1 persistent" one) while wget uses plain HTTP/1.0. We found our measurement results (Figure 6) using the new test port comparable with those of the mentioned applications. On the

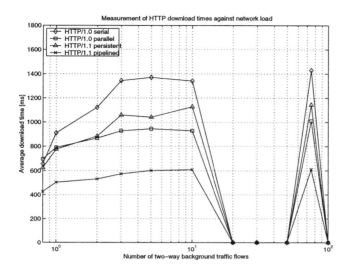

Figure 5. Measurement results using the performance test port of HTTP

one hand this proved that our test port and test cases are correct. On the other hand this gave the evidence that the PerfTTCN concept does really work and it is applicable for performance testing. Further investigation was only possible after these criteria has been fulfilled.

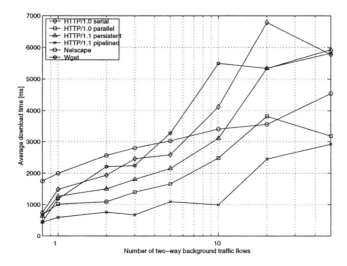

Figure 6. Measurement results on the second configuration (two segment)

Our last mentioned test case – simulating an HTTP/1.1 pipelining client – performed the best again. It is no wonder that it outperformed Netscape Navigator as we do not process the received data just discard it. However, wget was not agressive at all when it comes to download speed. Certainly, this is due to politeness criteria that has to be fulfilled by each robot application. The curves themselves look now as we expected. The mean download time is an exponential function of the number of parallel flows, which is proportional to the network utilization.

Our final measurement in Figure 7 shows what happened when we increased the amount of downloaded data. We designed a new test document consisting of a 2 kbyte HTML, a small 12 kbyte background and 4 pictures of length of about 100 kbytes each. The total size was 545 kbytes. This time we saw that HTTP/1.0 using multiple connections always outperformed the HTTP/1.1 pipelining implementation. We consider this as a consequence of the relatively small number of downloaded objects and their longer size. In the case of our previous test document, in which we had more objects of smaller size, we believe that the extreme background traffic filled up the router queues and pipelining could not work efficiently. This also justified our thought that "HTTP/1.0 parallel" can outperform "HTTP/1.1 pipeline" under certain circumstances. This is done, however, on the expense of more signalling overhead.

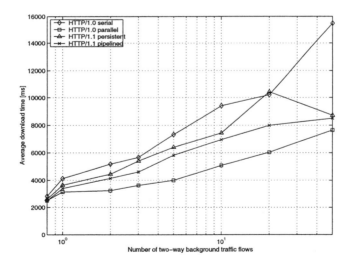

Figure 7.   Measurement results downloading the bigger test document (using the second configuration)

5. CONCLUSION

We have performed HTTP performance tests in a new way, we have run our comparison tests on top of loaded network. We have measured latency by the number of simultaneous TCP connections. We have managed to get similar results than many researchers before us, with only two remarks. First, we have run our tests on congested network, which has not been done before in any of the cited papers. Second, mainly our focus was on demonstrating that our environment based on PerfTTCN can cope with any other tool in real life measurements. Unlike other traffic engineers, as a result of our investigation we have found that there are situations when HTTP/1.0 using simultaneous connections achieves better performance than HTTP/1.1.

We have run our test cases on two different network configurations downloading two web pages having different characteristics using one type of three available traffic models. Future work can be, for instance, to repeat the measurements with other traffic models. The tests could also be repeated simultaneously, i.e. running all four types of data transfer in parallel and investigate their competition. Besides, the real-time evaluation of distributed measurements (when the start and end events are on different test components running on different machines) is a quite interesting (and difficult) problem, too.

References

[1] I. Schieferdecker, B. Stepien, A. Rennoch: PerfTTCN, a TTCN language extension for performance testing, Testing of Communicating Systems, Cheju Island, Korea, September 1997.

[2] R. Fielding, J. Gettys, J. Mogul, H. Frystyk, L. Masinter, P. Leach, T. Berners-Lee: Hypertext Transfer Protocol – HTTP/1.1, RFC 2616, IETF, Network Working Group, June 1999.

[3] T. Berners-Lee, R. Fielding, H. Frystyk: Hypertext Transfer Protocol – HTTP/1.0, RFC 1945, IETF Network Working Group, May 1996.

[4] OSI - Open System Interconnection, Conformance testing methodology and framework, ISO/IEC 9646, 1997.

[5] B. Baumgarten, A. Giessler: OSI conformance testing methodology and TTCN, North Holland, 1994.

[6] H. F. Nielsen, J. Gettys, A. Baird-Smith, E. Prud'hommeaux, H. W. Lie, C. Lilley: Network performance effects of HTTP/1.1, CSS1 and PNG, http://www.w3.org/Protocols/HTTP/Performance/Pipeline.html, June 1997.

[7] J. Heidemann: Performance interactions between P-HTTP and TCP implementations, ACM Computer Communications Review, pp. 65-73, April 1997.

[8] J. C. Mogul: The case for persistent-connection HTTP, Western Research Laboratory Research Report 95/4, http://www.research.digital.com/wrl/publications/abstracts/95.4.html, May 1995.

[9] V. N. Padmanabhan, J. C. Mogul: Improving HTTP latency, Computer Networks and ISDN Systems, Vol. 28., pp. 25-35, December 1995.

[10] S. E. Spero: Analysis of HTTP performance problems, http://www.w3.org/Protocols/HTTP/1.0/HTTPPerformance.html, July 1994.

[11] R. Gecse: Conformance testing methodology of Internet protocols, Testing of Communicating Systems, Tomsk, Russia, September 1998.

[12] R. Gecse, P. Krémer: Automated test of TCP congestion control algorithms, Testing of Communicating Systems, Budapest, Hungary, September 1999.

[13] R. Jonkman: NetSpec User Manual, http://www.ittc.ukans.edu/netspec, April 1999.

[14] J. Touch, J. Heidemann, K. Obraczka: Analysis of HTTPPerformance, USC/Information Sciences Institute, http://www.isi.edu/lsam/publications/http-perf/, Aug. 16, 1996.

[15] R. Braden: Requirements for Internet Hosts – Communication Layers, STD 3, RFC 1122, October 1989.

[16] V. Visweswaraiah, J. Heidemann: Improving Restart of Idle TCP Connections, USC TR 97-661, http://www.isi.edu/ johnh/PAPERS/Visweswaraiah97b.html, November 10, 1997.

[17] T. Ogishi, A. Idoue, T. Kato and K. Suzuki: Intelligent protocol analyzer for WWW server accesses with exception handling function, Testing of Communicating Systems, Tomsk, Russia, September 1998.

[18] M. Allman, V. Paxson, W. Stevens: TCP Congestion Control, RFC 2581, April 1999.

12

CONFORMANCE TESTING OF
CORBA SERVICES USING TTCN

Alexey Mednonogov[1], Hannu H. Kari[1], Olli Martikainen[1], Jari Malinen[2]

[1] *Telecommunications Software and Multimedia Laboratory*

[2] *Laboratory of Information Processing Science*

[1,2] *Helsinki University of Technology, P.O. Box 5400 , FIN-02015 Espoo, FINLAND*

E-mail: mednonog@lut.fi, hhk@cs.hut.fi, Olli.Martikainen@hut.fi, jtm@cs.hut.fi

Abstract   This paper presents a formal approach to conformance testing of CORBA-based distributed services using TTCN (Tree and Tabular Combined Notation) framework. It discusses mapping of CORBA IDL to TTCN, concentrating on the obstacles and the design issues to be considered. The paper overviews the architecture of the CORBA/TTCN gateway, which acts as an intermediary between test environment and system under test (SUT). It goes through an example of test session and distinguishes the typical stages of dynamic behaviour. The results of the study indicate that TTCN framework especially facilitates testing of active service components, although conformance testing of reactive parts is also possible.

Keywords: CORBA, TTCN, conformance testing, distributed systems

1.   INTRODUCTION

The introduction of Internet-based services suggests a major change in the future telecommunications service provisioning. The amount of Internet traffic will outweigh the amount of traditional telephone traffic, and new services and service architectures will evolve. The transition from the existing telecommunications services to Internet based ones will change the service infrastructure as well as service creation and management structures [3]. New solutions for service creation, testing and management will be a key success factor for distributed Internet based service provisioning. The distribution of services has been particularly fuelled by the appearance of

Common Object Request Broker Architecture (CORBA) in 1991, a technology that addresses the need for interoperable, object-oriented, robust and high-performance distributed solutions in the era of global networks and communications [4].

Surprisingly, the issue of conformance testing of CORBA-based distributed services remains relatively underdeveloped. To our knowledge, application of TTCN (Tree and Tabular Combined Notation) to testing CORBA interfaces has been addressed only in few papers (among them, [10] considers testing of TINA service components and [11] describes application of TTCN in the context of test case derivation from SDL models created within a framework of TOSCA project). At the same time, TTCN as a part of ISO-9646 standard [2] is a mature framework, widely recognized for its flexibility and industrial strength. Covering this gap, we share our experiences in testing CORBA objects described in terms of IDL interfaces using TTCN language. We limit ourselves to conformance testing of application-level CORBA servers and clients without considering other issues like testing of GIOP/IIOP protocols, performance testing, inter-operability testing, and so on. For the purposes of the present study we assume that all other components of ORB are already tested and are functioning properly. The paper is organized as follows: In section 2, we present an overview of OMG Object Management Architecture and CORBA middleware architecture. In section 3, we define the mapping of CORBA Interface Definition Language (IDL) to TTCN, focusing on the obstacles of such conversion. In section 4, an overall architecture of the CORBA/TTCN gateway is presented and its capabilities are explained. Section 5 goes through a test session example and provides recommendations on test suite coverage. Finally, in section 6 our conclusions are presented.

2.   CORBA MIDDLEWARE ARCHITECTURE

The timely creation of interoperable, distributed and high-performance applications capable of communicating transparently in the heterogeneous network environment has grown to be an increasingly challenging task in the telecommunications industry. To alleviate the problem, Object Management Group (OMG), a consortium consisting of several hundreds of member organizations world-wide, has devised a CORBA middleware architecture, the first version of which has been adopted in 1991. CORBA seamlessly interconnects multi-vendor applications and services and let them communicate with each other irrespectively of their location in the network, the operating system they use or the programming language utilized at the implementation stage. CORBA is the communications heart of the Object Management Architecture (OMA) which acts as a higher level reference model (RM) for CORBA framework.

OMA RM classifies all system components into four categories as shown in Figure 1, namely: the Object Request Broker (ORB); CORBA Services (COS); CORBA Facilities; Application Objects. The ORB provides generic low-level communication services to all other OMA components and is responsible for transparent distribution and communication of objects in the network. CORBA Services define a set of system-level interfaces complementing the basic functionality of ORB, and CORBA Facilities (classified into horizontal and vertical ones) provide standard frameworks responsible for defining the universal rules of engagement for collaborating objects [9].

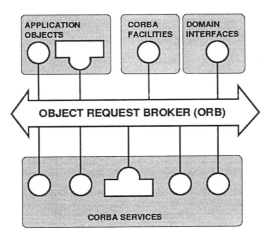

Figure 1. OMA Reference Model

CORBA specification constitutes a set of well-defined rules describing how OMA-oriented architecture shall be actually constructed. Its core part discusses the syntax and semantics of Interface Definition Language (IDL). Object declaration in terms of IDL acts as a contract between a concrete object implementation (server) and its potential clients which is independent from the programming language, platform and location of the contracted parties. IDL declaration provides the formal means for declaring the structure of CORBA object, abstracting from the implementation details. In particular, it defines object location within hierarchy of IDL modules and interfaces, operations accepted by the object, exceptions it raises and the data types of input and output parameters used in operation calls. By putting an IDL declaration into Interface Repository (IFR), the server advertises its capabilities to all interested clients which can introspect the contents of IFR and find out how invocation requests are handled by a specific server, thus turning CORBA into a self-described component architecture. Combined with Dynamic Invocation Interface (DII) and Dynamic Skeleton Interface (DSI), Interface Repository provides powerful and flexible means for dynamic invocation of server operations or dynamic processing of client requests. In case a distributed application or service platform does not need this flexibility, invocation requests can be processed directly using precompiled static IDL stubs and skeletons that act as a wrapper for low-level core part of the invocation request. CORBA also standardizes the way objects establish a communication path with each other by associating each server with an Interoperable Object Reference (IOR). As soon as the client finds out the exact contents of IOR, it may freely start communicating with the corresponding server. The overall architecture of CORBA ORB is shown in Figure 2.

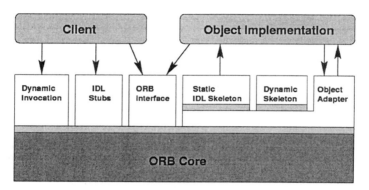

Figure 2. Common Object Request Broker Architecture

In TTCN-oriented conformance testing of CORBA clients and servers the power of DII, DSI and Interface Repository introspective capabilities is fully employed. Basically, TTCN ASP is converted to CORBA dynamic request or to dynamic response and vice versa, and appropriate generic clients and servants are created on the test environment side for that purpose. According to CORBA specification, a System Under Test (which is viewed as a collection of client and server objects) cannot distinguish whether a static or dynamic invocation scheme has been used, so the implementation details of the test environment are completely transparent from the SUT perspective.

3.   MAPPING OF CORBA IDL TO TTCN

The rules for mapping CORBA Interface Definition Language to TTCN have been presented in the paper of GMD FOKUS [10], although the related work of X/Open and Network Management Forum (NMF) conducted by the JIDM (Joint Inter Domain Management) working group shall be acknowledged, too [8]. However, to our knowledge JIDM mainly concentrates on the reverse task of mapping GDMO/ASN.1 to IDL, so JIDM mapping rules have had an indirect impact in the present paper; on the other hand, the mapping discussed in [10] has contributed most to our study. In particular, [10] defines the rules for mapping IDL operations, exceptions, constants, attributes and type definitions to TTCN, and we adhered to the same concepts as long as it has been practical. Yet, amendments and supplements to the approach discussed in [10] proved to be inevitable, either in cases when level of details in the original paper was insufficient or when alternative design of the mapping rules has been viewed as beneficial for improving the flexibility of the whole testing scheme. To avoid overlapping with [10], hereafter we provide a brief overview of the basic principles used in conversion of IDL to TTCN (mostly derived from [10]) and then concentrate on our findings in this area, as well as on the obstacles and the design issues to be considered.

The conversion rules are as follows: (a) one CORBA instance of IDL interface maps to one PCO declaration, which may be reused in different test cases to represent several CORBA objects; (b) declaration of one operation maps to a pair of ASPs, namely Call ASP and Reply ASP, the identifiers of which are prefixed by pCALL_ and pREPLY_ respectively; (c) one actual operation call maps to a pair of ASP constraints; (d) IDL data types are mapped to ASN.1 types; (e) IDL specification itself maps IDL attributes to normal operations, so additional arrangements are unnecessary; (f) one Exception ASP (identified as pRAISE) is defined for all possible CORBA exceptions, both system exceptions and user-defined, so that different constraints may be introduced as necessary for individual exceptions or for groups of them; (g) to facilitate debugging of an Abstract Test Suite (ATS) at design stage, CORBA/TTCN gateway introduces its own gateway-specific exceptions defined in terms of IDL to signal e.g. incorrectly constructed ASP and any other exceptional situations not directly related to SUT, and a separate namespace is reserved by the gateway for that purpose; (h) IDL-to-TTCN mapping implies an effective solution for IDL name resolution and inheritance mechanism without declaring that specifically, as both issues are more relevant to concrete language-dependent implementations of CORBA objects.

The mapping rules defined in our gateway specification ensure that every distinct entity of IDL language (interface, operation, type definition, identifier of enumerated type etc.) possesses a one-to-one mapping in TTCN namespace of identifiers. For that purpose, we use a symbolic chain "_i" as a separator and duplicate all occurrences of underscore characters ("_") in names of original IDL entities while mapping them to TTCN. For example, interface "IntC" defined within IDL module with scoped name "::ModA::ModB" according to these rules may be mapped to PCO identifier "PCO1_iModA_iModB_iIntC", and oneway operation "opD" defined within interface "::ModA_::_ModB::IntC" is mapped to Call ASP identifier "pCALL_iModA___i__ModB_iIntC_iopD". Apparently, we have to concatenate several IDL identifiers to form one TTCN identifier so that initial IDL entities can be reconstructed from the resulting TTCN aggregate, otherwise the conversion rules would not guarantee a one-to-one mapping, which can lead to possible collisions in TTCN namespace. To address this issue, we use a concept similar to "bit stuffing", a well-known pattern utilized in low-level telecommunication protocols to mark boundaries of frames sent over the communication medium. In our case, a similar idea is introduced to label margins of module, interface and operation names.

We also introduce an additional field CALL_ID of ASN.1 type INTEGER into every Call ASP, Reply ASP and Exception ASP to uniquely identify operation calls sent to CORBA servers or received from clients, which is necessary for support of concurrent invocations within one test case. It is required that concrete values of CALL_ID shall be unique within a test case, although set of CALL_ID values allocated for servers may overlap with values used by clients. Without having a CALL_ID field, it would be an ambiguous action to invoke the same operation while the previous operation call issued through the same PCO and within the same PTC is still pending, as in this case ATS would be unable to distinguish which operation completed its execution once one of them returns. Moreover, CALL_ID field is of vital importance for determining which operation call has thrown an exception if pRAISE ASP is received or sent. Alternatively, several parallel test components referring to the same PCO, or several PCOs referring to the same CORBA object could well be used instead of CALL_ID to resolve the ambiguous cases of concurrent invocations. Yet, in our view presence of CALL_ID brings more flexibility and determinism to concurrent operation calls. Moreover, concept of CALL_ID does not have an equally good substitution in the area of exception handling, taking into account mapping rules for IDL exceptions as they are presented in this paper.

Our specification of IDL/TTCN mapping intentionally avoids use of TTCN operations, although their use in concrete test suites is not restricted. One situation when TTCN operations could be used is related to associating PCO with a concrete CORBA object, and [10] seemingly uses operations for this purpose. However, this approach does not fit well with good TTCN design practice, as TTCN operations shall be normally limited to calculating the return value from the set of input parameters, and use of operation side effects for performing an association procedure does not strictly follows the formal part of ISO 9646-3 standard. Instead, we most naturally (and more formally) delegate the responsibility for associating PCO with a CORBA object to the CORBA/TTCN gateway by sending to it an ASP of a special format as needed.

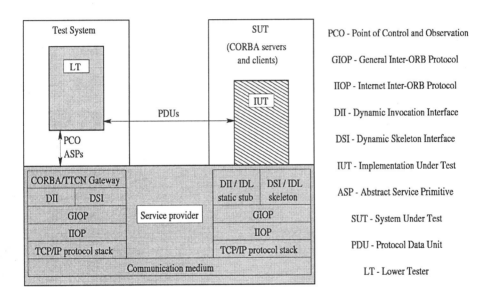

Figure 3. Position of CORBA/TTCN Gateway in Testing Architecture

For that purpose, we define several forms of Registration ASP which is sent to the gateway every time ATS is willing to bind PCO with a client or server object. For instance, ASP identified as pSREG_IOR sent through PCO in question will bind this PCO to the CORBA server advertised by IOR (Interoperable Object Reference); another form of Registration ASP identified as pCREG_NSERV is used to instruct the gateway to create a generic servant on its side and advertise its presence to all interested clients by putting this servant into object hierarchy in the Naming Service. Since then, all operation calls received from clients will be relayed to the PCO through which original pCREG_NSERV ASP has been issued. Gateway will inform ATS of whether the binding procedure failed or succeeded by sending a gateway-specific Exception ASP to the test environment. Gateway specification requires that all bindings mentioned above must be removed (and generic servants destroyed) without any special notification coming from ATS after test case completes its execution.

In our testing architecture the gateway is viewed as a part of Service Provider, as shown in Figure 3. Hence, we found it acceptable to introduce gateway-specific ASPs, thus following a gateway-aware approach, as long as even in this case the actual communication is still mediated by Service Provider (and the gateway being the part of it).

Our mapping of CORBA exceptions to TTCN allows catching group of exceptions in one TTCN constraint. As noted previously, we define one Exception ASP for all possible CORBA exceptions. This ASP contains three fields: (1) CALL_ID of the operation that has thrown an exception; (2) absolute scoped name of an exception thrown, as defined in IDL interface; (3) exception body encapsulated into separate PDU. Absolute scoped name is encapsulated into ASN.1 SEQUENCE OF IA5String, so that exception having scoped name "::ModA::IntB::ExcC" may be constrained by construct {"ModA", "IntB", "ExcC"}. At the same time, by defining construct {"ModA", "IntB", ?} we are capable of catching a group of exceptions defined within interface "::ModA::IntB". We assume that CORBA system exceptions are defined within interface "::CORBA::SystemException" and gateway-specific exceptions are defined within namespace "::GatewayException". The gateway-specific exceptions usually signal a receipt of structurally invalid ASP; inopportune ASPs may in some cases cause a gateway exception too, for instance when Call ASP has been issued before sending Registration ASP. However, the gateway will handle most of inopportune ASPs of all other kinds, as well as structurally correct ASPs containing invalid values.

It was already mentioned that IDL data types map almost naturally to ASN.1 types, as shown in Figure 4. Yet, there are two essential exceptions from this rule. One of them regards to IDL discriminated union. A straightforward solution implies mapping of IDL union directly to ASN.1 CHOICE. However, a direct conversion may lead to a loss of information about the exact value of the discriminator, since several values of the discriminator may refer to one branch of IDL union. On the other hand, CORBA language mappings allow a direct access to the discriminator, hence the result of such access is implementation-dependent. To eliminate this ambiguity, the final mapping for IDL union proposes encapsulation of discriminator field and ASN.1 CHOICE field into one ASN.1 SEQUENCE structure, where a field of type CHOICE contains the actual branch activated by union and an additional field carries the exact value of the discriminator.

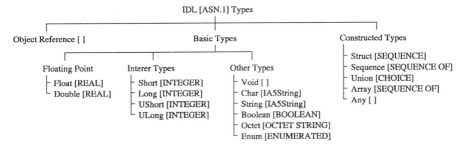

Figure 4. Mapping of IDL Data Types to ASN.1

More serious restrictions concern mapping of IDL Any type. Mapping of IDL to TTCN does not imply a one-to-one conversion from IDL data types to ASN.1 types, that is one ASN.1 type may correspond to several IDL types. For instance, ASN.1 INTEGER may correspond to IDL "unsigned short" or "long" or "unsigned long", and ASN.1 IA5String may correspond to IDL "char" or IDL "string". For this reason, CORBA/TTCN gateway internally uses Interface Repository (IFR) as its structural backbone, so every time it receives Call ASP going to server or Reply ASP or Exception ASP going to client, it performs a lookup of IFR to clarify how the structured data of the operation call must be actually processed. However, if the gateway meets the definition of Any type in IFR, it may not have any hint how to process the respective part of the received ASP, as the structural information extracted from the ASP is insufficient to provide a reliable guidance to the gateway, due to one-to-many nature of TTCN-to-IDL mapping of data types. If "any" value has been received from SUT, then its conversion to ASP is trivial, but in this case ATS shall contain multiple ASP definitions per one operation, what contradicts to the mapping rules described above. Due to all these complexities, current version of the gateway does not support mapping for Any type. Three possible solutions are anticipated, all three having their own advantages and drawbacks:

(1) Type information is explicitly inserted into corresponding identifiers of ASP fields in form of prefixes, e.g. "s" for strings, "c" for characters etc. However, this approach requires defining several pairs of ASPs per one operation call if the latter happens to contain "any". Moreover, since then ASP identifier itself must obey additional rules, because gateway must know exactly how ASP name is constructed at run-time, and if there are several

possible ASPs per response from the same operation, a kind of operation key is definitely needed. At the same time, we cannot encapsulate a variable of Any type into a separate PDU and then append prefixes only to what is inside (what would have eliminated the need for more than one pair of ASPs per operation), since ISO 9646-3 standard prohibits use of non-ASN.1 types inside ASN.1 data. Finally, it remains unclear how the whole scheme would address cases like "union" inside "any" or other challenges, so this approach shall be examined with caution.

(2) Values of "any" type are mapped to ASN.1 OCTET STRING type, having part of the marshalling buffer as a content filler. Although this approach seems to be the most natural solution from the gateway perspective, it puts a heavy burden on the designer of test suite who is since then responsible for manual construction of part of GIOP/ESIOP/IIOP stream, which might be practically unacceptable.

(3) The set of IDL interfaces describing SUT is supplemented with manually implemented decorators that do not contain "any", so that operation calls are routed through them. Decorator is a well-known design pattern, the main function of which is attaching additional responsibilities to the core system [1]. In our case, if we assume we know in advance all variations of typed information encapsulated into "any" variables, we can then make an equivalence transform of the operation containing "any" into several operations containing only conventional types. A set of these "wrapper" operations may be defined within one decorator object acting as a relay point between MOT and core SUT. This idea could be extended to mapping of one server operation containing "any" into two sets of decorating oneway operations, one set responsible for decorating a direct part of operation call and another one responsible for decorating operation return. This would help if operation call and operation return both contain "any" parameters. The similar idea could be also applied to operation calls coming from clients. The technique of decorators is a good trade-off between two previous approaches, although it requires additional effort to implement decorating CORBA objects.

4.   ARCHITECTURE OF CORBA/TTCN GATEWAY

The CORBA/TTCN gateway acts as an intermediary between the test environment and SUT. The gateway itself is a CORBA-based Java application, enjoying all the benefits of both CORBA distributed nature and Java platform independence. The ORB of our choice was ORBacus for Java from Object Oriented Concepts Inc [6], which fully implements the core functionality of CORBA 2.0 in accordance to the OMG standard. CORBA mapping to Java is the most natural and frequently used solution, as Java takes care of transparent object removal and has a comprehensive exception handling mechanism. Although Java is an interpreted language and is relatively slow, this issue is not of crucial importance as long as performance measurements in conformance testing are not usually involved. Java is not perfectly suitable for testing low-level protocols as Java API does not provide a direct access to protocol layers lower than TCP and UDP, but this could be solved by implementing a gateway logic on Java and a system-level part on C++. The two parts could be, again, combined using CORBA or Java native calls. Nevertheless, in case of testing CORBA services it is sufficient for the gateway to have indirect access to the protocol stack.

The gateway communicates with the test environment using CORBA ORB, thus making it possible to have both subsystems located on different machines. This architectural design was technically feasible, as we have used OpenTTCN from Open Environment Software Oy as our target test environment which internally uses ORB as its communication backbone [7]. The distributed nature of the means of testing (MOT) may significantly facilitate testing, especially if CORBA services are tested in conjunction with some other telecommunication software, what may require the physical distribution of multi-purpose gateways in the network. The interpreted nature of the test environment combined with a simple and developed-friendly API has noticeably reduced the time and budget spent on the gateway design, allowing the design process to be iterative, i.e. when the gateway components are designed and their functionality is tested immediately. The absence of the overhead related to compiling ATS into ETS (Executable Test Suite) was another speed-up factor at the design stage.

The overall structure of the gateway is shown in Figure 5. The core part of it consists of BuilderAdapter, VisitorAdapter, ClientCallBuilder, ServerReplyBuilder, ServerCallVisitor and ClientReplyVisitor. It performs a conversion from ASP format into format acceptable for making an operation call and vice versa. Here we have used three well-known design patterns, namely Builder, Visitor and Adapter [1]. Visitor performs introspection of ASP coming from the test environment, Builder constructs ASP corresponding either to client operation call or to server operation response and Adapter performs a technical conversion of information obtained from the Interface Repository into internal format acceptable for processing either by Builder or by Visitor.

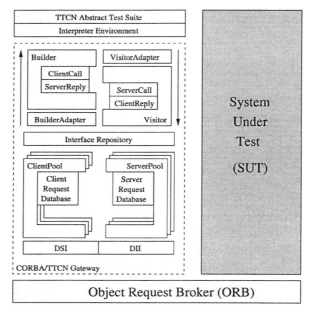

Figure 5. CORBA/TTCN Gateway Architecture

The gateway uses the asynchronous mode of DII invocation mechanism, what naturally correlates with asynchronous nature of TTCN operational semantics. The asynchronous mode enables the gateway to process new invocation requests coming from the test environment while other operation calls are still pending. This is also valid for concurrent invocations of one explicit operation attributed to the same CORBA object. This feature allows design of concurrent test suites with several parallel test components (PTC) or even simultaneous execution of several independent test suites served by one gateway component. In the latter case, the only restriction introduced by the gateway concerns prohibitive use of overlapping PCO identifiers in different test suites, as the TTCN test tool in use currently does not provide means for distinguishing between different test sessions as they are seen from the gateway perspective. Subsystem named "ServerRequestDatabase" is running in a dedicated thread and regularly polls for response from incomplete requests, accepting at the same time registration of new operation invocations issued by ServerPool. Client side of the gateway operates in a similar way, with "ClientRequestDatabase" subsystem designed to notify waiting threads in the ClientPool in case a response on client request has been issued by ATS. Client operation calls can be executed concurrently, too.

5.   TEST SUITE EXECUTION

For the purposes of verifying a gateway functionality, two sets of test cases have been developed. Each set contains nine test cases, one set for testing servers and another one for testing clients. This collection constitutes a minimum orthogonal set of tests of the gateway states and can therefore be recommended as a skeleton for developing more detailed test groups aimed at practical testing of ORB implementations. The test cases and their objectives are classified as follows: (1) *"prototype"* tests the very basic functionality of the gateway: it performs (or accepts) the only one operation call with "in" and "out" parameters of primitive IDL types; (2) *"primitive"* tests all IDL primitive data types as input and output parameters; (3) *"oneway"* tests correctness of executing oneway operations; (4) *"struct"* tests all IDL structured types (struct, sequence and array) except union; (5) *"choice"* tests IDL union, including recursive unions (i.e. union inside union); (6) *"context"* tests operations declared with context clause; (7) *"concurrent"* tests multiple invocation requests coming from clients or sent to servers; (8) *"exception"* tests handling of both system and user-defined exceptions; (9) finally, *"abnormal"* tests various invalid situations like incorrectly constructed ASP, most of which shall normally be observed on the ATS side in the form of gateway-specific Exception ASP.

A simplified example of a typical test session is shown in Figure 6. The session dynamic behaviour follows two stages of testing a CORBA service: (a) registering CORBA object, and (b) issuing (or accepting) actual operation calls. The third stage of deregistering CORBA object is implicitly performed by the gateway upon completion of test case and hence need not be reflected in test suite. The Interoperable Object Reference (IOR) used for registering CORBA server object is supplied to ATS in form of a PIXIT parameter.

It shall be acknowledged that the least formal part of CORBA services practical testing (requiring in some cases a manual control over SUT) relates to establishing a communication path between a CORBA object and the means of testing. In the presented example this is achieved by supplying an IOR of the object, but finding out the exact content of IOR may itself require knowledge about location of IOR file in e.g. file system or in the network. Moreover, an object may be located in the Naming Service or in the Trading Service what makes registration procedure even more complicated. The client side of SUT may be even harder to deal with, as ATS must explicitly instruct the gateway to create a generic servant and advertise its presence to the rest of the CORBA world. If for example SUT client is willing to immediately obtain a reference to the corresponding servant emulated by MOT before it becomes available, then such client shall be subject to manual control of the operator, and a corresponding IMPLICIT SEND statement shall be present in ATS.

Test Case					
Test Case ID:		tcINVOKE_COUNT			
Test Group Reference:		Operations/			
Test Case Purpose:		To test that after CORBA server object is registered by its IOR and "increment" operation is invoked, it returns the argument of the operation incremented by one.			
N	L	Behaviour Description	Constraint Ref	V	C
1		PCO1_iArithmetic_iCount ! pSREG_IOR START tSREG	cSREG_IOR (xIOR1_iArithmetic_iCount)		
2		PCO1_iArithmetic_iCount ? pRAISE	c1RAISE_iGatewayException _iRecoverable_iGeneral		
3		PCO1_iArithmetic_iCount ! pCALL_iArithmetic_iCount _iincrement START tREPLY	c1CALL1_iArithmetic_iCount _iincrement		
4		PCO1_iArithmetic_iCount ? pREPLY_iArithmetic_iCount _iincrement	c1REPLY1_iArithmetic_iCount _iincrement	P	
5		? TIMEOUT tREPLY		F	
6		PCO1_iArithmetic_iCount ? OTHERWISE		F	
7		? TIMEOUT tSREG		F	
8		PCO1_iArithmetic_iCount ? OTHERWISE		F	

Figure 6. Test Session Dynamic Behaviour Description

6.   DISCUSSION

This paper discusses the practical application of TTCN framework to conformance testing of CORBA services. An alternative approach to the same task would require a manual design of CORBA-based subsystem using language for which CORBA mapping specification exists, for instance C++, Java or Smalltalk. This manually implemented subsystem would then perform test invocations of SUT operations and in its turn respond to requests coming from SUT clients. Several issues shall be considered while making a choice between these two alternatives: (1) TTCN is a formal standardized framework specifically designed for reusable and automated testing of telecommunication systems; (2) TTCN facilitates practical testing of active components (servers) of CORBA services, as it clearly outperforms the above mentioned conventional technique due to ad-hoc and informal nature of the latter; (3) Yet, TTCN does not show up enough flexibility in emulating a CORBA servant, as TTCN framework may require implementation of TTCN operations for this purpose, what makes the processing of requests coming from reactive components of SUT (clients) look relatively awkward. To sum up, in case of testing CORBA clients TTCN obviously looses a comparison with its non-TTCN conventional counterpart. However, if SUT mostly contains active parts (which is normally the case in testing CORBA services), then a TTCN framework shall undoubtedly be preferred.

Several aspects constitute the grounds for the further research. First, the use of Modular TTCN for supplementing the existing IDL-to-TTCN mapping rules shall be investigated. The second aspect addresses the need for automating the process of test suite derivation from IDL interfaces and SDL models, as generating ASN.1 type definitions and ASP declarations from IDL is a relatively routine operation and is subject to automation. Finally, the third edition of TTCN, although not officially published yet, promises to be an innovative continuation of the TTCN standard which may bring a fresh breath to conformance testing of CORBA services and distributed applications. All these issues are planned to be addressed in our future work.

References

[1] E. Gamma, R. Helm, R. Johnson, J. Vlissides, *Design Patterns: Elements of Reusable Object-Oriented Software*, Addison-Wesley Publishing Company, 1995.

[2] ISO/IEC 9646-3, *Information technology - Open Systems Interconnection - Conformance testing methodology and framework - Part 3: The Tree and Tabular Combined Notation (TTCN)*, International Standard, second edition, 1994.

[3] O. Martikainen, J. Karvo, Internet Based Service Development, *Fifth International Conference on Intelligence in Networks (Smartnet '99)*, Thailand, November 22nd-26th, 1999.

[4] Object Management Group, *The Common Object Request Broker: Architecture and Specification, Revision 2.3.1*, OMG Document 99-10-08, October 1999.

[5] Object Management Group, *IDL-Java Language Mapping*, OMG Document 99-07-53, June 1999.

[6] Object Oriented Concepts, *ORBacus Manual*, http://www.ooc.com.

[7] Open Environment Software, *OpenTTCN Tester*, http://www.oes.fi.

[8]   Open Group, *JIDM Specification Translation*, Technical Document P509, February 1997.

[9]   R. Orfali, D. Harkey, *Client/Server Programming with Java and CORBA,* John Wiley & Sons, Inc., 1998.

[10]   I. Schieferdecker, M. Li, A. Hoffmann, Conformance Testing of TINA Service Components - the TTCN/CORBA Gateway, *Fifth International Conference on Intelligence in Services and Networks IS&N'98*, Antwerp, Belgium, May 25th - 28th, 1998.

[11]   R. Sinnott, M. Kolberg, Creating Telecommunication Services based on Object-Oriented Frameworks and SDL, *Proceedings of the Second IEEE International Symposium on Object-Oriented Real-Time Distributed Computing (ISORC'99)*, Saint Malo, France, 1999.

PART V

TEST AUTOMATION AND
INDUSTRIAL TESTING EXPERIENCE

13

FORMAL TEST AUTOMATION: THE CONFERENCE PROTOCOL WITH PHACT

Lex Heerink
Philips Research Laboratories, Eindhoven, The Netherlands
lex.heerink@philips.com

Jan Feenstra and Jan Tretmans*
University of Twente[†], Enschede, The Netherlands
{ feenstra,tretmans } @cs.utwente.nl

Abstract   We discuss a case study of automatic test generation and test execution based on formal methods. The case is the Conference Protocol, a simple, chatbox-like protocol, for which (formal) specifications and multiple implementations are publicly available and which is also used in other case study experiments. The tool used for test generation and test execution is PHACT, the PHilips Automated Conformance Tester. The formal method is (Extented) Finite State Machines which is the input language for PHACT. The experiment consists of developing a Finite State Machine specification for the Conference Protocol, generating 82 tests in TTCN with PHACT, and executing these tests against 28 different implementations of the Conference Protocol, both correct and erroneous ones. The result is that some erroneous implementations are not detected by the test cases. These results are analysed, the merits of Extented Finite State Machines for specification are discussed, and the achievements of PHACT are assessed. Moreover, the results are compared with a previous experiment in which the same 28 implementations were tested based on specifications in LOTOS and PROMELA.

Keywords: Conformance testing, case study, formal methods, finite state machines, test generation, test execution.

*Corresponding author: Jan Tretmans, University of Twente, Faculty of Computer Science, Formal Methods & Tools research group, P.O. Box 217, 7500 AE Enschede, The Netherlands; tretmans@cs.utwente.nl.
[†] This research is supported by the Dutch Technology Foundation STW under project STW TIF.4111: *Côte de Resyste* – COnformance TEsting of REactive SYSTEms; http://fmt.cs.utwente.nl/CdR.

1.   INTRODUCTION

In this note we describe a case study of automatic test generation and execution based on formal methods. The case is the *Conference Protocol*, a simple, chatbox-like protocol, for which (formal) specifications and multiple implementations are publicly available. The test tool used is PHACT. The formal method is (Extended) Finite State Machines – (E)FSM – which is the input language for PHACT.

The experiment consists of developing an EFSM specification for the Conference Protocol, generating tests in TTCN with the Conformance Kit – which is a part of PHACT – and executing these tests, also with PHACT, against 28 different implementations, both correct and erroneous ones. The goal of this paper is to present the experiment, to analyse the results, assess the merits of (E)FSM's and of PHACT, and to compare with a previous experiment in which the same 28 implementations were tested. That experiment is described in [1] and this paper is based on it.

2.   CONFERENCE PROTOCOL

The Conference Protocol provides a multicast service, i.e., a `chatbox' service, to its users. A group of users constitutes a conference. Each user in a conference can exchange messages with all other users in that conference. A conference can change dynamically: users can join or leave a conference at any time.

The Conference Protocol is a relatively simple protocol, yet, it contains many aspects of more realistic protocols. It has been used in other case studies, in particular, in [1] we used it with the test tool TORX and formal specifications in LOTOS, PROMELA and SDL. Hence, the Conference Protocol serves as a vehicle for testing and comparison of test tools. The web page [7] provides descriptions of the protocol, its service, formal specifications, and 28 different implementations, one of which is (assumed to be) correct, while the others are mutants in which (small) errors have been introduced.

3.   PHACT

PHACT (PHilips Automated Conformance Tester) is a set of tools that has been used within Philips for some time to generate tests automatically and to execute these tests against concrete implementations [4, 8]. The tool set consists of two major parts: a *generator* and an *executor*, see Figure 1.

The *generator* mainly consists of the Conformance Kit developed at KPN Research [2, 5]. It takes a formal specification in Extended Finite State Machine format (EFSM, Mealy Machine with state variables) as its input. This EFSM must be *input complete* and *deterministic*, i.e., for every state and input there

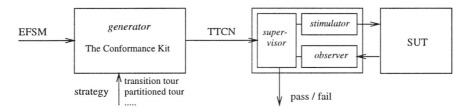

Figure 1. PHACT

shall be exactly one transition that specifies the corresponding output and destination state. The first step of the *generator* is to transform this EFSM into an equivalent Finite State Machine (FSM) by expanding all (necessarily finite-domain) variables. This FSM is minimized into an equivalent minimal FSM, and then tests are generated according to one of the standard, FSM-based algorithms (random sequences, transition tour, partitioned tour, i.e., UIO-based test sequences; see [6] for an overview). The tests are expressed in a subset of TTCN.

The *executor* part provides an environment for the execution of tests against a *System Under Test* (SUT). It consists of a *supervisor*, a *stimulator* and an *observer*. The *stimulator* and *observer* are application specific and must be developed, from a template, for each SUT separately. They provide the functionality for encoding and stimulating the SUT with a single input test event and for observing and decoding an output test event from the SUT. The *supervisor* is a generic kind of test engine. It reads the generated test cases and performs actions and test events given in the test case by using the *stimulator* and *observer* functions.

4.   TEST ARCHITECTURE

The test architecture for testing Conference Protocol implementations is the same as the one used in [1], see Figure 2. One Conference Protocol Entity (CPE) is tested; it is the IUT (*Implementation Under Test*). This CPE has one local conference user, denoted A, and two remote users, denoted B and C. The CPE communicates with its environment at the IAP's (*Implementation Access Points*) which are the CSAP (Conference Service Access Point) and the USAP (UDP Service Access Point). The test context is formed by the underlying layer, which is assumed to be a reliable and order-preserving UDP layer, sockets used for UDP communication, and pipes for communication at the CSAP. The SUT consists of the IUT and the test context. The PCO's (*Points of Control and Observation*) are the CSAP, which is the Upper Tester PCO (UT) and the remote USAP's which constitute the Lower Tester PCO (LT). The tester is instantiated with the PHACT *executor* (see Section 3.). The *stimulator* and *observer* provide the inputs and observe the outputs, respectively, at the UT and LT PCO's.

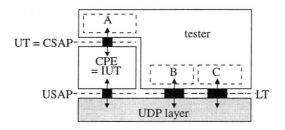

Figure 2. Test architecture

5. FINITE STATE MACHINE SPECIFICATION

The first step in the test experiment was developing an EFSM specification of the Conference Protocol. Starting point was the description in [7]. This resulted in an EFSM for one CPE with two conferences, one local user at the upper interface, and two remote users. It is written in the EFSM format of the Conformance Kit and it has 2 states, 3 variables, 18 inputs, 17 outputs and 36 explicitly specified transitions. The FSM expanded from this EFSM contains 9 states.

The restrictions on the (E)FSM imposed by the tool have implications for the way the SUT was modelled. We discuss a few aspects.

In EFSM's there is for every input at most one output. The Conference Protocol has multicast aspects. When user A in Figure 2 wants to join a conference, initiated by a *join-SP* (*join-Service-Primitive*) at CSAP, the CPE has to send a *join-PDU* (*join-Protocol-Data-Unit*) to all potential conference partners, i.e., to B and C. So, the input *join-SP* leads to two outputs: a *join-PDU* to B and one to C. Moreover, these outputs may arrive in any order – nondeterministically – at B and C. Neither two outputs in one transition, nor nondeterminism can be modelled directly in EFSM's. We solved this by using a single output action *join-A-to-BC-PDU*, which models the two SUT outputs occurring in any order (cf. [5]).

When the action *join-A-to-BC-PDU* occurs in a test case the *observer* has to take care to implement this action correctly, i.e., to observe two PDU's in arbitrary order. This is implemented by starting a timer before every observation. As long as this timer runs the implementation may perform its outputs. On expiration the *observer* reads all outputs from all PCO's, interprets them together, and tries to map them onto the corresponding single EFSM output action. So, any observation of outputs lasts at least the time required for this timer to expire. Note that this way of observation does not correspond with the standard TTCN snapshot semantics. A consequence of this strategy is that some orderings of actions are not tested, e.g., in the example above, an input action between the two *join-PDU*'s will never be considered for testing.

Alternation between inputs and outputs is always required in EFSM's. This also implies that a sequence of multiple inputs with delayed outputs is not considered, and hence not tested. In fact, EFSM's assume a kind of *synchrony hypothesis*.

Compositionality is not straightforward in EFSM's. This involves the test context. In our previous experiments with LOTOS and PROMELA we could easily combine the CPE model with the context models. For our EFSM model this turned out to be more complex. Actually, we do not take it into account at all, so we make the assumption that the test context, i.e., the queues which model the CSAP pipes and the lower layer UDP, does not influence the observable behaviour of the IUT. Whereas for queues this might seem a reasonable assumption it causes problems if the reliability and ordering assumptions on UDP are released, i.e., if UDP has to be modelled as a lossy bag.

The last important restriction of EFSM's in the Conformance Kit and PHACT is that they do not allow data parameters in inputs and outputs. Whereas in LOTOS and PROMELA different conferences, users and UDP addresses were easily modelled with data parameters, in EFSM's these have to be coded explicitly. This means that simplifications and abstractions have to be made with respect to the data aspects which will be tested in order to avoid an enormous set of explicitly coded input and output actions. Consequently, for all abstracted data aspects there will be no test cases generated. As an example, the input action *join-B-to-A-1-PDU* represents a *join-PDU* with parameters B as source address, A as destination address, and 1 as conference identifier. This explains the large number of actions necessary in our EFSM model.

6.   THE TEST EXPERIMENT

With the Conformance Kit 82 tests were generated from the EFSM specification with the partitioned tour method. For each explicitly specified transition of the FSM, obtained from the EFSM, a test case was generated; for transitions only added to make the EFSM input complete, no test cases were generated. A test case following the partitioned tour method consists of (*i*) a *synchronizing sequence* that brings the system from any arbitrary state into its initial state; (*ii*) a *transferring sequence* that brings the system from the initial state to the source state of the transition to be tested; (*iii*) a *transition test*, which performs the input and checks the output of the transition to be tested; and (*iv*) a *Simple Input/Output Sequence* (also known as Unique Input/Output (UIO) sequence), which verifies whether the correct destination state is reached. The length of the test cases varies from 6 to 16 test events: the synchronizing sequence has minimum 2 and maximum 4 test events, the 9 transferring sequences have a length from 0 to 6, every transition test consists of 2 test events, and the 9 UIO sequences have a length from 2 to 4 test events (not counting *start-timer* events).

The 82 test cases were successively applied, using the *executor* part of PHACT, to the 28 Conference Protocol implementations. They were executed for each implementation without resetting and re-initializing the implementation between the different test cases, which is possible due to the synchronizing sequence.

The results of the test experiments are that 21 obtained the verdict *fail*. These 21 implementations were correctly detected: they are known to be erroneous mutants. To 6 implementations the verdict *pass* was assigned, among which there was the correct one. Moreover, the following (erroneous) mutants got a *pass* verdict (using the mutant numbers of our internal identification scheme): 289, 293, 398, 444, and 666. One implementation (mutant 749) led to an abnormal termination (`core dump').

In our previous experiment with the Conference Protocol we tested the same 28 implementations with the test tool TORX [1]. Those experiments led to the detection of 25 out of the 28 implementations: the correct implementation and the mutants 444 and 666 got the verdict *pass*.

7.   ANALYSIS

While analysing the test results, the first observation was that some of the mutants were only detected because the 82 test cases were executed consecutively without resetting the implementation, i.e., the 82 test cases were actually executed as one large concatenated test case. This led to the situation where an error is triggered in one test case, without causing an incorrect observation, while the subsequently executed test case then led to the failure in terms of an incorrect observation. This situation becomes apparent, for instance, in the mutants numbered 332 and 345. These mutants implement the *leave-PDU* incorrectly, so that a conference partner which leaves the conference is not removed from the implementation's internal set of current conference partners. When one test case is finished and the synchronizing sequence of the next test case has been executed, the specification FSM is in its initial state again. But, since the SUT is not re-initialized between the two test cases, the SUT is, erroneously, not in its initial state: the conference partner that left is still in the set of current conference partners. This situation then leads to an incorrect observation with *fail* verdict in the subsequent test case, e.g., when PDU's are sent to the partners in the set.

The second point of analysis concerns mutant 749 that led to an abnormal termination. Analysis shows that this mutant generates an invalid PDU which is not correctly recognized by the *observer* and, consequently, cannot be mapped onto an EFSM output action. It turns out that our developed *observer* of PHACT is not robust for unrecognized PDU's, which is an anomaly of our *observer*. Mutant 749 was detected by TORX since the observer part of TORX maps every message which it cannot recognize onto a non-existing PDU. Since, of course, such a PDU does not occur in the specification this is subsequently considered as an incorrect response leading to the verdict *fail*. An analogous solution could have been implemented for PHACT.

The third point of analysis concerns the mutants not detected by PHACT.

Mutants 444 and 666:   Analysis in [1] showed that these mutants accept PDU's from any source, not only from (potential) partners. In the LOTOS and PROMELA specifications the responses to such PDU's are not explicitly specified, which means that, following TORX' correctness criterion **ioco** [1], any response from the SUT is allowed. Consequently, 444 and 666 are **ioco**-correct and pass TORX.

With PHACT we only generated test cases for explicitly specified transitions and not for those that were only added to make the FSM input complete (Section 6.). Consequently, it is no surprise that also with PHACT the mutants 444 and 666 were not detected. However, whereas with TORX these mutants can principally not be detected – the necessary stimuli are never generated by TORX – they might have been detected with PHACT – they can occur in a synchronizing sequence or UIO sequence. Moreover, in order to minimize the number of data parameters (which have to be coded explicitly; see Section 5.), inputs from non-existent partners were not considered in our EFSM. Hence, such inputs are outside the considered input set and, like for TORX, the necessary stimuli to trigger the faults are never generated.

Mutants 289, 293 and 398:   These erroneous mutants got the verdict *pass*, whereas they were detected in the TORX experiment. We now try to analyse why PHACT does not detect them by comparison with the TORX experiment. The failure trace-logs of TORX, i.e., the traces of inputs and outputs leading to the detection by TORX, were analysed. Then we projected these failure traces onto the EFSM and analysed the relevant EFSM behaviour and the PHACT test suite with respect to these failure traces.

Mutant 293:   Mutant 293 uses a bag instead of a set to administer its conference partners. With TORX the following failure trace detects this mutant.

1	input:	UT ! A ! join-SP(A,conf-1)	5	input:	LT ! C ! answer-PDU(C,conf-1)
2	output:	LT ! C ! join-PDU(A,conf-1)	6	input:	UT ! A ! datareq-SP(m1)
3	output:	LT ! B ! join-PDU(A,conf-1)	7	output:	LT ! C ! data-PDU(A,m1)
4	input:	LT ! C ! answer-PDU(C,conf-1)	8	output:	LT ! C ! data-PDU(A,m1) : fail

In step 1 the local user *A* wants to start a conference with name *conf-1* via a *join-Service-Primitive* at the upper test interface *UT*. In steps 2 and 3 the SUT informs the potential partners *B* and *C* at the lower test interface *LT* by sending two *join-PDU*'s. Then, at *LT*, user *C* sends an *answer-PDU* to the

SUT, twice. The SUT should react by simply neglecting the second *answer-PDU*, however, the SUT does store the second one in its bag. In step 6 the SUT receives a request from its local user A to send message *m1*. Then, in steps 7 and 8 the SUT sends the message to user C, also twice, which is incorrect: the SUT should send it only once. So, in step 8 no output was expected according to the specification (`quiescence'): *fail*.

Analysis of this sequence of events in the EFSM specification gives:

Steps	Condition	Input	Output	Action	Dest.
(1-3)	*TRUE*	*join-A-1-PDU*	*join-A-to-BC-1-PDU*	*(A-conf:=1)*	*Conn*
(4)	*(A-conf=1)*	*answer-C-to-A-1-PDU*	*(none)*	*(C-part:=TRUE)*	*Conn*
(5)	*(A-conf=1)*	*answer-C-to-A-1-PDU*	*(none)*	*(C-part:=TRUE)*	*Conn*
(6,7)	*notbc*	*datareq-SP*	*data-A-to-C-PDU*	*(none)*	*Conn*

notbc == (NOT(B-part) AND C-part)

In the first EFSM transition, corresponding to the steps *(1–3)* of the TORX trace, user A joins conference 1 and sends two *join-PDU*'s to users B and C, respectively. (This is modelled as one output, see Section 5.). The EFSM goes to state *Conn* and conference 1 is active: *(A-conf:=1)*. We see in steps *(4)* and *(5)* that user C is allowed to send an *answer-PDU*, twice. The second one does not change the state of the EFSM and, consequently, should have no effect. If in *(6,7)* the local user A issues a *datareq-SP* the corresponding *data-PDU* is sent only once. This state of the EFSM is never reached by the mutant: after transition *(5)* the mutant goes to a state where the set (bag) of current conference partners contains two entries for partner C. This state does not exist in the EFSM specification. Using the partioned tour method no test is generated for such a state and, consequently, the error is not detected. Typically, errors in implementation states which do not have a corresponding state in the specification are not always found by the UIO-based partitioned tour method.

Mutant 289:   Mutant 289 does not update its internal set of current conference partners correctly when an *answer-PDU* is received. With an analogous, although somewhat more complex analysis as for mutant 293 we concluded that also this failure of detection was caused by an additional state in the implementation which did not have a corresponding state in the EFSM specification.

Mutant 398:   Mutant 398 does not check the conference identifier when partners send *answer-PDU*'s. With TORX we constructed the following failure trace.

1	input:	*UT ! A ! join-SP(A,conf-1)*	4	input:	*LT ! C ! answer-PDU(C,conf-2)*
2	output:	*LT ! C ! join-PDU(A,conf-1)*	5	input:	*LT ! C ! data-PDU(C,m1)*
3	output:	*LT ! B ! join-PDU(A,conf-1)*	6	output:	*UT ! A ! dataind-SP(C,m1) : fail*

User A joins conference *conf-1* in steps 1-3. Then user C sends an *answer-PDU* to the SUT to join another conference: *conf-2*. The SUT should ignore this, but it does not: it erroneously adds C to its set of current conference partners. Then, when it receives in step 5 a *data-PDU* with message *m1* from *C*, the SUT passes the message *m1* erroneously as *dataind-SP* to local user *A*. Transposing this failure trace to the EFSM we obtain the following.

Steps	Condition	Input	Output	Action	Dest.
(1-3)	*TRUE*	*join-A-1-SP*	*join-A-to-BC-1-PDU*	*(A-conf:=1)*	*Conn*
(4)	*(A-conf=1)*	*answer-C-to-A-2-PDU*	*(none)*	*(none)*	*Conn*
(5,6)	*c1notbnotc*	*data-C-to-A-PDU*	*join-A-to-C-1-PDU*	*(none)*	*Conn*

c1notbnotc == ((A-conf=1) AND NOT(B-part) AND NOT(C-part))

Transition *(4)* is a transition which is added in the EFSM only to make the EFSM input complete. The EFSM specifies simply to neglect the incoming *answer-C-to-A-2-PDU*. Since for such transitions no test cases are derived the mutant is not detected. Probably, mutant 398 would have been detected if test cases had been generated for all transitions including the ones intended to make the EFSM complete, which is a possibility in PHACT. After all, the mutant makes an erroneous transition to a state known in the specification, i.e., the state in which *C* is a partner in the set of current conference partners. We did not perform this additional experiment yet.

8.   CONCLUDING REMARKS

If the Conference Protocol experiments were a match between TORX and PHACT the result would be 25 : 21 in favour of TORX. PHACT did not detect 4 mutants which TORX did. One of these caused a `core dump'; a simple improvement in the *observer* of PHACT would detect this mutant. Another mutant was not detected because we only tested the explicitly specified transitions. A PHACT test suite which tests all transitions would probably be able to detect this one. What remains are two implementations which are not detected because they clearly have states which do not exist in the EFSM specification. Such non-detected errors are typical for the partitioned tour method which is used by PHACT. In addition, we should note that some of the mutants were only detected since all 82 test cases were executed successively without resetting the implementations, see Section 7..

With respect to PHACT, the test tool is usable and successfully detected most of the faulty implementations. Most of its disadvantages are related to the restrictions imposed on the EFSM's it uses. They were discussed in Section 5.: the required alternation between single input and output actions, determinism, lack of compositionality, and the inability to cope with data parameters.

The Conference Protocol turned out to be a nice, useful and interesting case study: it is simple enough to be understood but not so simple that it is trivial, multiple specifications and implementations are publicly available, and the number of experiments with it increases so that interesting comparisons are possible [3]. As far as we know this is the first real comparison, based on actually detected erroneous implementations, of an FSM-based technique – PHACT – with an LTS-based technique – TORX.

Many additional experiments can be envisaged. First, the experiments mentioned above to detect the two not yet detected mutants with PHACT can be performed. Second, other orderings of test cases or independent test case execution could be considered. Third, other strategies of PHACT can be used: random sequences, transition tours. Fourth, more, and more tricky erroneous implementations can be developed to extend the comparison. Fifth, other test tools can be applied to the Conference Protocol and other case studies can be used to compare PHACT and TORX. Finally, apart from counting detected mutants, other comparison criteria should be investigated, such as total effort and cost of testing, cost per detected mutant, and ease of use.

Acknowledgements   The authors would like to thank the members of *Côte de Resyste*, in particular Axel Belinfante, René de Vries and Ron Koymans, and the anonymous reviewers for their comments and support. Erik Kwast from KPN Research is acknowledged for supplying the Conformance Kit and developing the first version of the Conference Protocol EFSM.

References

[1] A. Belinfante, J. Feenstra, R.G. de Vries, J. Tretmans, N. Goga, L. Feijs, S. Mauw, and L. Heerink. Formal Test Automation: A Simple Experiment. In G. Csopaki et al. (eds.), *Testing of Communicating Systems 12*, pp. 179–196. Kluwer Ac. Publ., 1999.

[2] S.P. van de Burgt, J. Kroon, E. Kwast, and H.J. Wilts. The RNL Conformance Kit. In J. de Meer et al. (eds.), *Protocol Test Systems 2*, pp. 279–294. North-Holland, 1990.

[3] L. Du Bousquet, S. Ramangalshy, C. Viho, A. Belinfante, and R.G. de Vries. Formal Test Automation: The Conference Protocol with TGV/TORX. In *TestCom 2000*. Kluwer Ac. Publ., 2000. This issue.

[4] L.M.G. Feijs, F.A.C. Meijs, J.R. Moonen, and J.J. Wamel. Conformance Testing of a Multimedia System using PHACT. In A. Petrenko and N. Yevtushenko (eds.), *Testing of Communicating Systems 11*, pp. 193–210. Kluwer Ac. Publ., 1998.

[5] E. Kwast, H. Wilts, H. Kloosterman, and J. Kroon. User Manual of the Conformance Kit. Version 2.2, PTT Research Neher Labs, Leidschendam, The Netherlands, Oct. 23 1991.

[6] D. Lee and M. Yannakakis. Principles and Methods for Testing Finite State Machines – A Survey. *The Procs. of the IEEE*, 84, Aug. 1996.

[7] Project Consortium Côte de Resyste. Conference Protocol Case Study. URL: http://fmt.cs.utwente.nl/ConfCase.

[8] T.I.P. Trew, B. Lanaspre, M. Hollenberg, J. Springintveld, and T.J. Harosia. Delivering High Definition TV to the USA – Testing Subcontracted Embedded Real-Time Software. In 16th *Conf. on Testing Computer Software (TCS'99)*, Washington D.C., June 1999.

14

FORMAL TEST AUTOMATION:
THE CONFERENCE PROTOCOL WITH TGV/TORX

Lydie Du Bousquet, Solofo Ramangalahy, Séverine Simon, César Viho[†]
IRISA, Campus de Beaulieu, F35042 Rennes, France
{ ldubousq, solofo, ssimon, viho } @irisa.fr

Axel Belinfante, René G. de Vries
Department of Computer Science – University of Twente
P.O. box 217, 7500 AE Enschede, the Netherlands
{ belinfan, rdevries } @cs.utwente.nl

Abstract   We present an experiment of automated formal conformance testing of the Conference Protocol Entity as reported in [2]. Our approach differs from other experiments, since it investigates the combination of the tools TGV for abstract test generation and TORX for test execution.

Keywords: Conformance testing, test generation and execution, TGV, TORX

1.   INTRODUCTION

As conformance testing theories and tools mature (i.e., leading to standards [5, 6]), it is interesting to compare different approaches by practical experiments. In [2] a comparison among several tools, specification formalisms and test execution paradigms was made for an implementation of a simple chat box protocol—called the *conference protocol*. In this paper we describe a similar

---

[*]This research was partly supported by the NWO in Van Gogh program for French-Dutch cooperation, project "Automatic Generation of Conformance tests", VGP 62-480 - 99.031 and by the Dutch Technology Foundation STW under project STW TIF.4111: *Côte de Resyste* – COnformance TEsting of REactive SYSTEms.
[†]First and third author with INRIA, second with DYADE/INRIA, fourth with IFSIC/Université de Rennes I

study using the TGV test generation tool [7], and the TORX tool environment [2] for test execution (see Fig. 1.(a)).

We use two different methods to generate the test suites: one based on manually-designed test purposes, and the other based on randomly-generated test purposes. Our goal is to evaluate advantages and drawbacks of the tools used during this experiment, and to investigate these two approaches of test purpose design.

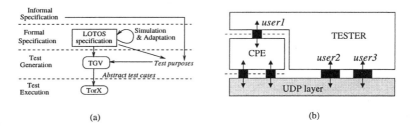

(a)   (b)

Figure 1.   Architecture: (a) steps and tools of the experiments; (b) the test architecture

2.   THE EXPERIMENT

2.1   Description

The experiment addresses testing of a multicast protocol implementation, facilitating a service similar to a `chatbox' to users participating in a conference. A conference is a session in which a group of users can participate by exchanging messages with other users (called *partners*). The partners involved in a conference can change dynamically. The Implementation Under Test (IUT) is the *Conference Protocol Entity* (CPE) which implements the protocol at a conference user site. The test architecture of the experiment is depicted in Fig. 1.(b). Since the IUT communicates using the User Datagram Protocol (UDP) underlying service, we need to incorporate it into the System Under Test (SUT). The specification of this protocol is described using the formal language LOTOS.

The experiment consists of testing 28 different implementations of the CPE—of which 27 are incorrect—and detecting the incorrect implementations (i.e., mutants [11]). In each mutant a single error has been introduced. There are three types of errors: *No outputs*, *No internal checks* and *No internal updates* (see [2]). There are respectively 6, 4, and 17 mutants for each type.

We evaluate the quality of the generated test suites by checking how many of these mutants are declared non-conforming with respect to the specification. See [2, 3] for more details about the experiment, the protocol and specifications.

2.2   Test Suite Generation

Description of TGV.   TGV [7] is a tool dedicated to the automatic generation of conformance tests based on a formal specification. Given a formal specification of a system to be tested and a formal description of a test purpose, TGV generates an abstract test case. The test purposes are used as test selection criteria and are formalized using automata. An abstract test case is a directed graph in which each path represents a test sequence with associated verdicts which indicate whether the SUT conforms to the specification. We use the conformance relation **ioco** [12], which is a correction notion for conformance between the specification and an implementation. The main characteristic of TGV is that it produces test cases "on-the-fly", i.e. the generation is done in a "lazy" way, so that the specification state space is not completely stored. "On-the-fly" techniques are one of the solutions proposed for the state-space explosion problem commonly encountered in verification techniques.

Formal specification.   We used the LOTOS specification, which is freely provided by the Côte de Resyste project [3]. As shown in Fig. 1.(b), the core of the LOTOS specification is a (state-oriented) description of the CPE behaviour. The CPE behaviour is parameterized with the potential conference partners. The instantiation of the CPE with concrete values for these parameters is part of the specification. We made some small modifications on the specification, to make processing by TGV more efficient. These adaptations are similar to optimizations of the PROMELA specification reported in [2].

Test purpose design.   In addition to a specification, we need test purposes [5]. In TGV, test purposes are given in the form of an automaton. They serve as a "guide" to the state-space exploration which is performed on the product between the specification and the test purpose. We followed two approaches to obtain them:

- In the first approach, the test purposes are designed manually, based on the informal requirements of the conference protocol [3].

- The second approach consists in automatic random generation of test purposes.

We began by designing 18 basic test purposes for basic protocol functionalities: joining and leaving the conference and data transfer. From these, we composed 8 more complex test purposes. We designed the test purposes to fulfill the informal requirements for a CPE. With the 8 complex test purposes and 11 basic test purposes, we generated a test suite with TGV. We select only 11 of the 18 basic test purposes to produce (basic) test cases with TGV because of the following testing equivalence hypothesis. Since the specification indicates that *user2* and *user3* behave equivalently towards *user1* (see Fig. 1.(b)), the verdict for a test case with *user1* and *user2* (noted *TC(user1,user2)*) should be the same as the verdict for *TC(user1,user3)*, i.e., the same test case replacing *user2* by *user3*. The time effort spent on designing and writing these 19 test purposes and generating tests was 4 hours. After execution of the test suite as described above, we designed 7 new test purposes, since the generated test suite was not able to detect one last mutant. To find the last mutant, we relied on the fault model used for the mutant generation. For each expected error, we designed a specific test purpose for which an implementation with this error would behave incorrectly. Those test purposes were more difficult to design than the previous ones. We decided to stop the test purpose design after ten hours of work.

For the second approach, we used the CADP [4] simulator to simulate the specification randomly. With this tool, we produced and saved 200 traces of 200 steps. We translated those traces into test purposes (with a script), and we used TGV to produce the associated test cases.

2.3   Test Execution

TorX and adaptation.   For the execution of the tests generated by TGV we used TORX [2] (configured for on-the-fly test generation and execution) and gave it the tests generated by TGV as "specifications". Because TORX can handle nondeterministic graphs, we were able to execute not only tests cases, but also uncontrollable tests graphs [7]. With TORX, we could reuse ADAPTER, the component that connects the tester to the SUT, as it was configured for the previous conference protocol experiments [2]. Minor changes were necessary for the components PRIMER, which implements the test derivation algorithm, DRIVER, which controls the testing, and EXPLORER, which explores the transition graph. PRIMER had to be changed because it initially only implemented the **ioco** test generation algorithm, while for this experiment it has to compute the traces of its input, i.e., test cases in TGV output format. In addition, it has to recognize the special events in the tests that encode quiescence [12] and verdicts and pass the verdicts to the DRIVER. The DRIVER had to be extended to accept verdicts from the PRIMER (in addition to computing the verdicts itself). The CADP libraries [4] allowed us to replace the LOTOS-specific EXPLORER by one that read the test case in BCG format.

Results.   Since TGV is based on the **ioco**-conformance relation, it can potentially detect **ioco**-incorrect implementations, but it cannot detect **ioco**-correct erroneous implementations. Among the 28 implementations, 25 were **ioco**-incorrect.

We first executed the test cases generated from the manually-written test purposes. In total 34 test cases were derived from 19 test purposes (11 basic

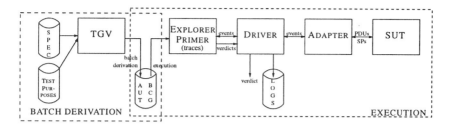

Figure 2. TORX tool architecture instantiated with TGV

and 8 complex). The test cases detected 24 **ioco**-incorrect mutants. One mutant was detected by all the test cases, 6 mutants were detected by 50% of the test cases, 7 mutants were detected by less than 10% of the test cases. The test suite generated from the 7 test purposes especially designed to find the last undetected mutant, was not able to detect it.

We then executed the test cases produced from the randomly generated test purposes. All 25 **ioco**-incorrect mutants were detected. One was detected by all the test cases, 3 were detected by 50% of the test cases, 3 were detected by less than 10% of the test cases.

3. CONCLUSION AND PERSPECTIVES

Tools. We were able to easily interface TGV and TORX, so that we could execute the generated tests. As a result of this experiment, we noted two ideas for the improvement of TGV, both of which have been implemented. First, the restriction that a test purpose should be deterministic has been removed, to facilitate the writing of a test purpose. For example, it eases the writing of wild cards in the transitions of a test purpose by avoiding the manual computation of the complement of a regular expression. The other adaptation of TGV addresses the controllability of a generated test case. In TGV, priority between inputs and outputs (of the specification) was given to the inputs. It can be advantageous to give less control to the environment (e.g., regarding quiescence [12]).

Specification of test purposes. Manually-written test purposes were more general than those produced by the random generation method: wild cards in transitions labels were used whereas they did not appear in random tests purpose coming from traces. To derive a test case by TGV, we began with a general test purpose and subsequently, restricted it (i.e., instantiated some wild cards). This iterative approach was needed, since the generation either took too much time or TGV did not produce a test case because of incoherence between the specification and the test purpose. That is, a test purpose is coherent with a specification if the behaviors it describes are included in those of the specification, see [7].

This experiment with TGV has underlined the difficulties that users have when translating a test purpose into TGV input format with which test cases are quickly generated. A balance has to be found between the expressiveness of test purposes and the amount of computation needed to generate a test case. This expressiveness comes from the ability to use wild cards in the labels. When test case generation takes too long, the expressiveness of the test purpose has to be reduced (e.g., by expanding the wild cards labels or by considering the different options of TGV). To succeed in a test campaign with manually-written test purposes, the user has to know the algorithms described in [7] to understand how TGV works in order to know what to do to obtain the expected test case.

From the experience gained in the use of TGV with the manually-written test purposes we chose to have the random test purposes as constrained as possible to ensure fast generation of test cases. This shows that it is possible to choose the level of automation of test generation: completely automated (as with random test purpose) or with TestComposer [8] or in a more interactive way (more control over the test generation process).

Manual vs. random test purpose design.   One should notice that the goal of this paper is not to compare the efficiency of the two test purpose design approaches. The fact that one method is better than another one cannot be established with a single experiment. We present our observations on the two design approaches. With the random generation of test purposes approach, all **ioco**-incorrect mutants were found. It is not clear why we did not manage to detect one **ioco**-incorrect mutant with the manual approach. The fact that all the mutants were detected by the random approach is not surprising considering that this method is essentially equivalent to the on-the-fly method of [2]. The efficiency of the random walk method for protocol (specification) testing has been known for a long time [13] even though few formal explanations exist regarding its efficiency [10]. Efficiency is transferred from specification testing to conformance testing when execution is combined with simulation. Another reason for the efficiency of both methods could be that the fault model we considered is restricted. Only "functional" mutants (in contrast to the usual syntactic mutants) were considered. The high-level fault model (no outputs, no internal checks, no internal updates), when applied at the source-code level of the implementation, give less choice in mutations than usual mutation operators [1] applied directly on the code of the implementation.

Analysis.   As in the previous experiment [2], the analysis of test case execution was the least automated part. We did not go much further than verdict checking, i.e., we did not diagnose the errors. We found that analysis of random-generated test purposes is quite difficult. In fact, many events in the test execution trace were unnecessary to trigger the error. The analysis of the trace produced from designed test purposes seems easier since these test purposes usually give a more precise idea of what is supposed to be tested (although obtaining a fail verdict is not necessary related to a fail on the "property" targeted by the test purpose). To diagnose a fail verdict on a mutant, we used an iterative approach to analyze the result of the test case execution, i.e., analyzing the test run. From this test run, we made a new test purpose by suppressing irrelevant events and derived a new test case. After execution of this test case, we iterated again to converge to a "minimal" test purpose. From a reduced test purpose, we hoped to interpret the detected error more easily. Although this methodological approach eases human diagnosis, it is not systematic and did not help for the last undetected mutant with manual test purposes[1]. Here, automation is lacking for fault diagnosis (which is a more complex problem than conformance testing, see [9] p. 1119).

Comparison with previous experiments.   Both approaches for test purpose development, resulted in a good mutation score (96% and 100% respectively). These results should be somewhat mitigated by the fact that mutant population is low and maybe not representative of common errors encountered when writing C code. We consider this experiment to be a case study. It is not possible to really compare this experiment with a previous one, i.e., benchmark it. The only conclusion to be drawn is that TGV and TORX (the full tool, not just the ADAPTER) have the same fault detecting power on this case study, which is not surprising since they use the same conformance relation. To do real benchmarking you need, besides benchmarking criteria, equivalent specifications in order to make a fair comparison. On the other hand, specifications may be adapted so that a tool is able to perform better. Therefore strict equivalence among specifications is hard to maintain.

Perspectives.   The experiment detailed here is the first part of a larger one using other tools. We intend to use TGV and TestComposer with the SDL specification of the CPE, and we have already produced some test cases. Currently we are working on the execution of these test cases. We want to extend the set of mutants by automatically generating more mutants. We are currently studying possible approaches towards this. We believe that more work is needed on the notion of a test purpose both at the theoretical and method-

---

[1]Note that since we obtained traces leading to **fail** with random test purposes, we could have used them for the other approach. But since we did not get the "meaning" of these test purposes, they cannot be considered as manual test purposes.

ological level. There are several informal definitions of what a test purpose is. TGV uses one specific, formal and constructive notion of a test purpose. One can consider that the test purpose description in TGV is too restrictive (for instance, it is not possible to specify quiescence in test purposes). It remains an open problem, to find the correct notion of test purpose. Such a definition should be as general as the informal one in [5] or the formal definition in [6]. In addition, we would still like to use the definition as a basis for test generation. Finally, tool support for writing test purposes is needed.

References

[1] H. Agrawal, R. A. DeMillo, B. Hataway, W. Hsu, W. Hsu, E. W. Krauser, R. J. Martin, and E. Spafford. Design of mutant operators for the C programming language. Technical Report TR-41-P, SERC, 1989.

[2] A. Belinfante, J. Feenstra, R.G. de Vries, J. Tretmans, N. Goga, L. Feijs, S. Mauw, and L. Heerink. Formal test automation: A simple experiment. In *IWTCS'99*, pages 179–196. Kluwer, 1999.

[3] Côte de Resyste. Conference protocol case study. http://fmt.cs.utwente.nl/ConfCase, 1999.

[4] H. Garavel. Open/cæsar: An open software architecture for verification, simulation, and testing. In *TACAS'98*, LNCS 1384. Springer, 1998.

[5] ISO. *International Standard IS-9646*. 1991.

[6] ITU-T recommendation Z-500: Framework on formal methods in conformance testing, 1997.

[7] T. Jéron and P. Morel. Test generation derived from model-checking. In *CAV'99*, LNCS 1633. Springer, 1999.

[8] A. Kerbrat, T. Jéron, and R. Groz. Automated test generation from SDL specifications. In *SDL'99*, pages 135–151. Elsevier, 1999.

[9] D. Lee and M. Yannakakis. Principles and methods of testing finite state machines–a survey. *Proceedings of the IEEE*, 84(8):1090–1123, 1996.

[10] M. Mihail and C. H. Papadimitriou. On the random walk method for protocol testing. In *CAV'94*, LNCS 818, pages 132–141. Springer, 1994.

[11] R. J. Lipton R. A. DeMillo and F. G. Sayward. Hints on test data selection: Help for the practicing programmer. *IEEE Computer*, 11(4):34–43, 1978.

[12] J. Tretmans. Test generation with inputs, outputs and repetitive quiescence. *Software— Concepts and Tools*, 17(3):103–120, 1996.

[13] C. H. West. Protocol validation by random state exploration. In *PSTV'86*, pages 7.1–7.12, 1986.

15

FUNCTIONAL TESTING GPRS SUPPORT NODES USING TTCN

Endre Horváth
Conformance Lab, Ericsson Hungary Ltd.
Laborc u. 1, H-1037 Budapest, Hungary
Endre.Horvath@eth.ericsson.se

Axel Manthey
Ericsson Eurolab Deutschland GmbH,
Ericssonallee 1, D-52134 Herzogenrath, Germany
Axel.Manthey@eed.ericsson.se

Abstract   Functional and conformance testing of new standardized services and protocols is a challenging task as standards may change during implementation and the test system has to be adapted to the system under test continuously. This paper describes how modular and concurrent TTCN was deployed for validation of the GPRS support nodes and reports on the experiences that were made.

Keywords: Functional Testing, GPRS, GPRS Support Node, Concurrent and Modular TTCN

1.   INTRODUCTION

The introduction of General Packet Radio Services brings with it a challenge to carry out functional and conformance tests. Never before such a large number of telecom and datacom protocols have been combined in a single network element. Demands to the test system are high; the system should be able to grow with the implementation, support a large variety of protocols in different national variants and be utilized in simulated and target environment.

Section 2 gives a brief introduction to General Packet Radio Services and presents the interfaces and protocols that need to be tested. The chosen solution is described in Section 3 with emphasis on the deployment of concurrent and modular TTCN. Section 4 summarizes the experiences that have been collected throughout the testing project.

2.   GENERAL PACKET RADIO SERVICES

General Packet Radio Services (GPRS) is a standardized extension to existing GSM networks that offers packet switched data services. Two new network elements will be added to the GSM network architecture: the Serving GPRS Support Node (SGSN) and the Gateway GPRS Support Node (GGSN). These nodes are interconnected by means of an IP based core network and have signaling connections to existing GSM network elements such as Home Location Registers (HLR), Mobile Services Switching Center/Visitor Location Registers (MSC/VLR), Base Station Controllers (BSC) or Short Message Service Gateway MSCs and Interworking MSCs (SMS-GMSC, SMS-IWMSC). The SGSN serves packet data users in a defined geographical area while the GGSN connects to external packet data networks.

From an end user's perspective GPRS offers permanent connectivity to IP networks, volume based charging and a higher bandwidth compared to existing GSM data services (up to 115 Kbps). Circuit switched GSM services and GPRS can coexist without disturbances, only one HLR based subscription is needed. Radio resources can be shared efficiently among several users. Horizontal applications (e.g. e-mail, FTP, HTTP) and vertical applications (such as telemetry, diagnostics, vending machines) can be offered. In addition SMS will be supported over GPRS. To connect the SGSN and GGSN to existing GSM network elements new protocols were standardized and existing protocols were modified to support the new services (*Figure 1*). The GPRS Tunneling Protocol (GTP) transports signaling and payload between GSNs. The User Datagram Protocol (UDP) over IP is the bearer for GTP Protocol Data Units (PDU). HLR, SMS-GMSC, SMS-IWMSC and Equipment Identity Register (EIR) use Mobile Application Part (MAP) PDUs over Transaction Capability Application Part (TCAP). The MSC/VLR is connected via the Base Station System Application Part + (BSSAP+) protocol using Signaling Connection Control Part (SCCP) Abstract Service Primitives (ASP). The interface to the BSC uses GPRS Mobility Management (GMM), Session Management (SM) and SMS PDUs over Logical Link Control (LLC) ASPs on the signaling plane and the Subnetwork Dependent Convergence Protocol (SNDCP) over LLC for the transmission of packet data. For configuration and monitoring purposes a Command Line Interface (CLI) exists in addition to graphical user interfaces.

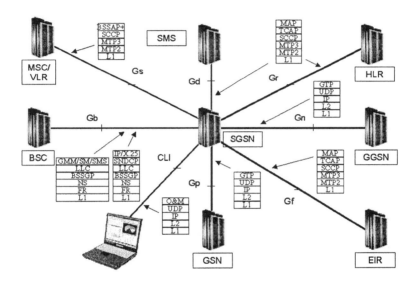

Figure 1. Interfaces to the SGSN and involved protocols

The enhancement of the GSM network requires software updates in the HLR and MSC/VLR, hard- and software updates in the BSC and the design of the two new nodes SGSN/GGSN. All interfaces to the GSNs have to be simulated in order to test the newly implemented protocols. Furthermore the protocols were affected by changes in standardization during design, thus the test system had to be upgraded permanently.

3.   FUNCTIONAL TESTING WITH TTCN

3.1   The Challenge

There are many protocols and interfaces in the GPRS network. Different protocols are used on different interfaces and all nodes interfacing the SGSN/GGSN have to be simulated, therefore testing a GPRS support node requires a complex test configuration simulating the other nodes adjacent to this entity. These expectations can be met by using *concurrent TTCN*.

Not all the interfaces are available from the first design increment on: The test system is becoming more and more sophisticated as the design grows and even the protocols may change during the design due to standardization changes. *Modular TTCN* provides the possibility to upgrade the test suites easily, handles large sized test suite production and facilitates parallel work.

The same test cases shall be executable in simulated and target environment as well as for different design increments that need to be maintained. This emphasizes the need to enable or disable different functionality. Testers can write easily adaptable test cases for different use cases by parameterized *TTCN test suites*.

3.2   Usage of Modular TTCN

Modular TTCN facilitates work-distribution (parallel work) and it increases the reusability of the test suites. It is the best possible way to handle large sized test suite production.

The key to modular test suite production is the architecture of modularization. In the literature there are three main modularization levels: Functional, Organizational and Language level. Functional level modularization describes the partition of test suites into several functional parts, based on the tree organization of the test cases/groups. Organizational level modularization is set up according to the ATS (Abstract Test Suite) writer's point of view. This level allows a design where modules are shared between ATSs (Global module, Common Module, Specific Module). Language level modularization is based on the classification of TTCN objects by the parts or sub-parts they are belonging to (i.e.: constants, test suite variables, PCOs, Timers, ASPs, PDUs, etc.).The GPRS project is based on a combination of the above mentioned levels. Protocol definitions are located in separate modules. Commonly used test steps and constraints are dealt within separate modules according to the different protocols/network elements to be simulated. The MAP protocol definitions are extracted from the ASN.1 definitions of GSM TS 09.02 utilizing the possibility to import ASN.1 definitions in TTCN test suites. *Figure 2* shows the architecture of modularization applied in GPRS function test.

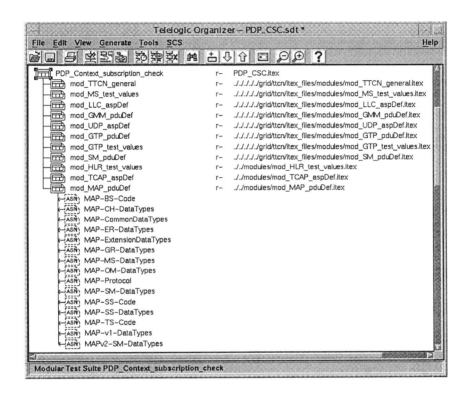

Figure 2. Architecture for modularization

It is important to realize that the test suite production within the GPRS project was carried out in two parts: The first part was to develop a test suite template i.e. to prepare the modules containing the protocol definitions, the common constraints and test steps. The second part was to build the actual test suite including the test cases. Tested protocols and functions are divided into test objects and a test suite is built for each of them using the modules of the test suite template. The test cases are located in these modular test suites, named according to the scope of the different test objects and importing the needed definitions from the connected modules (*Figure 2*). In this way each tester is able to work on his own test module having the appropriate import declarations from other modules.

The protocol modules are updated centrally and as more protocols are implemented by design, those can be added to the test environment in the same incremental approach. To handle the versions and the storage of modules the well-known version-control system ClearCase was used. Via this system each user has his own view (virtual workspace), which enables him to work on the files in ClearCase as if he was in regular Unix directories with regular tools on regular files.

3.3   Usage of Concurrent TTCN

There are many network elements that need to be connected to the Implementation Under Test (IUT), the SGSN or GGSN in this project. Concurrent TTCN offers the possibility to handle the parallel communication in an efficient way.

Connected network elements are simulated using parallel test components (PTCs) with Points of Control and Observation (PCO) associated. All parallel test components are controlled by one master test component (MTC). The MTC controls and synchronizes the PTCs: the body part of the test cases consist exclusively of CREATE/ ?DONE statement pairs controlling the parallel test components and assigning local behavior trees (local steps) to each PTC (*Figure 3*). Advantages of this strategy are that the test cases are easy to read and the configuration for new interface types is very simple. It is not difficult to follow the signaling on the different interfaces using local behavior trees for the PTCs. In order to avoid too long test cases, test steps are used for logically connected events to simplify their structure (*Figure 3*) and to increase the re-usage of TTCN code.

Each PTC described in the test cases is tagged with a qualifier containing a test suite parameter (*Figure 3*). This method allows enabling or disabling of the behavior trees: simulated network elements can easily be added or removed by using test suite parameters. When the network configuration changes (e.g. simulated/target environment), the test steps and constraints used in the test cases do not need to be changed - only the test suite parameters in the test suite configuration file have to be updated. This allows replacing the simulated network elements partly by the real environment.

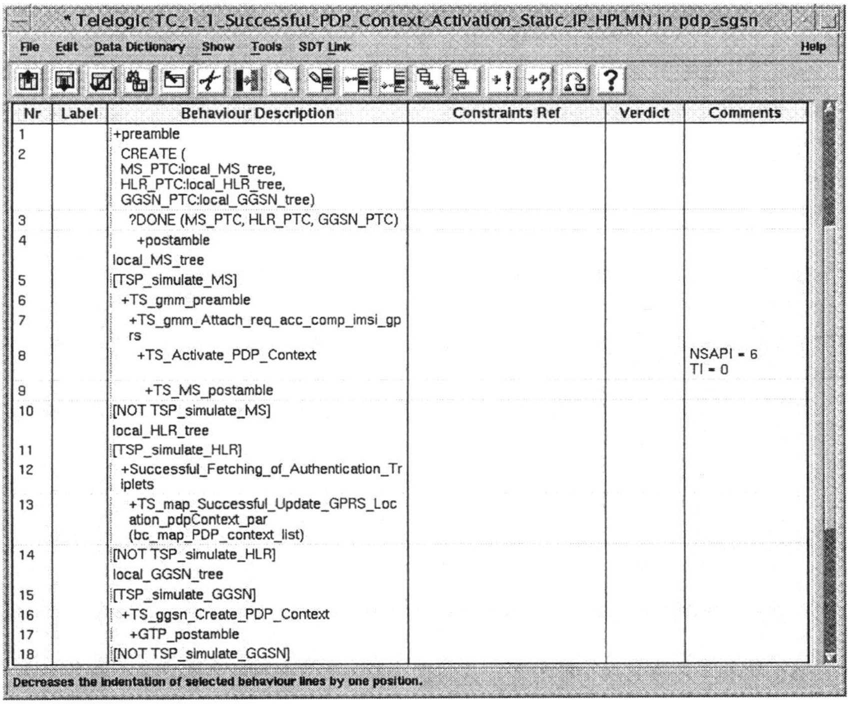
* Telelogic TC_1_1_Successful_PDP_Context_Activation_Static_IP_HPLMN in pdp_sgsn
File Edit Data Dictionary Show Tools SDT Link Help
Nr Label Behaviour Description Constraints Ref Verdict Comments
1 +preamble
2 CREATE (MS_PTC:local_MS_tree, HLR_PTC:local_HLR_tree, GGSN_PTC:local_GGSN_tree)
3 ?DONE (MS_PTC, HLR_PTC, GGSN_PTC)
4 +postamble
local_MS_tree
5 [TSP_simulate_MS]
6 +TS_gmm_preamble
7 +TS_gmm_Attach_req_acc_comp_imsi_gprs
8 +TS_Activate_PDP_Context NSAPI = 6 TI = 0
9 +TS_MS_postamble
10 [NOT TSP_simulate_MS]
local_HLR_tree
11 [TSP_simulate_HLR]
12 +Successful_Fetching_of_Authentication_Triplets
13 +TS_map_Successful_Update_GPRS_Location_pdpContext_par (bc_map_PDP_context_list)
14 [NOT TSP_simulate_HLR]
local_GGSN_tree
15 [TSP_simulate_GGSN]
16 +TS_ggsn_Create_PDP_Context
17 +GTP_postamble
18 [NOT TSP_simulate_GGSN]
Decreases the indentation of selected behaviour lines by one position.

Figure 3. Test case example

The parameterization of test suites has a remarkable effect on reusability and adaptability. Not only network configuration aspects can be parameterized in the test suites: address information (e.g. node addresses), subscriber data (e.g. International Mobile Subscriber Identity, IMSI) in constraints and other configuration related data such as execution control parameters/qualifiers are handled by using test suite parameters. The proper and reasonable usage of constraints together with test suite parameterization ensures a good portability, reusability and maintainability of test suites.

3.4   Test Configuration

The Abstract Test Configuration (ATC) is the representation of the testing environment within the TTCN test suites. It contains a description about how test events have to be carried out by means of interfaces and communicating channels. Within the test suites the ATC is specified on a logical level and the executing environment interprets it towards the IUT by the help of parallel communicating programs representing the test components.

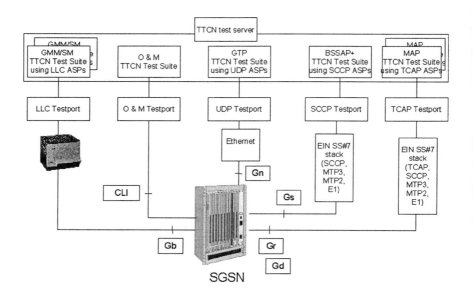

Figure 4. Test configuration for testing the SGSN

The executing tool for GPRS function test is the System Certification System (SCS) developed by Ericsson. The SCS is able to execute test suites in standardized TTCN.MP format towards the system under test (SUT) via implementation specific adapters. These so called test ports are dynamically linked to the SCS when the test session is started. For each type of interface a test port had to be developed. *Figure 4* shows the test configuration used for testing the SGSN with appropriate test ports applied on different interfaces.

The test port takes care of the communication between the SCS and the interface of the implementation under test (IUT). A test port has to be assigned to each PCO in order to execute the test suites. The assigned test port has to offer exactly the same set of ASPs that are specified in the test suite for the given PCO, e.g. LLC, TCAP or SCCP ASPs that are provided to the higher protocol layers.

Each test port in the SGSN test configuration has a direct connection to the IUT via real protocol stacks except from the LLC test port (*Figure 4*): the BSS (Base Station Subsystem) is simulated by a protocol simulator. Besides using the complete protocol stack it is also possible to interface the IUT on a higher protocol layer. This example shows the flexibility of the test port concept in the SCS: the test suite can be executed in simulated as well as in target environment.

4.   EXPERIENCES WITH TTCN, CONCLUSIONS

TTCN test suites are independent of test methods, layers, protocols or test tools, so TTCN is applicable under many conditions. It enables the handling of different protocols and multiple interfaces in one test suite and supports testing of complicated systems like the GPRS support nodes combining telecom and datacom protocols. TTCN provides powerful means of modularization and parameterization of test suites.

Writing of TTCN test suites is in many aspects similar to software development. A well-structured setup paves the way to a flexible and maintainable test system. A huge amount of tables in the test suites and a lot of code and definitions need to be written before testing can start. TTCN offers a lot of flexibility that brings with it many pitfalls to make mistakes.

Central configuration management is essential for success, without it the parallel testing of different protocol versions would be nearly impossible. A well-configured executing environment provides a lot of possibilities: interactive, automatic and distributed test execution, simulation of non-existent components and excellent logging/tracing facilities.

A total of approximately 2500 test cases distributed over ca. 50 test objects has been produced and executed during the GSN verification project using TTCN and SCS. A large subset of these test cases has been repeated in all design increments (regression test) and helped to secure the product quality with a minimum of test case redesign.

16

THE TEST SUB GROUP (TSG)
A Cooperative Approach to Improve the Release Quality before Type Acceptance

Giulio Maggiore✦, Silvio Valeau✦✦, Josè Pons✦✦✦
✦ *CSELT, Centro Studi e Laboratori Telecomunicazioni*
via Reiss Romoli 274 10148 Turin ITALY
Giulio.Maggiore@cselt.it

✦✦ *TIM, Telecom Italia Mobile*
largo Tassoni 323 Rome ITALY
svaleau@mail.tim.it

✦✦✦ *Ericsson Eurolab Deutschland GmbH*
Ericssonallee 1 52134 Herzogenrath GERMANY
Jose.Pons@eed.ericsson.se

Abstract   The development of telecommunication market has increased dramatically during the last ten years, forcing mobile operators to introduce new services for the customers. To do so, the time available for testing decreases drastically, forcing the operators to request always more quality to the manufacturers providing their own network elements. In this scenario, a group called Test Sub Group (TSG) has been created among the mobile operators using ERICSSON CME 20 Switching Systems, with the purpose to bring the operators' point of view inside Ericsson System Integration activities. In this phase, in fact, the possible improvements have lower costs for both, the supplier and the operators, and high effects on release software quality. In this paper, TSG and its way to work inside the Ericsson process is described, highlighting the benefits for Ericsson and the Operators. The paper continues addressing the methodology used and its application in the context of the Switching Systems R8 release with an example related to GPRS testing. A set of measurements to evaluate process and product quality is described. Some considerations about the validity of the "TSG" model are made, and its applicability to various market situations is analysed.

Keywords: CME 20, End-to-End testing, TSG, GSM, GPRS

1.   INTRODUCTION

In the last decade the development of the telecommunications had a tremendous growth. This development has resulted in new market opportunities and reduction in costs for the final users. The deregulation of the market has resulted in the entry of new players, as network operators and service providers.

The cellular communication arrived in the early 1980s marked a turning point in telecommunications. GSM and DCS 1800 are undoubtedly the major achievements in modern cellular solutions. Millions of customers use nowadays those voice services and look forward to adding non-voice services to their service portfolio. In fact with the new data services like WAP, GPRS, the idea of "Wireless office" is a reality. Wireless means you can get the information any time and any place.

In this scenario, GSM network operators need to put in service high quality network elements, sooner than in the past.

In this paper, the Ericsson CME 20 TSG Switching System[1] (SS) mobile operators' experience is described. Inside CME 20 User Group, a subgroup called Test Sub Group (TSG) was formed in January 1999 to support Ericsson in increasing the quality of its SS equipment's, and consequently to have an earlier roll out for the operators. This is a work in progress paper so no results have been reported yet. The paper is structured as follows. The section 2 explains what is TSG, how it works in the Ericsson industrialisation process, its reasons and goals, why it represents a win-win situation between Ericsson and the Operators. The section 3 gives the TSG SS methodology. The section 4 reports the current project organisation, named "Tornado", referred to R8 release of the Switching System's nodes and the measurements proposed to evaluate R8 process and product quality. In the section 5 the validity of the TSG model is addressed.

---

[1] In the TSG Switching System are included MSC/VLR, HLR, SSF, SCP and in this specific case the GPRS packet nodes.

2.   THE TEST SUB GROUP (TSG) SWITCHING SYSTEMS (SS)

Starting from the positive experience of the TSG BSS, made inside CME 20 for the BSS R7, in January 1999 a group of operators, Mannesmann Mobilfunk, Swisscom Mobile, Telenor Mobil, Telia Mobile, TIM, Vodafone UK, Libertel, and later on in the project Pacbell Wireless, proposed to Ericsson to launch, under the hat of CME 20 User Group, a TSG for the next SS release. Ericsson saw this new activity as an opportunity to improve its own systems. In this section what is the TSG and how it works in the Ericsson process is reported, what the TSG reasons and goals are, and why it is considered as "win-win" situation between Ericsson and the Operators.

2.1   What is TSG

The TSG is a group operating on behalf of the CME 20 User Group, during the Ericsson system integration phase of a release, in order to have a faster acceptance and rollout and, consequently to obtain a high quality release delivered at the end of the industrialisation process. The purpose of the group is to provide feedback, from the operator's point of view, to the Ericsson system integration phase. The TSG does not provide any requirements that Ericsson has to fulfil, and any constraints, which can cause some delay in the delivery date. It will be an Ericsson choice to implement or not the comments received according to its own policy. Quality is meant as the service perception seen from the operators' point of view, which means the ability to use a service, to manage the network elements and to be able to collect all the statistics, measurements and billing information.

2.2   The TSG SS in the Ericsson Process

The Ericsson's test process, consists of two main activities:
- Global Application System (GAS) testing activity, under the responsibility of a central Product Unit, in which is made the testing of all the world common parts of the new release;
- Market Application System (MAS), under the responsibility of the Ericsson's local company, in which the testing of all the market adaptations is made.

Both Ericsson and the CME UG (User Group) have identified a need for Generic Switching System Testing which is performed on the GAS part.

TSG Generic testing has the objective to check the common part of the release (GAS). By doing so, the operators share resources in analysing the

GAS testing, and Ericsson has an early feedback, which in the past was supposed to be received only during the type acceptance[2] activities.

A higher level of quality is achieved earlier in the life span of the release and the individual Ericsson local offices execute less repetitive testing for the respective customers.

The scope of a generic testing phase is limited to core CME 20 and common functionality across the CME UG operators. It remains the responsibility of the local Ericsson companies to prove national or customer specific software package and features.

The testing is carried out on a generic network environment. This environment includes as many customer specific settings as possible. TSG aims, together with Ericsson, to reach an agreement on identifying a suitable environment. The test environment is as close as possible to an end-to-end environment so that *the whole product* can be verified.

The parameter settings for the generic network environment are as close to the customer recommended setting as possible with as many functions as possible switched on. This recommendation is done jointly by Ericsson and TSG. The TSG has access to the parameter settings to enable individual members to assess the impact of their own parameter requirements on the generic test results.

TSG does not want to change Ericsson's way of working: Ericsson is not expected to produce extra documentation because of TSG more than what is produced anyhow. TSG just simply analyses, and gives feedback to the Ericsson work procedures. It is Ericsson's choice to make changes into the methodology using the TSG suggestions. It is in the interest of Ericsson to incorporate the TSG suggestions to increase the quality of the Ericsson products.

2.3   The New "Supplier-Customer" Process

The "Supplier-Customer" process that TSG is supposed to modify has two separated phases:
-    The Supplier's phase, in which the new product is designed and integrated;
-    The Customer's phase in which the product is tested against the internal requirements and then if accepted rolled out in the network.

In this way, any problem or a not completely satisfactory feature in the new product, is found when the Supplier's design and integration process is

---

[2] The type acceptance activity is made by all the testing aspects of a new product/release prototype, whose acceptance implies the acceptance for all the instances of the same product/release.

finished, and the same product has been delivered to various customers. The possible modifications on the product has to be done on several packages or postponed to a next release. From the customer's point of view this means as the best option, to have a delay in the roll out phase or even worse to postpone the delivering of the new service to a next product's release.

The new "Supplier-Customer" process due to the TSG interaction, see the Customers involved since the Supplier's System Integration phase, giving early feedback to the Supplier, that can possibly modify the product ones and for all the customers. This early involvement allows to the operators to get more product's information that can help reducing type acceptance and roll out and consequently the whole time to market.

2.4   The Win-Win Situation

The TSG activity is a joint venture between Ericsson and the CME User Group. The basis for the joint venture is built upon the expectation of reaching a mutual benefit, which results in a win-win situation for both parties. For example it is expected that members of the Test Sub Group perform the First Office Application (FOA). In other words, FOA operators are expected to be interested to participate actively in TSG. This is only one example of possible win-win situation, in which operators have an early roll out and Ericsson a better knowledge of the FOA site.

Both parties define in detail in the Joint Project Specification for each release what are the minimum benefits for the process to be undertaken. These are specified as measurable goals jointly valid for Ericsson and TSG members.

2.4.1   The TSG benefits as seen by the Operators

A successful TSG may result in a reduction of the Type Acceptance and roll out time for each operator, with an increased release quality. This results in a measurable cost saving for the operator.

At General Availability, the date at which the products are available for all customers, the operators receive products from Ericsson that have been tested according to the operator's point of view, prior to releasing the product for customer acceptance.

By doing this, the operators are in the position to accept Ericsson products sooner and doing this reduces trouble shooting on the new release. The customers also have the possibility to join efforts and resources among themselves, since they have been co-operating from early stages of the project already.

The TSG operators actively strive to find synergies among themselves by identifying common Test Objects that will result in a reduction of the time used to verify the different Market Application Systems.

By providing early information to the operators the rollout can be prepared early.

2.4.2   The TSG benefits as seen by Ericsson

A successful TSG may result in a reduction of Ericsson Time to Customer, allowing to use the results of the report that is written by TSG at the end of the project, towards customers outside the CME User Group.

Ericsson has an early involvement of the Supply organisations for those operators that are members of TSG during the verification phase, by creating an early awareness of the products within TSG operators and the corresponding Ericsson Local Offices.

A successful TSG activity will reduce the time and activities for Type Acceptance and roll out for the total Ericsson markets. This will be checked as resulting from the normal Type Acceptance activities and specially for TSG customers. This is measured on a project basis together with TSG customers. Ericsson delivers products with better quality since they have been tested early following the operator's point of view.

Ericsson staff (testers etc.) involved in the TSG activities meet the customer on a technical level. By that, they get an understanding of how the product quality impacts the customer and by that how important their own performance is. Additionally they have access to the verification competence available with the operator's technicians [1].

Realistic "customer-like" configurations are used when testing. This has a positive effect on the rollout as more "configuration-specific" faults are found before the Roll Out.

The test and verification staff gains a more customer oriented attitude which results in a more customer oriented way of working at an early stage of the production process then as it was the case before. The TSG effort and results are documented in agreement with Ericsson.

No results are available on both Ericsson's and Operators' side, because this is a work in progress paper.

3. THE TSG SS METHODOLOGY

The TSG SS methodology is based on the End to End methodology defined in the EURESCOM P412 [2] and P613[3] project [3] applied to the operator's operational environment. The application of the End-to-End methodology allows to check the service, since the very early stages of the products availability, checking not only network elements (NEs) functionality in the new release, but also the interactions with the NEs in the previous release.

All the aspects of the service functionality are checked together with network operability aspects, e.g. O&M, billing and statistics and measurements verifications. All these verifications, performed together, allow to check properly the services functionality as well as operators do in their own network. All this verification will increase the release quality, allowing speeding up the type acceptance and the roll out activities.

3.1   The End-to-End Methodology

In general, the object of the End to End methodology is the set of two Network Elements connected directly (or possibly through a third transit network element, or a sequence of transit NEs), as described in figure 1. The abstract testing methods used is Multi Party Testing Method as described in [4].

The networks are observed and controlled through some interfaces (user-network or network-network) which are called PCOs (Point of Control and Observation) [5]. There are at least two PCOs for the control and observation at the external borders (minimum mandatory configuration) and, as an optional feature, one or more additional PCOs only for observation (i.e. in order to monitor the internal network behaviour).

The actual number and location of the PCOs depends on the characteristics of the networks and on the functionality to be tested. The physical testers shown in figure 1 are logically located at such Points of Control and Observation. Such "points", in real life testing, may actually be close to each other or may possibly be very far: actually End to End may be applied to geographically distributed testing.

The monitoring point, indicated in the figure 2, is optional. The two other PCOs must always exist; as a particular case, the functions of one tester

---

[3] EURESCOM P613 project defined methodology and End to End test specifications written in Concurrent TTCN [TTCNV2_99] for the GSM-ISDN-PSTN Network Integration Testing.   Projects   results   are   publicly   available   on http://www.eurescom.de/public/Projects/P600-series/P613/P613.HTM

located in one of such PCOs might actually be replaced by equivalent actions performed by the corresponding node.

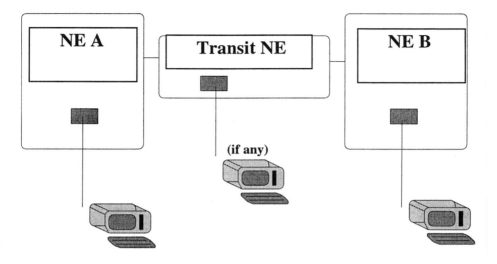

Figure 1. Generic configuration for an End to End testing session, including distributed Testers

In End to End, the network is tested as it is seen from the user's Terminal Equipment (TE), i.e. taking the user-network interfaces as PCOs (Point of Control and Observation). For example, A = GSM Access protocol, B = an ISDN Basic or Primary Access protocol.

The "End to End Network Under Test" is composed of all the parts that contribute to perform the expected network functionality, i.e. connection and transport of data and signals between the external gates (A, B, [M]):

❑   protocols that manage the external entities connected to the System Under Test (SUT) (access protocols in the case of End-to-End testing);

❑   each network component or function involved in a call between the A-side and the B-side (e.g. all call-control functions in all crossed NEs).

EtE is concerned only with the external behaviour of the network. For a complete description of the methodology refer to [1].

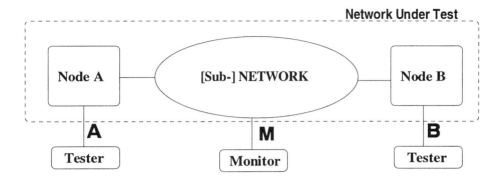

Figure 2. Example of a System Under Test (SUT), from the End to End point of view

3.2 Examples of Application of End-to-End Methodology

Examples of the aspects listed above are the worlds of ISDN, IN, GSM. For example in the case of a data call between a GSM user belonging to a PVN and an ISDN user, several network features are involved:
- switching functions (LE, TE, MSC);
- network databases (HLR);
- nodes containing service logic (SCP).

When each network element is tested separately, there is no confidence on the correct behaviour of the whole network. This is particularly valid for those services, whose implementation requires the interaction between NE based on telecommunication (TLC) technology with NE based on information technology (IT). This is exactly the case of the new GPRS service, whose deployment requires TLC NEs (BSS, MSC, HLR, and so on) and IT NEs (SGSN and GGSN). In fact as shown in figure 3, where is represented the ETSI GPRS functional architecture, there are quite a lot of interactions between TLC and IT NEs. Using the End-to-End methodology, in order to have a clear picture of the overall GPRS functionality, Terminal Equipment (TE) will be simulated on radio interface (Um) and on data network (Gi), and monitor points will be placed on the all the G interfaces.

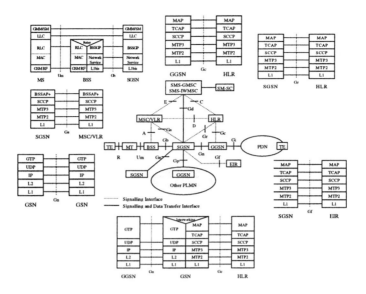

Figure 3. Overview of the GPRS Logical Architecture and protocol stacks

3.3 TSG SS R8 End to End Application

The application of the End to End methodology considered inside the TSG SS R8 can be summarised with the following items:

❑ Simulate access features, using protocol analyser, real terminal or network elements which provide the access facilities;

❑ Monitor network features, using monitor facilities, by applying End to End stimulus to use the service, as seen in the previous point.

❑ Check within each test case executed not only the telecommunication service from the protocols point of view, but also all the other aspects relevant for the operators like:

    – Charging;
    – Statistic measurements (STS, routes, traffic dispersion measurement);
    – Operation and Maintenance.
    – Only when all these aspects are correct, an End-to-End test case can be declared PASS.

4.   THE TORNADO PROJECT

Inside the R8 project, Ericsson decided to set-up a project addressing the TSG activities. This project was called "Tornado". The name was chosen to highlight the Ericsson's awareness of the customer's presence inside the project. In this section the project organisation will be described.

4.1   The Project Organisation and Phases

The project has been organised under Ericsson's responsibility, and follows Ericsson's system integration phases: features analysis, Test Description design, test case execution. These phases have been mapped into set-up phase, analysis phase and test execution phase inside the Tornado project.

4.1.1   Set-up phase

After the development of the project terms of reference together with the operators, Ericsson presented the new features to be developed and verified in the new release and the customers made a choice on the list of interesting features. A selection of feature was necessary because of a limitation of resources. For the remaining features not selected by TSG, Ericsson tried to apply the TSG methodology. A detailed time plan with deadline for the feedback was made, in order not to delay the ongoing project with late customer comments or suggestions. Agreements have been made with the Ericsson Local Companies to get them involved in the project as soon as possible. The Ericsson development and verification Switching System Project was informed about the direct customer involvement and its implications, since this was done for the first time for the Circuit Switching Systems Product Unit.

4.1.2   Analysis phase

In the analysis phase TSG operators provide feedback to the Test Descriptions (TDs) written by Ericsson for the release features verification. The specifications used as a reference for commenting the TDs are the ETSI standards and in case of proprietary services Ericsson Functional Specifications (FSs).

A common understanding is to have preliminary information provided by Ericsson. In this phase early information on TDs allow to TSG operators to give feedback that can be endorsed by Ericsson with low impact on the process.

The suggestions made by the operators are considered by Ericsson case by case and incorporated into the documentation.

The experience done, highlight minimal delays due to the customer involvement and valuable customer comments. The usage of Internet as the media to exchange documentation and information in Tornado Project optimise the project activity and reduce the project disturbance.

Face to face meetings are arranged when needed.

Ericsson provides to TSG operators, also documents relevant for the roll out, such as parameter information, information on changes in the Operation and Maintenance documentation. To all the documents TSG operators give feedback, highlighting the aspects to be improved and the relevant aspects from the operators' point of view.

4.1.2.1   The TSG SS feedback

As previously described, TSG is a "Win-Win" situation between Ericsson and the operators. This means that TSG doesn't put constraints to the Ericsson plan, in order not to delay the product delivery and consequently operator's time to market. The TSG SS analyses test architectures and specifications, and all the documents useful for the roll out of the new release, and gives feedback to Ericsson. It will be Ericsson's choice to endorse or not the comments received, on the basis of its own development policy, strategy and so on.

4.1.3   Test execution inspection

The test case execution phase starts after the analysis phase and the production of the final TDs on the Ericsson side. This phase intends to give to the TSG operators an early feeling of the services in the new release and the evidence of the System Integration process executed according with the reviewed documents. The position of TSG is being an observer. The feeling perceived in this phase together with the measurements taken during the project will help TSG operators in writing the final report, expressing opinion on the process and product quality.

4.1.4   Final report

All the activities performed by TSG operators together with Ericsson will be written in a final report. This TSG document represents the experience of the TSG operators related to the release quality and the Ericsson process applied for it. The final TSG opinion is built up on the bases of the measurements collected during the project and the service perception of the TSG operators during Ericsson test cases execution. The impact of TSG feedback on the Ericsson work is also evaluated.

This report is supposed to be delivered by TSG in preliminary version two weeks after the Ready for Service Date. A final version will be available two weeks after General availability Date. At the end of the roll out activity, the TSG operators will once more review the report. All these versions will be available inside the CME 20 user group.

4.2   The Measurements Proposed

In order to collect some figures to be used together with the feeling perceived during test cases executions, TSG and Ericsson agreed on a set of measurements. In some cases comparison between R8 and R7 will be attempted, in order to highlight positive or negative trends between the two releases. The proposed measurements are mainly taken from EIRUS[4] [6], but also other measurements are taken.

4.2.1   What TSG SS is going to measure

During R8 project, TSG SS is going to measure:
❑ Ericsson Process established for R8;
❑ Ericsson R8 Products.
Measuring Ericsson process TSG would like to keep track of the information flow, which is going to set-up the R8 products. Measuring R8 products TSG is going to measure the quality perceived, in terms of "trouble

---

[4] EIRUS is E-IPQM and E-RQMS User Group (http://www.eurescom.de/~public-webspace/P300-series/P307/eirus/rules.htm). The objectives of the group are:
❑   The implementation of E-IPQM and E-RQMS;
❑   Discussion in order to find best solutions and practices;
❑   Learn from the experiences of other companies;
❑   Formulate a consensus on change requests to E-IPQM and E-RQMS in order to preserve uniformity;
❑   Provide input to system user groups on the issue of quality measurements;
❑   Maintain contacts between PNOs and Suppliers;
❑   Establish a common attitude towards quality;
❑   Promote the knowledge about E-IPQM and E-RQMS.
❑   Communicate with various institutions and organisations in order to exchange ideas.
❑   Decide on actions which improve the internal organisations of EIRUS and its activities in general.

report", during the Ericsson test executions and afterwards during type acceptance and roll out.

4.2.2 How TSG wants to measure process and products

For the Process measurements will be considered the following indicators:
 ❑ Milestone delay (EIRUS);
 ❑ Requirements and Design Stability (EIRUS);
 ❑ Test Tracking (EIRUS);
 ❑ Action Points Delay (to TSG);
 ❑ Documentation Delay (to TSG);
For the Product measurement will be considered:
 ❑ Trouble report per service;
 ❑ Trouble Reports comparison per test phase;
 ❑ Trouble Reports comparison per life cycle.
 In the following subsections each measurement will be described.

4.2.2.1 Milestone Delay

Milestone status metrics can help the suppliers in assessing the robustness of their software process by monitoring key milestone schedules. Frequent occurrences of milestone slip may indicate planning, resource allocation, testing and other software quality related problems. These metrics can help the suppliers to better understand and control the Ericsson's milestone delays that are visible to the customers through customer interactions.

Data in input to TSG:
 ❑ List of INDUS R8 project milestones;
 ❑ Planned completion for each milestone;
 ❑ Actual completion date for each milestone.
 The Milestone Delay will be calculated as:

Milestone Delay = (Actual date – Planned Date) for each milestone

---

4.2.2.2 Requirements and Design Stability

Excessive number of changes, additions and deletion to requirements, design, code and test cases can adversely impact release schedules and software quality. Minimising and controlling these changes, the supplier can improve software quality and customer satisfaction. The TSG application for R8 will be related only to addition and deletion of features.

Data in input to TSG:

Original number of Features originally planned;
- ❏ Total number of deleted features;
- ❏ Total number of added features.

So the Requirements and Design Stability can be calculated as

$$StabilityIndex = \frac{FeatureChanges(addition, deletion)}{FeatureChanges + FeaturesInitialPlanned}\%$$

4.2.2.3   Test Tracking

Software testing plays a very important role in improving software reliability by removing defects and verifying performance before the release of a software package to the customers. These metrics will provide the suppliers with data by which they can assess the progress of their test planning and test activities. The information is used to track testing activities during integration, regression and system testing life cycle phases.

Data in input to TSG:
- Total number of Test Cases Planned (P);
- Total number of Test Cases Executed (E);
- Total number of Test Cases Passed (PA).

The test tracking metrics are the following:

- Percentage of Planned Test Execution (PTE):

$$PTE = \frac{E}{P}\%$$

- Percentage of Yield of Executed Test (YET):

$$YET = \frac{PA}{E}\%$$

- Pass percentage based on Tests Planned (PTP):

$$PTP = \frac{PA}{P}\%$$

The thresholds fixed by EIRUS for these parameters are:

❑ PTE

$$PTE \geq 97\% \Leftrightarrow GreenCondition$$

$$90\% \leq PTE \leq 97\% \Leftrightarrow YellowCondition$$

$$PTE \leq 90\% \Leftrightarrow \operatorname{Re}dCondition$$

❑ YET

$$YET \geq 95\% \Leftrightarrow GreenCondition$$
$$85\% \leq YET \leq 95\% \Leftrightarrow YellowCondition$$

$$YET \leq 85\% \Leftrightarrow \operatorname{Re}dCondition$$

❑ PTP

$$PTP \geq 92\% \Leftrightarrow GreenCondition$$
$$77\% \leq PTP \leq 92\% \Leftrightarrow YellowCondition$$

$$PTP \leq 77\% \Leftrightarrow \operatorname{Re}dCondition$$

4.2.2.4   Action Points Delay

Frequent occurrences of action points slip might be the indication of problems in the release progress, or in the supplier process.

Data in input to TSG:
- Action points planned dates;
- Action points Actual dates.

$$APD = Actual - Planned$$

4.2.2.5   Documentation Delay (to TSG)

As indicated also for the Action Points Delay, frequent occurrences of documentation delay might be the indication of problems in the release progress, or in the supplier process. The documentation considered is described in the project specification and the Documentation Delay will be measured as:

$$DD = Actual - Planned$$

4.2.2.6   Trouble report per service

Ericsson usually writes trouble reports (TRs) per software block. To be able to make a comparison between two releases (e.g. R7 and R8), TSG proposed to group TRs in order to create containers that can be compared. Each container can be a service or a group of services belonging to the same area (e.g. CCS#7, ISDN, etc.). For each service or group of services will be established a priority, which can be High, Medium, Low. The regression test will be considered as a big group, because of the importance it has for the operators. The services mentioned are evolution or improvements from the previous release. Examples are:

❑   Regression Test (High);
❑   Service 1 (High priority);
❑   Service 2 (Medium priority);
❑   Service 3 (Low priority).

This grouping can allow the comparison as showed in the table 1

Table 1. TR per serrvice comparison between R7 and R8

	R7	R8
Regression (High)	TR (A,B,C)	TR (A,B,C)
Service 1 (High)	Phase 1 TR(A,B,C)	Phase 2 TR(A,B,C)
Service 2 (Medium)	TR(A,B,C)	TR(A,B,C)

In order to have a fair comparison between the two releases, for each service or group of services will be considered the number of the included blocks, in order to make an average that will be used for the comparison between the two releases.

4.2.2.7   Trouble Reports comparison per test phase

With this measurement, for each test phase inside the system integration process, will be made a comparison between the release under test and the previous one. Trouble Reports are normalised using the test cases executed. The Delta value will give the indication about the increasing or reducing of the percentage of Trouble Reports per test cases. The figures in the table 2 show that for phase 1 and 2 the new release has better values than the previous one. This is not the case of the phase 3 that has to be analysed.

Table 2. TR/TC per phase comparison between Release N and Release N+1

Project	TR/TC (%) Phase 1	TR/TC (%) Phase 2	TR/TC (%) Phase 3
Release N	2	10	1.5
Release N+1	1.5	8	2
Delta	-0.5	-2	0.5

4.2.2.8   Trouble Reports comparison per life cycle

With this measurement, a projection outside the system integration phase is attempted. In fact Trouble Report found during system integration, are compared with those found during the First Office Application (FOA), and after 6 months rollout.

A comparison is also made between the last and the previous release (table 3).

Table 3. TR per life cycle comparison between R7 and R8

Project	TRs System Integration	TRs FOA	TRs GA + 6 months*
R7	TR (A,B,C)	TR (A,B,C)	TR (A,B,C)
R8.0	TR (A,B,C)	TR (A,B,C)	TR (A,B,C)
Delta	TR (A,B,C)	TR (A,B,C)	TR (A,B,C)

4.3   The Present Status of the Project

The TSG SS R8 project launched in January 1999 has reached at the moment half of its planned activities. The methodology chosen has been applied to provide feedback on test descriptions, and half of the comments provided to Ericsson have been implemented.

The future actions will be the test cases execution monitoring phase, collection of TRs and the reporting of the final consideration in the TSG Final Report. A clear picture of the overall benefits of the TSG approach will be seen only after the rollout of the R8 release, when product quality will be evident and the ratio quality/costs will be drawn for both the supplier and the operators.

5.   VALIDITY OF THE TSG MODEL

The TSG SS model applied inside Ericsson CME 20 User Group, can also be valid for those operators, whose network or most of it, is provided by one supplier. In this case this model is fully applicable and the operators can get more confidence on the new release, with the exception of the verification of the market customisations. In fact the custom solution cannot be proved for obvious reasons in a co-operative environment.

The TSG model is also applicable when such a NE typology (e.g. all the MSC/VLR, HLR, SCP etc.) belong to the same supplier. In this case the operators can be confident of the new release with the exception of the compatibility tests between the NE from one supplier and NE from another supplier. So in this case a special compatibility testing session will be required during the type acceptance phase.

There is a third case in which the model is hardly applicable. It is the case when different suppliers provide the same NE type. In this case it could be very difficult to establish a co-operative approach for confidentiality reasons. Even if the model is applied with one supplier, a full Network Integration Testing Campaign should be issued to make sure everything works together. In any case this point needs to be experimented.

6.   CONCLUSION

Customers are fundamentally changing the dynamics of the market place. The TSG is one example of a market become a forum in which customers play an active role in creating and competing for value. The distinguishing feature of this new marketplace is that customers become a new source of competence. The competence that customers bring is a function of the knowledge and skill they possess, their willingness to learn and experiment, and their ability to engage in an active dialogue. This will carry the market to develop a product shaped to the user needs, and not to the other way round. The products has to evolve in a way that enables future modifications and extension based both on customers' changing needs and companies' changing capabilities. Whether or not TSG has accomplished to his task will be discovered at the end of the R8 roll out, when pros and cons will be highlight for Ericsson and the Operators.

References

[1] C.K. Prahalad, V. Ramaswamy "Co-opting Customer Competence" Harvard Business Review January-February 2000.

[2] EURESCOM P.412 Deliverable 3: "Guidelines for NIT session management: volume 2: Guidelines and proformas".

[3] EURESCOM P613 Deliverable 3 "Test Specification for narrowband services (GSM, ISDN, PSTN)"

[4] ETSI ETR 193 "Methods for testing and Specification (MTS); Network Integration testing (NIT); Methodology aspects; Test Co-ordination Procedure (TCP) style guide".

[5] ISO/IEC 9646 (1994): "Information technology - Open Systems Interconnection - Conformance testing methodology and framework

[6] E-IPQM and E-RQMS User Group (http://www.eurescom.de/~public-webspace/P300-series/P307/eirus/rules.htm).

ABBREVIATIONS

CME 20	Ericsson GSM Network product
DCS	Digital Communication System
EtE	End to End
ETSI	European Telecommunication Standard Institute
FOA	First Office Application
FS	Functional Specification
GAS	Global Application System
GGSN	Gatway GPRS Support Node
GPRS	General Packet Radio Services
GSM	Global System for Mobile communication
HLR	Home Location Register
IN	Intelligent Network
ISDN	Integrated Services Digital Network
ISUP	ISDN User Part
IT	Information technology
LE	Local Exchange
MAS	Market Application System
MSC	Mobile Switching Centre
NE	Network Element
NtN	Node to Node
O&M	Operation and Maintenance
PCO	Point of Control and Observation
PVN	Private Virtual Network
SCF	Service Control Function
SCP	Service Control Point
SGSN	Serving Gprs Service Node
SS	Swithing System
SSF	Service Switching Function
STS	Statisctic and Traffic measurement Subsystem
SUT	System Under Test
TD	Test Description
TE	Terminal Equipment
TLC	Telecommunication
TR	Trouble Report
TSG	Test Sub Group
VLR	Visitor Location Register
WAP	Wireless Application Protocol

PART VI

REAL-TIME TESTING

17

VERIFICATION AND TESTING OF CONCURRENT SYSTEMS WITH ACTION RACES

Alex Petrenko[a] and Andreas Ulrich[b]

[a] *Centre de Recherche Informatique de Montreal, 550 Sherbrooke West, Suite 100, Montreal, H3A 1B9, Canada*
E-mail: petrenko@crim.ca

[b] *Siemens AG, Corporate Technology, ZT SE 1, 81730 Munich, Germany*
E-mail: andreas.ulrich@mchp.siemens.de

Abstract   We propose a categorization of action race conditions which occur in the specification of a concurrent system given as a collection of communicating labeled transition systems. We present several conditions to detect action races in the concurrent system operating in three different environments: a sequential-slow environment that allows execution of external actions only in stable states, a sequential-fast environment that supports the sequential execution of an action sequence before the next stable state is reached, and a concurrent environment that performs independent actions simultaneously. Race analysis facilitates the design of testers that can execute test actions without races at a highest possible speed.

Keywords: Concurrent systems, communicating LTS, action race analysis, race-free testing, test execution.

1.   INTRODUCTION

It is understood that races, besides interleaving of concurrent actions and internal nondeterminism, are a form of nondeterminism that is often inherent to concurrent systems. Depending on the execution order of actions, the global state of the system reached after a race is ambiguous.

In general, races are a source of software faults that are difficult to detect during test execution. They require repeated test runs of the same test case with varying execution speeds of test events and an analysis of the test result after each test run. As an alternative to testing, possible races can be analyzed based on a formal specification of the concurrent system within the design phase. However, general verification methods do not pay particular attention to race conditions and other types of nondeterminism since they evaluate all possible execution sequences of the system in a systematic manner [5]. Therefore, analysis techniques to detect race conditions are needed.

Work on race analysis started first in the context of sequential circuits [3]. Here, a distinction between race and race condition was introduced: a race condition results into a race if a circuit reaches different destination states during execution. The existence of races is explained from the specification of the circuits only.

Race analysis of concurrent software systems has mostly concentrated on an analysis of traces observed during the execution of the concurrent system. The approach in [11] assumes no specification of the concurrent system. Instead it tries to deduce possible race conditions from the mere observation of traces. The test engineer has to decide whether a certain constellation of message exchanges really constitutes a race. The trace analysis is based on a suitable definition of a happen-before relation between messages [4]. Similar work is presented in [9]. Another approach tries to detect races on-the-fly during the execution of a concurrent program using probing software [2]. Its drawback however is that races remain undiscovered if the suitable trace is not executed. A more systematic approach to trace analysis is presented in [1]. It discusses possible races based on an analysis of the system's behavior using different communication schemas like asynchronous communication with or without message queues.

In this paper we use another approach to detect races. We provide a characterization of races based on available formal specifications of the components of a concurrent system. We show that races occur due to conflicts between internal actions used for internal communication between the components and external actions that the system uses to communicate with its environment. Depending on the speed at which the environment can execute actions with the system, different race conditions occur. To distinguish this type of races from message or data races that are caused due to the underlying communication infrastructure, we call them *action races*.

The paper is organized as follows. Section 2 introduces basic notations. Section 3 presents definitions of races in the context of different environments. Section 4 applies the findings to construct test sequences that guarantee a race-free test execution. Section 5 concludes the paper.

2.   ASSUMPTIONS AND DEFINITIONS

A concurrent system comprises a number of independent components communicating with each other via message passing. Inputs and outputs of a component are not distinguished. Furthermore, there are no spontaneous transitions within a component. Each component has ports to communicate synchronously with the environment or with other components. The concurrent system is open, i.e., it is reactive with its environment.

More formally, a *concurrent system* $\Im$ is modeled by a system of n finite labeled transition systems (LTSs) communicating synchronously (rendezvous communication). Both, two-way rendezvous and multi-rendezvous, are considered. Each LTS defines a component of the concurrent system. Transmitting messages and their receipt through ports are referred to *actions* in an LTS.

Definition 1. A *labeled transition system* (LTS) M is defined by a quadruple $(S, A \cup \{\tau\}, \rightarrow, s_0)$, where S is a finite set of states; A is a finite set of actions; $\rightarrow \subseteq S \times (A \cup \{\tau\}) \times S$ is a transition relation; and $s_0 \in S$ is the initial state. Symbol τ denotes an unobservable action of M.

A transition $(s_1, a, s_2) \in \rightarrow$ is also written as $s_1-a\rightarrow s_2$. Let $s=a\Rightarrow s'$ denote the sequence $s-\tau\rightarrow\ldots-a\rightarrow\ldots-\tau\rightarrow s'$. Furthermore, let $s=\sigma\Rightarrow s'$ express the sequence $s=a_1\Rightarrow s_1=a_2\Rightarrow s_2=\ldots\Rightarrow s'$, where $\sigma = a_1a_2\ldots$. We assume that each component LTS is initially connected, i.e. for each state s' there exists a trace σ such that $s_0=\sigma\Rightarrow s'$. Since an LTS is in general nondeterministic, the state s' reached after execution of trace σ might be not unique. Let therefore (s **after** σ) denote the set $\{s' \mid s=\sigma\Rightarrow s'\}$.

Definition 2. An LTS M is *deterministic* if $|(s$ **after** $\sigma)| \leq 1$ for all $\sigma \in A^*$ and $s \in S$.

We further consider concurrent systems composed of deterministic LTSs without the unobservable action τ (rigid LTSs). Their behavior is characterized in terms of traces. The set of traces of state s is defined as the set $\{\sigma \in A^* \mid \exists s' \in S: s=\sigma\Rightarrow s'\} = Tr(s)$. The set of traces of the LTS M is $Tr(s_0)$.

The composition of two deterministic LTSs P and Q defined over the action sets A_P and A_Q, respectively, to a composite machine is expressed using the composition operator $\|$: $LTS(A_P) \times LTS(A_Q) \rightarrow LTS(A_P \cup A_Q)$, where $LTS(A)$ denotes the set of all possible LTSs over action set A.

Definition 3. Given two deterministic LTSs $P = (S_P, A_P, \rightarrow_P, s_P)$ and $Q = (S_Q, A_Q, \rightarrow_Q, s_Q)$, the parallel *composition* operator ‖ is defined by the following inference rules:

1. If $P -a\rightarrow P'$, $a \notin A_Q$ then $(P \parallel Q) -a\rightarrow (P' \parallel Q)$.
2. If $Q -a\rightarrow Q'$, $a \notin A_P$ then $(P \parallel Q) -a\rightarrow (P \parallel Q')$.
3. If $P -a\rightarrow P'$, $Q -a\rightarrow Q'$ then $(P \parallel Q) -a\rightarrow (P' \parallel Q')$.

Applied to a given concurrent system, this operator defines a composite machine. States of this machine are called *global* opposed to *local* states of component LTSs.

Definition 4. A *composite machine* $C_\Im = M_1 \parallel \ldots \parallel M_n$ of a given concurrent system $\Im$ of n LTSs $M_i = (S_i, A_i, \rightarrow_i, s_{0i})$ is the LTS $(S_\Im, A_\Im, \rightarrow_\Im, s_{0\Im})$, where $s_{0\Im} = (s_{01}, \ldots, s_{0n})$ is the initial global state; $S_\Im \subseteq S_1 \times \ldots \times S_n$, $A_\Im \subseteq A_1 \cup \ldots \cup A_n$, and $\rightarrow_\Im \subseteq S_\Im \times A_\Im \times S_\Im$ are the smallest sets obtained by application of the parallel composition operator ‖.

For simplicity, we assume that $A_\Im = A_1 \cup \ldots \cup A_n$. Let N be the set of indices $\{1, \ldots, n\}$ of the LTSs in $\Im$. Given action $a \in A_\Im$, $id(a)$ denotes the *occurrence* set $\{i \in N \mid a \in A_i\}$ of a. We assume that in a given system $\Im$ there are no isolated components, i.e., each component has at least an action a such that $|id(a)| > 1$. Given a subset of actions $A_{ext} \subseteq A_\Im$, actions in A_{ext} are called *external* actions, if $A_{ext} \supseteq \{a \in A_\Im \mid |id(a)| = 1\}$. We do not consider here *autonomous* systems, which have the empty set of external actions A_{ext} and we always assume $A_{ext} \neq \varnothing$. So, a concurrent system as defined here is an open system: It needs an environment to execute external actions. The set of actions of $\Im$ that do not belong to A_{ext} is the set of *internal* actions $A_{int} = A_\Im \setminus A_{ext}$ of system $\Im$. We shall refer to a (local or global) transition as *internal* or *external* if it is defined by an internal or external action, respectively.

Let $enabled(s) = \{a \mid \exists s' \in S_\Im: (s, a, s') \in \rightarrow_\Im\}$ denote the set of actions that belong to global transitions enabled in global state s. A *stable* global state is a state s such that $enabled(s) \subseteq A_{ext}$; a *transient* global state is a state s such that $enabled(s) \neq \varnothing$ and $enabled(s) \not\subset A_{ext}$. A *deadlock* state s is defined as a global state with no enabled actions, $enabled(s) = \varnothing$. We assume in this paper that a concurrent system has at least one stable global state (the initial one) in order to exclude "oscillating" systems.

EXAMPLE

Figure 1 shows the topology of an example system $\Im$ comprising three components communicating via the internal actions x, y, and z. The whole system communicates via the external actions a and b with environment E. Figure 2 depicts the behavior of each component in $\Im$ as a LTS. The composite machine defining the global behavior of the system $\Im$ is shown in Figure 3. In this figure, the bold circles refer to stable states.

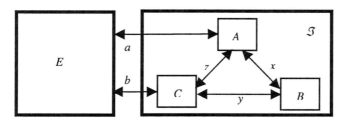

Figure 1. Example system $\Im = \{A, B, C\}$ and environment E.

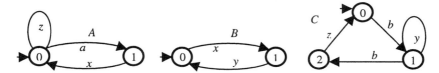

Figure 2. Component LTSs A, B, and C.

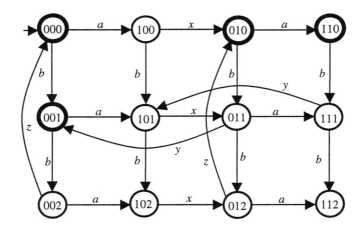

Figure 3. The composite machine $C_\Im$ of concurrent system $\Im$.

3.   RACES IN CONCURRENT SPECIFICATIONS

3.1   Race Conditions Categorization

Intuitively in our model of synchronous communication, an *action race condition* arises when in a given global state several actions are enabled. A race condition leads to a *race* when, as a result, the system may reach different destination (stable) states. Note that this phenomenon is pretty much the same as (critical) races in sequential circuits [3].

Depending on types of actions creating race conditions, we distinguish three types of race conditions:

- internal actions race conditions or *i-race* conditions;
- external actions race conditions or *e-race* conditions;
- mixed (internal and external) race conditions or *m-race* conditions.

Autonomous (closed) systems may have only i-race conditions. A given (open) concurrent system may or may not have a particular type of race conditions if it is put into different environments. Accordingly, we define three basic types of environments depending on its ability to offer either single or several concurrent actions at a time when the system is in a stable or in any (stable or transient) state.

- An environment that submits a single external action at a time only when the concurrent system is in a stable state is called a *sequential-slow* environment. Here only i-race conditions can occur.
- An environment that may submit external actions sequentially even before the concurrent system has reached the next stable state is called a *sequential-fast* environment. M-race conditions (in addition to i-race conditions) occur in a system within such an environment.
- Finally, if the environment can offer simultaneously several external actions starting from a stable state, it is called a *concurrent* environment. Note that since these external actions can be executed in any possible interleaving, a concurrent environment is also a sequential-fast environment. E-race conditions may arise within such an environment.

In this section, we formally define races that can occur in a given concurrent system placed within various environments and discuss the ways to detect them.

3.2   Sequential-slow Environment

A sequential-slow environment waits until a concurrent system completes the execution of internal actions before it offers a next external action. The behavior of a concurrent system that is controlled by a sequential-slow environment can be represented by a proper submachine of a composite machine. To define such a machine, we first introduce a modified composition operator for two LTSs, P and Q, with a fixed set of external actions $A_{ext} \subseteq A_P \cup A_Q$, as the mapping $\rfloor A_{ext}\llcorner : LTS(A_P) \times LTS(A_Q) \to LTS(A_P \cup A_Q)$.

Compared to the usual parallel composition operator $\parallel$, the new operator prohibits external actions from execution whenever the system can execute an internal action.

Definition 5. Given two LTSs $P = (S_P, A_P, \to_P, s_P)$ and $Q = (S_Q, A_Q, \to_Q, s_Q)$ and the sets A_{ext} and A_{int} of external and internal actions, respectively, $A_{ext} \cup A_{int} = A_P \cup A_Q$, the A_{ext}-*slow* composition operator $\rfloor A_{ext}\llcorner$ is defined by the following inference rule:

If there exists $a \in A_{int}$ such that $P -a\to P', Q -a\to Q'$
then $(P \rfloor A_{ext}\llcorner Q) -a\to (P' \rfloor A_{ext}\llcorner Q')$ **otherwise**
1. **If** $P -a\to P', a \notin A_Q$ **then** $(P \rfloor A_{ext}\llcorner Q) -a\to (P' \rfloor A_{ext}\llcorner Q)$.
2. **If** $Q -a\to Q', a \notin A_P$ **then** $(P \rfloor A_{ext}\llcorner Q) -a\to (P \rfloor A_{ext}\llcorner Q')$.
3. **If** $P -a\to P', Q -a\to Q'$ **then** $(P \rfloor A_{ext}\llcorner Q) -a\to (P' \rfloor A_{ext}\llcorner Q')$.

Applied to a given concurrent system, this operator defines the following composition.

Definition 6. Given the concurrent system $\mathfrak{S}$ of n LTSs $M_i = (S_i, A_i, \to_i, s_{0i})$ with the set of external actions A_{ext}, a *slow-composite* machine $Sc_{\mathfrak{S}} = M_1 \rfloor A_{ext}\llcorner \ldots \rfloor A_{ext}\llcorner M_n$ of $\mathfrak{S}$ is the LTS $(S_{Sc}, A_{Sc}, \to_{Sc}, s_{0Sc})$, where $s_{0Sc} = s_{0\mathfrak{S}}$ is the initial stable state; $S_{Sc} \subseteq S_1 \times \ldots \times S_n, A_{Sc} \subseteq A_1 \cup \ldots \cup A_n$, and $\to_{Sc} \subseteq S_{\mathfrak{S}} \times A_{\mathfrak{S}} \times S_{\mathfrak{S}}$ are the smallest sets obtained by application of the A_{ext}-slow composition operator $\rfloor A_{ext}\llcorner$.

For simplicity, we assume that $A_{ext} \subseteq A_{Sc}$. When no confusion arises, we omit the set A_{ext} from the operator to use $\rfloor\llcorner$. The relationships between the two types of composite machines are stated as follows.

Proposition 1. Given a concurrent system $\mathfrak{S}$ and its composite machine $C_{\mathfrak{S}}$, the slow-composite machine $Sc_{\mathfrak{S}}$ is a submachine of $C_{\mathfrak{S}}$ for any set of external actions A_{ext}. If $A_{ext} = A_{\mathfrak{S}}$, i.e., if the system has no internal actions, then the machines $Sc_{\mathfrak{S}}$ and $C_{\mathfrak{S}}$ coincide.

Let $(s$ *after-ext* $a)$ denote the set of all possible stable global states that can be reached by $\mathfrak{S}$ when external action a is executed in stable global state s, i.e., $(s$ *after-ext* $a) = \{s' \mid \exists \sigma \in A_{int}^{*}: s' \in (s \text{ *after* } a\sigma) \wedge \text{ *enabled*}(s') \subseteq A_{ext}\}$.

Definition 7. Given stable global state s and external action $a \in$ ***enabled***(s),
- system $\mathfrak{I}$ is *divergent*, i.e., it has a *livelock* for a in s if $(s$ ***after-ext*** $a) = \varnothing$;
- the action a creates a *race* in s if $|(s$ ***after-ext*** $a)| > 1$;
- the action a is *race-free* in s if $|(s$ ***after-ext*** $a)| = 1$.

The system $\mathfrak{I}$ is said to be *race-free* within a sequential-slow environment if in each stable state s every $a \in$ ***enabled***(s) is race-free.

In the sequel we consider only convergent concurrent systems, i.e. systems without livelock. All races in such systems within a sequential-slow environment can be characterized in a so-called stable composite machine defined to be as follows.

Definition 8. Given a concurrent system $\mathfrak{I}$ with the slow-composite machine $Sc_\mathfrak{I}$, a *stable* composite machine is an LTS $(S_{St}, A_{St}, \rightarrow_{St}, s_{0St})$, denoted $St_\mathfrak{I}$, where $s_{0St} = s_{0Sc}$; $S_{St} \subseteq S_{Sc}$, $A_{St} \subseteq A_{ext}$, and $\rightarrow_{St} \subseteq S_{St} \times A_{ext} \times S_{St}$ are the smallest sets obtained by application of the following rule:
$(s, a, s') \in \rightarrow_{St}$ iff $s' \in (s$ ***after-ext*** $a)$.

For simplicity, we assume that $A_{St} = A_{ext}$. A stable composite machine $St_\mathfrak{I}$ can be obtained from the slow composite machine by replacing all internal actions with the unobservable action τ and subsequently hiding τ in the obtained LTS by performing the failure-equivalent transformation as defined in [8]. Then, a nondeterministic stable composite machine indicates the presence of races.

Proposition 2. Concurrent system $\mathfrak{I}$ is race-free within a sequential-slow environment iff its stable composite machine $St_\mathfrak{I}$ is deterministic.

Any action sequence of a deterministic stable composite machine is race-free when it is executed within a sequential-slow environment.

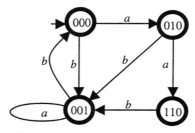

Figure 4. Stable composite machine $St_\mathfrak{I}$ of system $\mathfrak{I}$.

Paper on action races:

- May be used in our
  UM e testing for some
  special functionality, e.g.
  fail-over can be considered
  as sequential-slow case.

EXAMPLE

Figure 4 shows the stable composite machine for our working example that is deterministic. This means that the given concurrent system is race-free within a sequential-slow environment. Moreover, as one can see from the composite machine (Figure 3), at most one internal action is enabled in each global state. I-race conditions do not occur in such a system.

The number of states of a stable composite machine may reach that of a composite machine in the worst case situation. The question arises whether races can be detected without actually constructing the full (stable) composite machine. To check i-race conditions there is no need to store the whole stable composite machine, as each combination of external action and stable state can be checked separately.

Some sufficient conditions for the absence of i-race conditions can be established based on the following observation. Internal actions simultaneously enabled in a global state do not create a race if they belong to different local states, as the final state does not depend on the order of their execution. Only internal actions enabled at the same local state may create a race, thus we have the following sufficient condition.

Proposition 3. Concurrent system $\mathfrak{I}$ is race-free within a sequential-slow environment if there exists at most a single internal transition from every local state.

As follows from the definition, this type of action races is completely out of control of the environment in spite of its slow sequential behavior. If a given specification has races within a sequential-slow environment then we may conclude that either there exists a design error (in other words, it has unintended nondeterminism [11]) or the specification is, in fact, an abstraction of a future design and has to be refined further.

3.3   Sequential-fast Environment

Assume now that a concurrent system $\mathfrak{I}$ interacts with a sequential-fast environment that may submit external actions sequentially before the system has reached the next stable state. In this case, a sequence of external actions may create races even if each individual action is race-free in a corresponding stable state. An external transition can be enabled in some transient global state where an internal transition is also enabled. This may create a race.

Let σ be a sequence of k external actions, $\sigma = a_1 \ldots a_k \in A_{\text{ext}}^*$. Let $Tr_{St_\mathfrak{I}}(s)$ denote the set of traces of stable state s of stable composite machine $St_\mathfrak{I}$.

Given stable global state s and a sequence σ of k external actions, $\sigma = a_1 \ldots a_k$ $\in A_{ext}^*$, such that $\sigma \in Tr_{St_3}(s)$, we define (s **after-ext** σ) to be the set of all possible stable global states that can be reached in C_3 when the external trace σ is executed from s within a sequential-fast environment, i.e., (s **after-ext** σ) $= \{s' \mid \exists \beta_1, \ldots, \beta_k \in A_{int}^* : s' \in (s \text{ after } a_1\beta_1 \ldots a_k\beta_k) \wedge enabled(s') \subseteq A_{ext}\}$.

Definition 9. Given stable state s and external trace $\sigma \in Tr_{St_3}(s)$,
• the trace σ creates a *race* in s if $|(s \text{ } after\text{-}ext \text{ } \sigma)| > 1$;
• σ is *race-free* in s if $|(s \text{ } after\text{-}ext \text{ } \sigma)| = 1$.
The system $\mathfrak{S}$ is said to be *race-free* within a sequential-fast environment if in each stable state s every $\sigma \in Tr_{St_3}(s)$ is race-free.

If an external trace is race-free in some state, then it can be executed at this stable state in a speed-independent manner (see Section 4.2). The final stable state is always the same regardless of the speed at which the actions are consecutively executed. Similar to a sequential-slow environment (cf. Proposition 3), there exists a simple sufficient condition for the absence of races in the case of a sequential-fast environment.

Proposition 4. Concurrent system $\mathfrak{S}$ is race-free within a sequential-fast environment if in each local state all outgoing transitions are external or there is a single internal outgoing transition.

The stable composite machine does not represent m-race conditions, and the presence of races within a sequential-fast environment can be verified by constructing a part of the composite machine as follows.

Proposition 5. Given a stable state $s = (s_1, \ldots, s_n)$, let $\sigma = a_1 \ldots a_k$ be an external trace $\sigma \in Tr_{St_3}(s)$ and $E(\sigma)$ denote the linear LTS such that it has $k+1$ states and a single completed trace σ. The trace σ is race-free in s iff the system $E(\sigma) \parallel (M_1(s_1) \parallel \ldots \parallel M_n(s_n))$ has a single deadlock state.

Here, the notation $M(s)$ refers to LTS M initialized in state s. As this statement indicates, the problem of race detection is computationally as difficult as the classical reachability problem.

EXAMPLE
Consider the system $E(bb) \parallel A(1) \parallel B(1) \parallel C(0)$, i.e. execution of the external trace bb in the stable global state 110 (Figure 5). In the stable composite machine, i.e., under the sequential-slow environment assumption, the trace takes the system into the unique stable state 000. Not so within a sequential-

fast environment since three possible final states can be reached. Hence the trace *bb* creates a race in state 110 under the sequential-fast environment assumption. Consider another external trace *ab* in state 000, the system $E(ab)$ ‖ $A(0)$ ‖ $B(0)$ ‖ $C(0)$ has only one deadlock state in stable state 001 (cf. Figure 3), therefore *ab* is race-free in state 000.

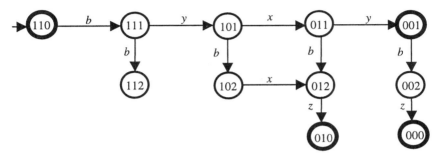

Figure 5. A race in state 110 for the sequence *bb*.

By definition, an m-race condition arises when an external and an internal action caused by a previous external action can be both executed in some (transient) global state. It is intuitively clear that such a condition does not lead to a race if external actions excite internal actions in different components. This observation leads us to the following sufficient conditions for the absence of races in external action sequences executed within a sequential-fast environment.

Let (s **act-ext** a) denote the set of all components that execute internal transitions in response to an external action a applied at stable state s, i.e., (s **act-ext** a) = $\{i \in id(b) \mid b \in A_{int} \wedge \exists \beta \in A_{int}^{*}: a\beta b \in Tr_{Sc_{\mathfrak{I}}}(s)\}$, where $Tr_{Sc_{\mathfrak{I}}}(s)$ is the set of traces of a slow-composite machine $Sc_{\mathfrak{I}}$.

Proposition 6. Given a stable state s and external trace $\sigma = a_1...a_k \in A_{ext}^{*}$, let $s_1...s_k s_{k+1}$ be the corresponding sequence of traversed stable states, where $s = s_1$. The trace σ is race-free in s if for all i, $1 \leq i \leq k$, the action a_i is race-free in state s_i and (s_i **act-ext** a_i) $\cap \bigcup_{j < i} (s_j$ **act-ext** $a_j) = \varnothing$.

3.4   Concurrent Environment

We now consider a concurrent environment that can simultaneously execute several external actions of different components, however only one action at a time for a particular component. Assuming that the concurrent system comprises n components, the concurrent environment can simultaneously execute at most n external actions with the system as a single *concurrent action.* To be consistent with the interleaving semantics, we represent a concurrent action with k elementary actions, $k \le n$, by an LTS with 2^k states (a hypercube), each completed trace of it is one possible ordering (permutation) of length k of all the constituent external actions.

To constitute a concurrent action, it is required that two actions a, $b \in A_{ext}$ have disjoint occurrence sets, i.e., $\mathbf{id}(a) \cap \mathbf{id}(b) = \varnothing$ (the actions are independent). This is a syntactic sufficient condition for concurrent transitions used in [5] that can be applied also in our context: two independent external actions a and b can be executed as a concurrent action if in a given stable state neither a nor b enable subsequent internal actions. However if the concurrent system has internal actions, it is not guaranteed that two independent external actions executed concurrently will result in the same final stable state. Therefore, we need a stronger requirement to define concurrent actions.

Definition 10. Given a deterministic stable composite machine St_3, two external actions a, $b \in A_{ext}$ are *potentially concurrent* in stable state $s \in S_{St}$ if :
1. $a, b \in \mathbf{enabled}(s)$, $b \in \mathbf{enabled}(s \ \mathbf{after}_{St} \ a)$, $a \in \mathbf{enabled}(s \ \mathbf{after}_{St} \ b)$ and
2. $s{=}ab{\Rightarrow}_{St} s'$, $s{=}ba{\Rightarrow}_{St} s''$, $s' = s''$.

The above two conditions, (1) enabledness and (2) commutativity, are similar to the conditions of independence relation given in [7], [5]. The difference is that we formulate them in terms of a stable composite machine and not, as it is usually the case in verification, of a composite machine. Moreover, in our context, to be truly concurrent, potentially concurrent actions should not create races. Given a subset of potentially concurrent external actions P, we use $[P]$ to denote the set of all possible linearization traces of the constituent actions, i.e. all of its permutations.

Definition 11. Given stable state s and a set P of external actions that are in pairs potentially concurrent, $P \subseteq \mathbf{enabled}(s)$,
- the set P creates a *race* in s if $|\bigcup_{\sigma \in [P]} (s \ \mathbf{after\text{-}ext} \ \sigma)| > 1$,
- P is *race-free* in s if $|\bigcup_{\sigma \in [P]} (s \ \mathbf{after\text{-}ext} \ \sigma)| = 1$.

If set P is race-free, then we also say that P is a *concurrent action* in state s.

Proposition 7. Given a stable state s and a set $P \subseteq \mathbf{enabled}(s)$ containing in pairs potentially concurrent external actions, the set P is a concurrent action in s iff all possible linearizations $\sigma \in [P]$ are race-free in s.

Proof. If the set P is a concurrent action in s then the union $\bigcup_{\sigma \in [P]} (s\ \textbf{\textit{after-ext}}\ \sigma)$ is a singleton. The latter implies that the set $(s\ \textbf{\textit{after-ext}}\ \sigma)$ is also a singleton for all $\sigma \in [P]$. We conclude that each $\sigma \in [P]$ is race-free in s.

To demonstrate the other part of the statement, we assume that there are two potentially concurrent external actions a, $b \in P$ such that ab are ba are race-free in s, but the set $<a, b>$ (and thus the set P) creates a race in s. We have $|(s\ \textbf{\textit{after-ext}}\ ab) \cup (s\ \textbf{\textit{after-ext}}\ ba)| > 1$, at the same time a and b are potentially concurrent in state s. The commutativity property of actions a and b of a deterministic system implies that $s=ab\Rightarrow_{St} s'$, $s=ba\Rightarrow_{St} s''$, $s' = s''$. Then $s' = ((s\ \textbf{\textit{after-ext}}\ a)\ \textbf{\textit{after-ext}}\ b) = ((s\ \textbf{\textit{after-ext}}\ b)\ \textbf{\textit{after-ext}}\ a) = s''$ holds. Taking into account that $|(s\ \textbf{\textit{after-ext}}\ ab) \cup (s\ \textbf{\textit{after-ext}}\ ba)| > 1$, we conclude that either $|(s\ \textbf{\textit{after-ext}}\ ab)| > 1$ or $|(s\ \textbf{\textit{after-ext}}\ ba)| > 1$. That means, trace ab or ba should create a race in s, which contradicts our assumption that ab and ba are race-free in s.

EXAMPLE
In our working example (cf. Figure 4), we have two external actions a and b. They cannot be potentially concurrent in stable state 110, since $a \notin \textbf{\textit{enabled}}(110)$, neither can they in stable state 001, for $001=ab\Rightarrow_{St} 000$ but $001=ba\Rightarrow_{St} 010$. On the other hand, it can be verified that they constitute a concurrent action $<a, b>$ in the states 000 and 010 of the stable composite machine.

A sufficient condition can be formulated for a pair of potentially concurrent external actions to form a concurrent action taking into account the components that participate in the execution of each action.

Proposition 8. Given a stable state s and external actions a, $b \in \textbf{\textit{enabled}}(s)$, $<a, b>$ is a concurrent action in state s if $(s\ \textbf{\textit{act-ext}}\ a) \cap (s\ \textbf{\textit{act-ext}}\ b) = \varnothing$.

E-race conditions can be verified by a direct construction of a part of the composite machine instead of verifying each linearization as suggested by Proposition 7. This fact is expressed by the following proposition.

Proposition 9. Given a stable state $s = (s_1, \ldots, s_n)$, let P be a set of external actions that are in pairs potentially concurrent in s and $E(P)$ denote an LTS such that it contains $2^{|P|}$ states and the set of all of its completed traces is $[P]$. The set P is a concurrent action in s iff the system $E(P) \parallel (M_1(s_1) \parallel \ldots \parallel M_n(s_n))$ has a single deadlock state.

Up to here we have established all necessary definitions and propositions needed to perform a race analysis for a concurrent system of communicating LTSs. The next section demonstrates how this knowledge helps deal with races during testing of such systems.

4.   TEST EXECUTION AVOIDING ACTION RACES

4.1   Test Architecture and Testing Goal

The test architecture to test concurrent system $\mathfrak{S}$ consists of the parallel composition of the system under test (SUT), the implementation of $\mathfrak{S}$, and a tester modeled in terms of one or several communicating LTSs. The tester is able to observe and control all actions of the SUT, which it has access to (synchronous communication). In opposition to [10] we ease the assumption that the tester participates in all internal and external actions of the SUT. Instead we assume that the SUT exposes only its external actions to the tester. Consequently, race conditions between external test actions and/or internal actions of the SUT might occur during a test run.

Testing implementations of a concurrent system specification with respect to races, we may intend to verify whether or not a test case of an external action trace that is race-free in the specification is also race-free in the implementation. Such a testing goal is interesting since opposed to a typical testing assumption, an implementation might be in this case "more nondeterministic" than its specification. Since races cause nondeterministic behavior, their guaranteed detection requires that a certain fairness assumption (often called *complete testing assumption*) is satisfied for the implementation. To obtain confidence that the SUT contains no additional races than given in its specification, we need to rerun the same test several times and to identify the global state reached of the SUT after each test run. These tests must be deterministic, i.e., they do not contain any known specification races, to clearly indicate deviating behavior and thus faults of the SUT. Besides the detection of further races in an implementation, deterministic, race-free tests greatly reduce test efforts since the test purpose of a test case can be definitely reached for a correct SUT, which is hardly possible for nondeterministic tests.

In this section, we consider how testers should be devised that execute test cases at the highest possible speed without triggering specification races in order to detect additional races in the SUT and possibly other faulty behavior.

4.2   Steady Testing

Given a *test case* as an external action sequence ω over A_{ext} for concurrent system $\mathfrak{S}$, the execution of the test case on the SUT of $\mathfrak{S}$ is usually modeled as a parallel composition with the $\|$ operator of the tester $T(\omega)$ implementing ω as a linear LTS and the SUT itself.

Deterministic test cases exists only when i-race conditions in the specification do not lead to races, i.e., the stable composite machine of the specification is deterministic (see Section 3.2). We assume that it is always the case. To avoid further specification races within a sequential-slow environment, the tester must guarantee that the SUT always returns into a stable state before the next external action of the test case is applied to the SUT. This behavior is achieved if the $\rfloor\lfloor$ operator is applied between tester and SUT instead of $\|$. We refer to this mode of testing as *steady testing*.

To implement the $\rfloor\lfloor$ operator, a tester implementing a test case has to use delays, denoted as Δ, to insure that the SUT reaches a stable state in response to a previous external test action[1]. Consider our example system and assume the test designer derives the test case $\omega = ababaabbbb$ (a transition tour through the stable composite machine from Figure 4), then the test case for system $\mathfrak{S}$ executed under the slow environment assumption is $\omega_{steady} = a \Delta b \Delta a \Delta b \Delta a \Delta a \Delta b \Delta b \Delta b \Delta b$.

4.3   Sequential-Fast Testing

To increase the execution speed of tests, one has to use a tester that behaves as a sequential-fast environment to the SUT provided that specification races do not occur. Formally speaking, we have to derive maximum race-free *test steps* (nonempty sub-sequences) $\omega_1, \omega_2, \ldots$ from the original test case ω, such that $\omega = \omega_1 \omega_2 \ldots$. Then delays have to be inserted only after each test step, $\omega_{seq} = \omega_1 \Delta \omega_2 \Delta \ldots$. The longer these test steps are the fewer delays have to be inserted to avoid specification races. The resulting test can be executed faster in most cases than the previous secure test in steady testing that requires inserting delays after each single test action.

Thus, given test case ω, we have to determine a minimum number of delays or, in other words, a minimum partitioning of ω in order to execute this

---

[1] Note that delays cannot be expressed in our LTS model quantitatively since it does not support time. Therefore the value of a delay must be determined by other means. The delay depends mainly on the execution time of messages exchanged between two participating components and on the length of the execution sequence of internal actions. The test engineer needs to adjust this time for any specific system.

sequence in the fastest sequential way. More formally, find a partitioning ω_1, ..., ω_r of ω such that:

1. $\omega_1 \ldots \omega_r = \omega$;
2. ω_1 is race-free in the initial stable state s_{03} and ω_i is race-free in stable state $(s_{03}$ *after-ext* $(\omega_1 \ldots \omega_{i-1}))$ for all $i > 1$; and
3. if there exist another partitioning ω'_1, ..., ω'_k such that (1) and (2) are satisfied, then $k \geq r$.

Algorithm 1.
1. The initial stable state is s_{03}.
2. Find a maximum race-free prefix of ω in the given initial stable state by analyzing the sequence of the first two consecutive actions to be race-free using Proposition 5 and then consecutively adding actions one by one until the sequence is no longer race-free.
3. Remove the obtained prefix from ω and do Step 1 starting from the stable state reached by the previous prefix until ω is finished.

Proposition 10. Algorithm 1 yields a minimum race-free partitioning of ω.

Proof. Let us assume that the sequence ω is partitioned into n race-free non-empty sequences, $\omega = \omega_1 \omega_2 \ldots \omega_n$. Suppose there exists a partitioning of $n-1$ race-free sequences only, $\omega = v_1 v_2 \ldots v_{n-1}$. In this case, there should exist two subsequences ω_i and v_j such that $\alpha\omega_i\beta = v_j$, where β is a non-empty sequence, otherwise the second partitioning cannot have fewer sequences. Consider the case when α is an empty sequence. Then ω_i cannot be extended by a non-empty sequence β to obtain a race-free sequence $\omega_i\beta$, for Algorithm 1 delivers the longest race-free sequence in stable state $(s_{03}$ *after-ext* $\omega_1 \omega_2 \ldots \omega_{i-1})$. If α is a non-empty sequence, then $v_j = \alpha\omega_i\beta$ must be race-free in stable state $(s_{03}$ *after-ext* $v_1 v_2 \ldots v_{j-1})$. This implies that $\omega_i\beta$ is race-free in stable state $(s_{03}$ *after-ext* $v_1 v_2 \ldots v_{j-1}\alpha)$. Furthermore, we have $(s_{03}$ *after-ext* $v_1 v_2 \ldots v_{j-1}\alpha) = (s_{03}$ *after-ext* $\omega_1 \omega_2 \ldots \omega_{i-1})$. This means that $\omega_i\beta$ is race-free in stable state $(s_{03}$ *after-ext* $\omega_1 \omega_2 \ldots \omega_{i-1})$, which is impossible since ω_i is the longest race-free sequence in stable state $(s_{03}$ *after-ext* $\omega_1 \omega_2 \ldots \omega_{i-1})$.

Once it is proven that a subsequence (a test step during testing) does not cause specification races, the execution time of the whole test case can be immediately reduced by avoiding all the delays between the consecutive actions within the subsequence.

A tester implementation of the partitioned test case $\omega_{seq} = \omega_1 \Delta \omega_2 \Delta \ldots$ is a deterministic global tester that has control to all external actions A_{ext} of SUT. The global tester executes a test action one after another and performs additional delays as indicated in ω_{seq} to ensure that the SUT reaches a stable state from where the test is resumed. Consider our example system again, it can be analyzed that the same test case $\omega = ababaabbbb$ can be partitioned into four test steps that are race-free in a sequential-fast environment. Therefore, it is sufficient to execute the test case $\omega_{seq} = a\ b\ a\ \Delta\ b\ a\ \Delta\ a\ b\ \Delta\ b\ b\ b$ to guarantee a deterministic sequential-fast test run without specification races. Instead of nine delays in ω_{steady} only three are needed in ω_{seq}.

4.4 Concurrent Testing

The results obtained in the previous section rely on sequential testers only. Further timesaving in the execution of tests is possible if concurrency is introduced in the tester. We discuss the interplay of concurrency and specification races first for the abstract notion of a global concurrent tester and next for the model of a distributed concurrent tester that can be the basis for practical tester implementations.

To start with, we consider a global concurrent tester that can simultaneously execute several external actions with different components, however only one action at a time for a particular component. That means, such a tester can execute several external actions with the system as a single concurrent action as discussed in Section 3.4.

We assume that a test case is partitioned into several consecutive test steps using the analysis method presented in Section 4.2. Additionally, each of the test steps may contain concurrent actions according to Section 3.4. If concurrent actions exist in a test step, we refer to it as a *concurrent test step*. Delays are inserted only between the test steps of the test case. A delay inserted between two test steps indicates that the tester must slow down to allow the concurrent system to stabilize in the next stable state in order to avoid specification races.

A concurrent test step for concurrent system $\Im$ is an acyclic LTS T with the initial state t_{0T} and a single sink state. Its set of actions coincides with A_{ext} of $\Im$. If the test step is a single concurrent action, the LTS T is a hypercube (see Section 3.4). E-race conditions occur if there is a state in T with more than one action enabled, i.e., if the test step contains concurrent actions. To verify race conditions between tester and specification, a part of their composite machine has to be constructed, similar to Proposition 9.

Proposition 11. Given a concurrent test step T for $\Im$ in stable state s, test step T is race-free in s iff the system $T(t_{0T}) \parallel (M_1(s_1) \parallel \dots \parallel M_n(s_n))$ has a single deadlock state.

When specification races are detected for a given concurrent test step, it has to be modified to ensure a deterministic test run. The composition $T(t_{0T})$ ‖ $(M_1(s_1)$ ‖ ... ‖ $M_n(s_n))$ helps identify traces of T that have to be serialized to avoid specification races. In the extreme situation, a concurrent test step might be transformed into a race-free test sequence.

EXAMPLE

Taking the initial partitioning of the (sequential) test case for our example system from Section 4.2 $\omega_{seq} = a\ b\ a\ \Delta\ b\ a\ \Delta\ a\ b\ \Delta\ b\ b\ b$, an analysis for concurrent actions reveals the concurrent test case $\omega_{conc} = <a,\ b>\ a\ \Delta\ b\ a\ \Delta\ <a, b>\ \Delta\ b\ b\ b.$[2] The savings to use a concurrent tester instead a sequential-fast one are two concurrent actions, i.e., the length of the test case (including delays) could be reduced from 19 test actions for a sequential-slow tester to 13 test actions for a sequential-fast tester to 11 test actions for the concurrent tester.

The global concurrent tester can be refined such that concurrent actions are executed truly concurrent instead of executing them in an arbitrary interleaving order. To obtain true concurrency, the tester must be distributed among several concurrent tester components. We assume that such a distributed tester comprises k tester components, where k is the number of external ports of system $\mathfrak{S}$. Each tester components controls and observes a subset of external actions from A_{ext}. That means, a distributed tester is another concurrent system T with the same alphabet of external action A_{ext} as concurrent system $\mathfrak{S}$.

To support collaboration among the distributed tester components, a synchronization mechanism must be implemented. If broadcast communication is available, a single synchronization action is used as an internal action of the distributed tester T. In case of multicast, the alphabet of internal actions of the system T consists of several synchronization actions for each necessary subset of tester components that needs to be synchronized. Each tester component is modeled as an LTS with the unobservable action τ that represents a delay needed to avoid races in a test run. The global behavior of the distributed tester is given by the parallel composition of its tester components, $T = T_1$ ‖ ... ‖ T_k.

Once a distributed tester is constructed for a given concurrent test case, it can be verified whether it contains race conditions using the results presented

---

[2]   Note that the partitioning *ba* after the first delay does not constitute a concurrent action, although *a* and *b* are independent. However, the two possible linearizations *ba* and *ab* starting from stable state 001 result in the two different stable states 010 and 000, respectively, which can be verified using Figure 4 (cf. Definition 10).

in previous sections. A distributed tester is usually required to behave deterministically (unless the test designer intentionally opts for nondeterministic testing). This means that its internal synchronization actions should not create races. A stable composite machine of the distributed tester has to be deterministic.

The approach in [6] requires, similar to our work, that the parallel composition of tester components (after hiding) is a given test case. It neglects however the possibility of races inside the tester and between tester and SUT. Races must be completely avoided to ensure a deterministic test run. One can say that a deterministic concurrent distributed tester exemplifies the design of a concurrent system without races.

5. CONCLUSIONS

The paper introduced the issue of action races as an inherent property of concurrent systems. Such races occur when internal and external actions of the system are enabled in certain global states that result in different end states. Three different assumptions on the environment of a concurrent system and their influence on race conditions were discussed. A very stringent assumption is the sequential-slow environment where only race conditions between internal actions occur. However the execution time of the concurrent system is artificially inflated.

A more realistic environment is a sequential-fast environment that can perform several external actions sequentially before the system reaches the next stable state. Here mixed race conditions between internal and external actions may occur, their absence has to be verified in the specification in order to guarantee a deterministic run of the concurrent system. Finally a concurrent environment was introduced that represents what seems to be the most universal environment. In addition to mixed race conditions, race conditions between external actions may occur due to the concurrent execution of these actions.

Moreover, the paper presented necessary and sufficient conditions to avoid races in a concurrent system. Still, races may be not completely avoided in the design of a concurrent system. In spite of this, races have to be identified to take preventive measures in the implementation of the system. The issue of races is of interest too when it comes to testing concurrent systems. We demonstrated how the analysis of races helps design testers that execute test actions without races at a highest possible speed.

The paper presented only initial results in the field of race analysis based on specifications. More work needs to be done. More subtle conditions should be established to detect race conditions in a concurrent system. It is also interesting to study the interplay of verification techniques based on partial orders and race analysis techniques proposed in this paper. Another interesting topic is to study the test generation process that avoids specification races. Last but not least, race analysis must be discussed in the context of communicating I/O-LTSs in an asynchronously communicating environment to address more realistic systems.

6.   ACKNOWLEDGEMENT

The first author acknowledges support of NSERC grant OGP0194381.

References

[1]   R. Alur, G. Holzmann, D. Peled: *An Analyzer for Message Sequence Charts*; 2nd Int'l Workshop TACAS'96, 1996; pp. 35–48.

[2]   K. Audenaert: *Maintaining Concurrency Information for On-the-fly Data Race Detection*; Parallel Computing 1997; Bonn, Germany, September 1997; pp. 19–22.

[3]   J. A. Brzozowski, C.-J. Seger: *Asynchronous Circuits*; Springer-Verlag, 1994; 404 p.

[4]   A. Bechini, K. C. Tai: *Timestamps for Programs Using Messages and Shared Variables*; 18th Inter'l Conference on Distributed Computing Systems; Amsterdam, The Netherlands, May 1998; pp. 266–273.

[5]   P. Godefroid: *Partial-order methods for the verification of concurrent systems*; LNCS 1032; Springer, 1996.

[6]   C. Jard, T. Jeron, H. Kahlouche, C. Viho: *Towards automatic distribution of testers for distributed conformance testing*; FORTE/PSTV'98; Paris, France, 1998.

[7]   S. Katz, D. Peled: *Defining conditional independence using collapses*. Theoretical Computer Science, vol. 101, 1992; pp. 337–359.

[8]   G. Luo, G. v. Bochmann, A. Das, Ch. Wu: *Failure-equivalent transformation of transition systems to avoid internal actions*; Information Processing Letters, vol. 44 (1992), pp. 333–343.

[9]   R. H. Netzer, B. P. Miller: *Optimal Tracing and Replay for Debugging Message-Passing Parallel Programs*; Conference on Supercomputing 1992; pp. 502–511.

[10] A. Petrenko, A. Ulrich, V. Chapenko: *Using partial-orders for detecting faults in concurrent systems*; IWTCS'98, Tomsk, Russia, 1998; pp. 175-190.

[11] K. C. Tai: *Race analysis of traces of asynchronous message-passing programs*. 17th Int'l Conference on Distributed Computing Systems (ICDCS'97); Baltimore, MD, USA, 1997; pp. 261–268.

18

AN APPROACH FOR TESTING REAL TIME PROTOCOL ENTITIES

Ahmed Khoumsi
Université de Sherbrooke, Dep. GEGI, Sherbrooke, CANADA
khoumsi@gel.usherb.ca

Mehdi Akalay, Rachida Dssouli, Abdeslam En-Nouaary
Université de Montréal, Dep. IRO, Montréal, CANADA
(akalay,dssouli,ennouaar)@iro.umontreal.ca

Louis Granger
École Polytechnique de Montréal, Dep. GEGI, Montréal, CANADA
louis.granger@mail.polymtl.ca

Abstract   We propose a two-step method and an architecture for testing real-time protocol entities. In the first step, the timed specification of the implementation under test is transformed into an equivalent untimed specification. Then in the second step, test methods for non-real-time systems may be used. In comparison with other test methods using similar approaches, our method avoids state explosion. A transformation algorithm of the first step and a test architecture are proposed.

Keywords: Test generation, Test architecture, Implementation under test, Real time protocol entity, Timed automata, Discrete time, Minimization.

1.   INTRODUCTION

Many formal methods are used to describe timed systems; real-time logic, timed Petri nets, and timed automata are among the most popular. Two approaches are known to represent the time : *continuous* time [2] and *discrete* time [15, 10]. With continuous time, time measures represent exact values of time. With discrete time, a *unit of clock time* (uct) is defined and time measures are integer values which are incremented after the passing of each uct. Various techniques and tools have been developed to ensure formal *verification* of timed systems, such as COSPAN, KRONOS, UPAAL and HYTECH. But few work has been done for *testing* timed systems; for example :

- In [12] timers and counters are used to guarantee a reliable transmission of messages in a given bounded time. The used model allows to describe constraints on delays separating *send*s and corresponding *receive*s.

- In [14] temporal logic formulas are extended with discrete time. Tests are generated from formulas written in that logic. Their method allows to describe simple formulas using a single variable.

- In [3] the authors generate tests from a *Constraint Graph* (CG). Their model allows to describe constraints on delays separating consecutive events.

- In [5] the authors propose a test method based on the transition tour of the entire region graph of the control part. A state is identified only by a single empirical value for each clock. The fault coverage of the method is limited.

- In [17] the authors propose a theoretical method for test cases generation based on a variant of the Timed Automata of [2]. This is the first approach inspired by methods for untimed systems.

- In [6] the authors continue in the same direction as [17] and provide a practical and complete method for test cases generation.

In this article[1], we propose a test method and a test architecture for timed systems. The proposed test method consists of two steps : (1) the timed specification of the $\mathcal{IUT}$ (Implementation Under Test) is transformed into an equivalent untimed specification; and (2) existing test methods for untimed systems may then be used to generate test sequences. Our work is inspired from [6] which uses a similar two-step approach. An important difference with [6] is that we use discrete time while in [6] continuous time is used. Discrete time does not allow to represent exactly the physical behaviour of the system. However, discrete time is sufficiently accurate in many instances, for instance when dealing with digital control systems [11].

In [6] after the transformation of the first step, the elapsing of each (uct) of each clock is represented by a transition ν. Therefore a state explosion problem arises. Our method avoids state explosion by representing only "relevant" time elapsing. For instance, if the possible behaviour of a system changes only after ten ucts, then only the elapsing of the tenth uct will be represented. Although this optimisation is inspired from [1] which uses a continuous time, it is of no use to [6]. In fact, in continuous time the relevant time elapsing cannot in

---

[1] [9] is a longer version with more details

general be related to any concrete aspect, while in discrete time the relevant time elapsing can be related to the expiration of a real timer. We propose an algorithm to realize the transformation of the first step. We also propose a test architecture applicable to the specifications obtained at the first step. Henceforth FSA denotes "Finite State Automaton".

The remaining of this article is structured as follows. Sect. 2 describes the model of timed automata (TA) and its corresponding fault model. In Sect. 3 we present our tick-FSA model and its minimization. In Sect. 4 we propose the se-FSA model which, in comparison to tick-FSA, avoids state explosion. And then we present an algorithm for transforming a TA into a se-FSA. Sect. 5 presents a test architecture applicable to specifications described by se-FSAs. In Sect. 6 we show how test cases are generated. And finally in Sect. 7 we conclude by discussing some future work.

2.   TIMED AUTOMATA AND FAULT MODEL

2.1.   Model of Time and Timers

We consider a digital clock [10] which generates a tick at a constant frequency; the delay between two consecutive ticks is called unit of clock time (uct). The time is modeled by a variable τ which is initially equal to zero and is incremented by one after the passing of each uct (Fig. 1.a). With such a model, there is an inaccuracy of one uct on the instants of events and an inaccuracy of two ucts on the delays separating two events [10] (Fig. 1.b). A *timer* is an integer variable which : (1) is automatically incremented after each tick and (2) may be set to zero at the occurrence of every event.

Figure 1.   (a) Time model; (b) Accuracy of the time model

2.2.   Timed Automata (TA)

We consider a set of timers $\mathcal{T} = \{t_1, \cdots, t_{N_t}\}$, and we define a *canonical Enabling Condition* (EC) as being any formula in the form "$t_i \sim k$", where $\sim \in \{<, >, \leq, \geq, =\}$. More generally, a EC may consist of a single canonical EC or of a conjunction of canonical ECs. A EC may also be the constant $True$. We also define $\mathcal{EC}_{\mathcal{T}}$ as being the set of ECs depending on timers of $\mathcal{T}$, and $\mathcal{P}_{\mathcal{T}}$ as being the set of subsets of $\mathcal{T}$.

A TA may be obtained from a FSA if we associate an EC and a set Z of timers to each transition Tr of the FSA. More formally, a TA may be defined by $(\mathcal{L}, \mathcal{E}, \mathcal{T}, \mathcal{T}r, l_0)$ [10] where : $\mathcal{L}$ is a finite set of locations, l_0 is the initial location, $\mathcal{E}$ is a finite set of events, $\mathcal{T}$ is a finite set of timers, and

$Tr \subseteq \mathcal{L} \times \mathcal{E} \times \mathcal{L} \times \mathcal{EC}_T \times \mathcal{P}_T$ is a transition relation. A transition is therefore defined by Tr = $\langle q; \sigma; r; EC; Z \rangle$ where : q and r are origin and destination locations, σ is the event of the transition, Tr may occur only if $EC = true$, and after the occurrence of Tr the timers in Z are set to zero. Z is called *reset* of Tr.

A TA allows to express constraints on the number of ticks between events. For example, to specify that there may be 1 to 3 ticks between transitions Tr1 and Tr2, we may use a timer t_1 as follows : the Z of Tr1 is $\{t_1\}$ and the EC of Tr2 is $(t_1 \geq 1) \wedge (t_1 \leq 3)$. Note that a state s of a system specified by a TA may be defined by (l, t), where l is a location and t is an N_t-tuple specifying the current value of each timer.

2.3.   Example of TA

We consider the simplified version of the sender of the CSMA/CD communication protocol [19] which is described by the TA of Fig. 2 where : (1) λ is the transmission time of a message and σ is the worst case propagation delay; (2) a single timer t is used; (3) $?u$(resp. $!u$) means "u is received (resp. sent)"; and (4) a transition Tr = $\langle q; \sigma; r; EC; \{t\} \rangle$ is labelled by "$EC; \sigma; t := 0$". The absence of EC or of timers to reset are indicated by "-". As described in Fig. 2 :

- The sender is initially at location *idle* where it waits for a message to send. When the latter is received ($?send$), the sender reaches location *ready*.

- When location *ready* is reached, the sender verifies the bus. If the latter is free ($?free$), the sender starts the transmission ($!begin$) and reaches location *transmit*; otherwise the sender reaches location *retry* either if the bus is busy ($?busy$) or if a collision is detected ($?cd$).

- When location *retry* is reached, the sender waits for the availability of the bus during twice the worst propagation delay ($t = 2\sigma$) before re-attempting the transmission of the message ($!begin$) and then reaching location *transmit*. If a collision is detected in location *retry* before the delay 2σ ($t < 2\sigma$), then the sender begins again to wait for the availability of the bus (selfloop).

- When location *transmit* is reached (from *ready* or *retry*), the transmission is started ($!begin$). If a collision is detected ($?cd$) before the worst propagation delay ($t < \sigma$), the sender goes to location *retry*; otherwise it terminates sending the message ($!end$) after exactly λ ucts ($t = \lambda$) and goes to location *idle*.

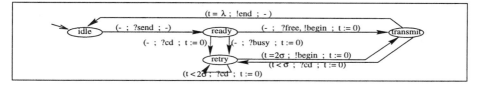

Figure 2.   CSMA-CD sender

2.4.   Fault Model

Let SP and IUT be two TAs describing a specification and its implementation, respectively. We may categorize faults that may arise in an implementation of a timed system by : (1) faults independent on timing constraints; and (2) timing faults [7]. For the first category, we consider the types of faults given in [13] : output faults, transfer faults and hybrid faults. The second category of faults are caused by the non respect by IUT of timing constraints associated to outputs of SP. During a testing process and for a given test sequence, the tester respects timing constraints of inputs and checks whether timing constraints of outputs are respected. An essential type of timing faults may be defined as follows : (1) SP requires that the $\mathcal{IUT}$ sends an output σ at an instant $t \in [t1, t2]$; and (2) IUT allows the sending of σ at certain instants which do not fall within $[t1, t2]$. In reality, with discrete time $t, t1$ and $t2$ are numbers of ticks. For example, in the specification of Fig. 2, when location *transmit* is reached, the output !*end* must occur after λ ticks. The $\mathcal{IUT}$ is faulty if !*end* occurs before the λth tick or after the $(\lambda + 1)$th tick.

3.   TICK-FSAS AND THEIR MINIMIZATION

We show here how a TA may be transformed into an equivalent FSA called tick-FSA. Then we show in an example how state explosion may be avoided by combining states of the tick-FSA.

3.1.   Transforming TA into Tick-FSA

A TA may be transformed into an equivalent FSA called tick-FSA where the event tick is represented by a transition [10]. A state of a tick-FSA may be defined by (l, t) where : (1) l is a location of the corresponding TA and (2) t is a N_t-tuple specifying the current value of each timer (N_t is the number of timers). The transformation from a TA into a tick-FSA can lead to a state explosion problem [10].

We consider the example of Fig. 2 where : the bus is 2^{20} bits/second, the propagation delay σ is 25 ucts, the length of each message is 2^{10} bytes, and 1 second = 10^6 ucts. We compute that the delay λ to transmit a message, including the propagation delay, is 806.25 ucts. Therefore, events !*begin* and

!*end* are separated by 806 ticks. The resulting tick-FSA which illustrates these results is represented on Fig. 3.

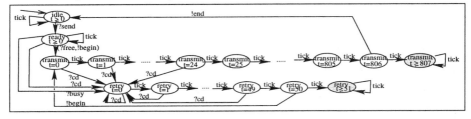

Figure 3. State explosion with a tick-FSA

3.2. Combining States of a Tick-FSA

For example, the 808 states of the tick-FSA of Fig. 3 which are associated to location *transmit* can be combined into :

Group 1 : $transmit, t \leq 24$: from which ?*cd* leads to $(retry, t = 0)$,
Group 2 : $transmit, 25 \leq t \leq 805$: from which no event $(\neq tick)$ occurs,
Group 3 : $transmit, t = 806$: from which !*end* leads to $(idle, t \geq 0)$,
Group 4 : $transmit, t \geq 807$: which is a deadlock.

Transitions between these groups are :

(Group 1 → Group 2) is the 25th tick from $(transmit, t = 0)$,
(Group 2 → Group 3) is the 806th tick from $(transmit, t = 0)$,
(Group 3 → Group 4) is the 807th tick from $(transmit, t = 0)$.

These combinations illustrate the possibility to avoid state explosion. In Sect. 4 we propose the se-FSA model which is the result of this kind of combinations, and we present an algorithm for transforming a TA into a se-FSA without using an intermediate tick-FSA.

4. TRANSFORMING TA INTO A SE-FSA

4.1. Two New Types of Events

?$Set(t_i, k)$ means : timer t_i is set to zero and will expire when its value is equal to k. ?$Set(t_i, k_1, k_2, \cdots, k_p)$ means that t_i is set to zero and will expire several times when its value is equal to $k_1, k_2, \cdots, k_p$.

!$Exp(t_i, k)$ means : timer t_i expires and its currewnt value is k.

Therefore, each event ?$Set(t_i, k)$ is followed (after a delay k) by !$Exp(t_i, k)$, and each event ?$Set(t_i, k_1, k_2, \cdots, k_p)$ is followed by the sequence : !$Exp(t_i, k_1)$, !$Exp(t_i, k_2)$, $\cdots$, !$Exp(t_i, k_p)$. An event Set (resp. Exp) is associated with Sign "?" (resp. "!") because, as we will see in Sect. 5, Set (resp.Exp) is sent (resp. received) by the tester and may therefore be conceptually considered as an input (resp. output) of the system under test (SUT).

In [4], events *Set* and *Exp* have been used differently than here, for *verifying* timed systems.

4.2. Transformation Algorithm

We propose here an algorithm for transforming a TA into a FSA with events *Set* and *Exp*, which we denote se-FSA. The algorithm consists of four steps which may be intuitively introduced as follows :

Step 1 : Rewrite the ECs by using only "$<$" and "$\geq$". This rewriting is convenient because the value (true or false) of a canonical EC "$t < k$" or "$t \geq k$" changes *once* exactly when $t = k$. After the rewriting, the only relevant information for checking "$t \sim k$", will be : (1) whether t has been set to zero and (2) whether its value has reached k.

Step 2 : Associate an event $?Set(t, k)$ to each transition which : (1) resets a timer t and (2) is followed (not necessarily in the next transition) by an EC "$t \sim k$".

Step 3 : For each location L :

    1 Associate an event $!Exp(t, k)$ to each "$t \sim k$" which is in an outgoing transition of L.

    2 Construct all the sequences consisting of all the determined events *Exp*. Each of these sequences, which we denote Exp-sequence, corresponds to an order of events *Exp*.

    3 Remove certain *impossible* Exp-sequences by using certain rules. For example, let $!Exp(t1, k_1)$ and $!Exp(t2, k_2)$ be contained in the set of constructed Exp-sequences, where $k_1 \leq k_2$. If in all paths which allow to reach location L, $?Set(t1, k_1)$ is *before* $?Set(t2, k_2)$, then any Exp-sequence where $!Exp(t1, k_1)$ is *after* $!Exp(t2, k_2)$ is removed.

    4 Simplify each constructed Exp-sequence *Seq* by removing certain "irrelevant" events *Exp* according to the following rule. If just after an event *Exp* the EC of a transition *Tr* becomes false, then remove all the following events *Exp* which are associated exclusively to ECs of *Tr*. The removed events are irrelevant because when a transition becomes false, it remains false in the whole remaining part of *Seq*. This simplification will be illustrated in the example of Fig. 6.

    5 For each constructed Exp-sequence, we construct a sequence of states $L_1, L_2, \cdots$ which are connected by this Exp-sequence.

    6 For each outgoing transition *Tr* of L leading to any L', we construct a transition executing the same event than *Tr* from every $L, L_1, L_2, \cdots$ where the EC of *Tr* is *true*. All the constructed transitions lead to L'.

Step 4 : The obtained FSA is determinized and minimized, and undesirable states are removed.

For the TA of Fig. 2 the result just before the removal of undesirable states is represented in Fig. 4. Here is an intuitive explanation of how this se-FSA is constructed : ($\lambda = 806$ and $\sigma = 25$)

1 "$t = 806$" is rewritten "$(t \geq 806) \wedge (t < 807)$", and

"$t = 50$" is rewritten "$(t \geq 50) \wedge (t < 51)$".

2 Events *Set* are determined as follows :

- "$t := 0$" of the two transitions leading to *transmit* are followed by comparisons :

  "$t < 25$" of the transition *transmit* $\rightarrow$ *retry*, and

  "$(t \geq 806) \wedge (t < 807)$" of the transition *transmit* $\rightarrow$ *idle*.

  Therefore $?Set(t, 25, 806, 807)$ is associated to the transitions leading to *transmit*.

- "$t := 0$" of the four transitions leading to *retry* are followed by comparisons :

  "$t < 50$" of the transition *retry* $\rightarrow$ *retry*, and

  "$(t \geq 50) \wedge (t < 51)$" of the transition *retry* $\rightarrow$ *transmit*.

  Therefore $?Set(t, 50, 51)$ is associated to the transitions leading to *retry*.

3 Events *Exp* are constructed as follows :

- Three events $!Exp(t, 25)$, $!Exp(t, 806)$ and $!Exp(t, 807)$ are consequences of $?Set(t, 25, 806, 807)$.

- Two events $!Exp(t, 50)$ and $!Exp(t, 51)$ are consequences of $?Set(t, 50, 51)$.

- The sequence $(!Exp(t, 25), !Exp(t, 806), !Exp(t, 807))$ "divides" *transmit* into four states which correspond to the four groups introduced in Sect. 3.2.

- The sequence $(!Exp(t, 50), !Exp(t, 51))$ divides *retry* into three states : $(retry, t \leq 49)$, $(retry, t = 50)$ and $(retry, t \geq 51)$.

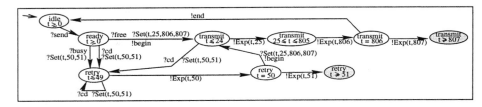

Figure 4. se-FSA obtained from the TA of Fig. 2. Undesirable states are marked.

4.3.   Operational Presentation of the Algorithm

Step 1	**For** every canonical EC : **If** the EC is "$t_i = k$" **then** rewrite it into "$(t_i \geq k) \wedge (t_i < k + 1)$", **If** the EC is "$t_i \leq k$" **then** rewrite it into "$t_i < k + 1$", **If** the EC is "$t_i > k$" **then** rewrite it into "$t_i \geq k + 1$". **End**

Step 2	**For** every location L of the TA :
	    For every outgoing transition Tr of L :
	      For every timer t which is reset by Tr :
	        Search ECs in transitions which are reachable from Tr without
	        crossing a transition which resets t
	      For every EC "$t \sim k$" which has been found :
	        Associate the event $?Set(t, k)$ to Tr
	      End
	    End
	    Events $?Set(t, k_1), ?Set(t, k_2), \cdots, ?Set(t, k_p)$ may be replaced by
	    $?Set(t, k_1, k_2, \cdots, k_p)$
	    End
	 End

Step 3	**For** every location L
	    For every EC "$t \sim k$" of an outgoing transition of L : (Point 1)
	      Associate an event $!Exp(t, k)$
	    End
	    For every timer t used in at least one of the events Exp :
	      $SeqOfExp_t := \epsilon$
	    End
	    For every event $!Exp(t, k)$:
	      Insert $!Exp(t, k)$ into $SeqOfExp_t$ in temporal order
	    End
	    Construct the $SetOfSeq$ containing all the sequences which may be
	      obtained by interleaving all the sequences $SeqOfExp_t$ (With the
	      preceding two loops : Point 2)
	    Remove certain impossible sequences of Exp from $SetOfSeq$ (Point 3)
	    For every $Seq \in SetOfSeq$:
	      For every $!Exp(t, k)$ of Seq which corresponds to "$t < k$" : (Point 4)
	        Remove all the following events Exp in Seq which are associated
	        exclusively to the same transition(s) than $!Exp(t, k)$
	      End
	      Construct states $L_1, L_2, \cdots$ such that $L, L_1, L_2, \cdots$ are connected by Seq
	      (Point 5)
	      For every outgoing transition $Tr = \langle L, \sigma, L', EC, Z \rangle$ of L : (Point 6)
	        For every state $s \in \{L, L_1, L_2, \cdots\}$
	          If s is (*before* the $Exps$ associated to canonical ECs of Tr using "$<$"
	          and after the $Exps$ associated to canonical ECs of Tr using "$\geq$")
	          then Add a transition leading to L' and executing σ and the events
	          Set associated to Tr
	        End
	        Remove Tr
	      End
	    End
	    L becomes a *state* instead of location
	 End

Step 4	Determinize and minimize the obtained FSA Remove the undesirable states

Let us illustrate the simplification of the sequences of Exp (Point 4 of Step 3) with the the TA of Fig. 5 which uses timers $t1$ and $t2$. The results of the four steps are represented in Fig. 6. In Fig. 6.c (resp. d) we show the result of Step 4 without (resp. with) the simplification. The latter may be explained as follows. The transition Tr executing ϕ has the EC "$(t1 < 4) \wedge (t2 < 3)$", and therefore it becomes disabled after the first of $!Exp(t1, 4)$ and $!Exp(t2, 3)$. For this reason, it is useless to represent the second of the two Exp. This implies that every simplified sequence contains *either* $!Exp(t1, 4)$ or $!Exp(t2, 3)$.

In general a timer t may expire at an instant τ and its expiration is either : (1) immediately relevant; or (2) irrelevant and will become relevant at a future instant ν ($\nu > \tau$); or (3) irrelevant and will never become relevant. Case (3) can be considered as a particular case of Case (2) when $\nu = \infty$. In our transformation algorithm, in Cases (2) and (3) we have conceptually *delayed* the expiration until instant ν. As an illustration of Case (3) in the example of Fig. 6, when ϕ occurs in State $L2_0$ and then leads to State $L0_0$, none of the events $!Exp(t_1, 2), !Exp(t_1, 4), !Exp(t_2, 3)$ and $!Exp(t_2, 5)$ is relevant in State $L0_0$. For this reason, although their occurrences remain possible in $L0_0$, these events have been conceptually delayed to the instant $\nu = \infty$. In Sect. 5.1.3, we will see how this delaying action can be realized physically.

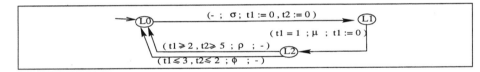

Figure 5.   Example of TA using two timers

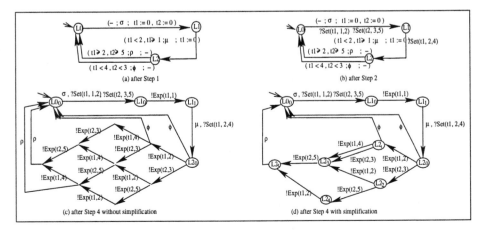

Figure 6.   Results of the algorithm when it is applied to the TA of Fig. 5

5. TEST ARCHITECTURE

We propose a test system ($\mathcal{TS}$) consisting of three modules called Test-Controller, Exp-Delayer and Timer-Handler, respectively. The interest of such a $\mathcal{TS}$ is that it guarantees the following equivalence.

TESTING EQUIVALENCE : Let S1 be a TA and S2 be the se-FSA obtained from S1 using the algorithm of Sect. 4. The following two points are equivalent :

1 The $\mathcal{IUT}$ is conformant to the specification S1.

2 The $\mathcal{SUT}$ (System Under Test) consisting of the set { $\mathcal{IUT}$, Exp-Delayer, Timer-Handler } is conformant to the specification S2.

5.1. Test System

The test system ($\mathcal{TS}$) consists of three modules : Test-Controller, Timer-Handler and Exp-Delayer.

5.1.1 Test-Controller. This module executes test sequences generated from the se-FSA generated itself from the TA describing the specification. Therefore Test-Controller : (1) *sends* inputs of the $\mathcal{IUT}$ and events Set; and (2) *receives* outputs of the $\mathcal{IUT}$ and events Exp.

5.1.2 Timer-Handler. This module consists of : (1) a process PC which generates the event tick at a constant frequency $=\frac{1}{uct}$; and (2) a process $PT_i(k)$ created with the reception of each $?Set(t_i, k)$. k ucts after its creation, $PT_i(k)$ generates $!Exp(t_i, k)$ and then terminates. If a $?Set(t_i, p)$ is generated before the termination of $PT_i(k)$, then the latter is killed and a new process $PT_i(p)$ is created.

5.1.3 Exp-Delayer. This module realizes the delaying action of events Exp which has been introduced at the end of Sect. 4. To achieve this delaying, Exp-Delayer :

1 "shows" the occurrence of an event $Exp(t, k)$ if the latter is allowed;

2 "hides and stores" the occurrence of an $Exp(t, k)$ if the latter is not allowed;

3 "removes" a stored event $Exp(t, k)$ when a $Set(t, *)$ is executed;

4 "generates" automatically and *immediately* a (non-removed) stored $Exp(t, k)$ when the latter becomes allowed.

For choosing between "shows", "hides and stores" and "generates", the Exp-Delayer needs to know the current state of the se-FSA. Therefore it needs to observe all the events (inputs, outputs, Set and Exp). For performing the "removes" action, the Exp-Delayer needs to observe *Set* events. A specification of the Exp-Delayer may therefore be obtained by modifying the se-FSA as follows :

- All events become inputs (Exp-Delayer observes all the events).

- Every input $?Exp(t, k)$ is replaced by :

  "$?Exp(t, k), !Exp(t, k)$" **if** $Exp(t, k)$ is not stored (Exp-Delayer "shows" the input $?Exp(t, k)$),

  "$!Exp(t, k)$" **if** $Exp(t, k)$ is stored (Exp-Delayer "generates" the stored $Exp(t, k)$).

- In every state q, add selfloops of inputs *Exps* which are not allowed in q and associate to them the action "hides and stores".

- To every input $?Set(t, k)$, associate the action "remove($!Exp(t, k)$)".

5.1.4   Test system ($\mathcal{TS}$) and System Under Test ($\mathcal{SUT}$).   Our test system ($\mathcal{TS}$) has therefore the structure represented in Fig. 7. Let S1 be a TA describing the specification and S2 be the se-FSA obtained from S1. The proposed structure allows to test the conformance of the set $\{\mathcal{IUT}$, Timer-Handler, Exp-Delayer$\}$ to the specification S2. From the testing equivalence (see beginning of Sect. 5), we deduce that this structure allows to test the conformance of the $\mathcal{IUT}$ to the specification S1. Testing equivalence can be intuitively explained as follows : the Timer-Handler and the Exp-Delayer are assumed correct, and therefore the detection of any error by the Test-Controller in the set $\{\mathcal{IUT}$, Timer-Handler, Exp-Delayer$\}$ implies that the $\mathcal{IUT}$ is faulty.

Note that certain transitions of the se-FSA may have two inputs : an event *Set* and an input for the $\mathcal{IUT}$ (see Fig. 4). This is not a problem because the two inputs are sent by the Test-Controller to different destinations ($\mathcal{IUT}$ versus $\{$ Timer-Handler, Exp-Delayer $\}$. Note that the action "generates" of the Exp-Delayer must be *immediate* because the testing equivalence is guaranteed if and only if no event of the $\mathcal{IUT}$ occurs between the instant when a stored $Exp(t, k)$ becomes allowed and the instant when the Exp-Delayer generates it. This assumption can be removed by using the following mechanism. Each event *Exp* is timestamped by the instant when the Timer-Handler generates it. When the Test-Controller receives an event (input or output) x of the $\mathcal{IUT}$ and an event *Exp* e from the Exp-Delayer, if x precedes e by a short delay then the Test-Controller determines the correct order of the two events from the timestamp of e. In the remaining of this article, for the sake of simplicity the use by the Test-Controller of such a mechanism is implicit.

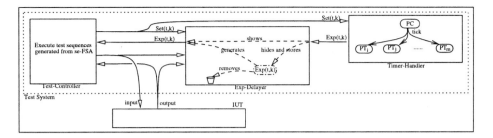

Figure 7.   Structure of the test system

5.2.   Model of $\mathcal{IUT}$

Since the specification of the $\mathcal{IUT}$ is initially described by a TA, we assume that the $\mathcal{IUT}$ may itself be described by a TA (possibly unknown). The proposed model of $\mathcal{IUT}$ is inspired from [6] and is represented in Fig. 8. The $\mathcal{IUT}$ consists of two parts :

A IUT-Controller which executes the TA modelling the $\mathcal{IUT}$

A Timer-Handler : (different than the Timer-Handler of the $\mathcal{TS}$)

- a process PC which is similar to the process PC of $\mathcal{TS}$; and
- a process PT_i is created with each $ResetTimer(t_i)$ which sets the value of t_i to zero. When the timer handler receives $PleaseValue(t_i)$ then it immediately sends $GetValue(t_i, k)$ where k is the current value of t_i.

With this model, we assume that the PC of the $\mathcal{TS}$ and the PC of the $\mathcal{IUT}$ are synchronized. In a first approach of synchronization, the two PCs have a direct access to a single time source. In a more usual approach, each of the PCs uses an internal hardware clock that gives an adequate approximation to the passage of time in the environment. With this approach, it is necessary to coordinate the two local clocks. There are many algorithms for doing this [16].

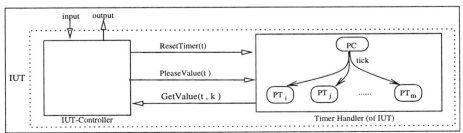

Figure 8.   Model of $\mathcal{IUT}$

6.   TEST CASE GENERATION

6.1.   Transforming se-FSA into io-FSA

We intend to use a test generation method which is applicable to input/output FSAs (denoted io-FSA). For this reason, the se-FSA obtained by the algorithm of Sect. 4 needs to be transformed into a io-FSA. For the se-FSA of Fig. 4, after the removal of undesirable states and the transformation into a io-FSA, we obtain the io-FSA of Fig. 9.

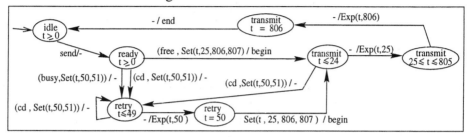

Figure 9.   Input/output FSA obtained from the se-FSA of Fig. 4

6.2.   Use of Wp-Method

For generating test sequences, we can use any software tool based on Wp-Method [8]. As an example, we have used TAG [18] (Test Automatic Generation). First of all TAG is used to construct, for each state q of the io-FSA : (where q_0 is the initial state of the io-FSA)

An Identification set W_q which is a set of input sequences which can be used to identify q.

A Preamble P_q which is an input sequence which brings the io-FSA from q_0 to q.

A Postamble R_q which is an input sequence which brings the io-FSA from q to q_0.

Then for each transition Tr : "$q \xrightarrow{\sigma/\rho} r$" of the io-FSA, TAG is used to construct a set of test cases, denoted $TestCase_{Tr}$, which allow to check Tr. Each test case is defined by : P_q, σ, seq ($\in W_r$) and R_s, where s is the state reached after the execution of seq from r. Therefore, all test cases of $TestCase_{Tr}$ are differenciated by the input sequence of W_r which is used to identify Tr. A test case is therefore executed into the following four steps :

1. Execute P_q to bring the io-FSA from its initial state to State q;
2. Execute σ and check whether the observed output(s) is(are) ρ;
3. Execute an input sequence seq of W_r and check the observed output in order to identify the destination state of Tr. Let s be the state reached after the execution of seq.
4. Execute R_s to bring the io-FSA from State s to its initial state.

6.3.   Test Hypotheses

Correctness of the results obtained from TAG (or any tool based on Wp-Method [8]) is guaranteed only if the following hypotheses hold.

Hypothesis 1 The TA describing a specification is deterministic, i.e., from any location of the TA we cannot have :

- several outgoing transitions labeled with the *same input* and the enabling conditions (EC) of which can be satisfied simultaneously;

- several outgoing transitions labeled by *outputs* and the EC of which can be satisfied simultaneously.

Hypothesis 2 The io-FSA derived from the TA describing the specification is deterministic, i.e., from any state of the io-FSA, we cannot have several transitions with the same (possibly empty) input.

The io-FSA must also satisfy the following other hypotheses :

1 It must contain no transition with an empty input. Since the io-FSA contains transitions "*-/Exp*", this hypothesis is never initially satisfied. A solution consists of replacing every "*-/x*" by "*w/x*", where w is an internal action which models "waiting the event x". This solution can be applied to the example of Fig. 9.

2 It must be completely specified over the set of inputs; otherwise, a solution consists of completing it by adding a selfloop "$\sigma/-$" in each state q where an input σ is not specified.

3 It must be minimal, i.e. it has no undistinguishable states. This hypothesis is always satisfied because the corresponding se-FSA generated by the algorithm of Sect 4 is minimal.

6.4.   Exp-Undeterministic io-FSAs

Let A and B be a TA and the corresponding io-FSA, respectively. Hypothesis 2 is equivalent to Hypothesis 1 in the case where, to each location of A corresponds a single sequence of "*-/Exp*" in B. As an example, the TA of Fig. 2 and its corresponding io-FSA of Fig. 9 are both deterministic. On the other hand, Hypothesis 2 is stronger than Hypothesis 1 in the case where, to a given location of A correspond several sequences of "*-/Exp*" in B. In this case, we say that the io-FSA is *Exp-Undeterministic*. As an example, the TA of Fig. 5 is deterministic and its corresponding io-FSA (which can be obtained from Fig. 6.d) is Exp-Undeterministic. Therefore, TAG cannot be used for this io-FSA.

Here is an approach which allows to use TAG for Exp-Undeterministic io-FSAs. If B is Exp-Undeterministic, then we generate several deterministic io-FSAs $B_1, B_2, \cdots$. The only difference between the B_is and B is that, for each set of sequences of Exp of B which correspond to the same location of A, each B_i will have only a single sequence. For example, for the io-FSA corresponding to the se-FSA of Fig. 6.d, we obtain four io-FSAs containing one of the four sequences of Exp, respectively. TAG can then be used to generate test cases for each of them.

But if we consider test *execution*, a tester does not know in advance which sequence, among a set of sequences of Exp corresponding to the same location, will be executed. To deal with this problem, we propose that each test purpose be : *To check any transition among a set of transitions which are reachable by paths : (1) with the same length; and (2) which diverge by transitions Exp.* For the example in Fig. 6.d, the following three sets of transitions constitute three test purposes :

- the three outgoing transitions of State $L2_0$;
- the five outgoing transitions of States $L2_1$ and $L2_2$;
- the two outgoing Exp of States $L2_3$ and $L2_4$.

6.5.   Procedure of Test Generation

Entry	TA
Step 0	**If** Hypothesis 2 is not satisfied **Then** : terminate **Else**
Step 1	Transform the TA into a se-FSA (see Sect. 4)
Step 2	Transform the se-FSA into a io-FSA (see Sect. 6.1)
Step 3	Remove empty inputs (see Point 1 in Sect. 6.3)
Step 4	Complete the io-FSA (see Point 2 in Sect. 6.3)
Step 5	**If** the obtained io-FSA is Exp-Undeterministic **Then** : Generate several deterministic io-FSAs $B_1, B_2, \cdots$ (see Sect. 6.4) **End-If**

Step 6	**For** each deterministic io-FSA :
	     Compute Identification Sets, Preambles, Postambles, and Test cases
	     Remove test cases containing selfloops which have been added at Step 4
	 End-For

	End-If

6.6.   Example

Let us consider the specification of Fig. 2. Hypothesis 2 is satisfied and after Step 2 of the test procedure, we obtain the io-FSA of Fig. 9. After the application of the whole test procedure, we obtain the following identification sets, preambles and postambles.

Identification sets :

$$W_{\langle idle, t \geq 0 \rangle} = W_{\langle retry, t=50 \rangle} = \{ w \cdot free\_Set(t, 25, 806, 807) \cdot Set(t, 25, 806, 807) \}$$

$$W_{\langle ready, t \geq 0 \rangle} = \{ w \cdot free\_Set(t, 25, 806, 807) \}$$

$$W_{\langle transmit, t \leq 24 \rangle} = W_{\langle transmit, 25 \leq t \leq 805 \rangle} = W_{\langle transmit, t=806 \rangle} = W_{\langle retry, t \leq 49 \rangle} = \{ w \}$$

Preambles :

$$P_{\langle idle, t \geq 0 \rangle} = \epsilon \text{ (i.e., empty); } P_{\langle ready, t \geq 0 \rangle} = send;$$

$$P_{\langle transmit,t\leq 24\rangle} = send \cdot free\_Set(t, 25, 806, 807);$$
$$P_{\langle transmit,25\leq t\leq 805\rangle} = send \cdot free\_Set(t, 25, 806, 807) \cdot w;$$
$$P_{\langle transmit,t=806\rangle} = send \cdot free\_Set(t, 25, 806, 807) \cdot w \cdot w;$$
$$P_{\langle retry,t\leq 49\rangle} = send \cdot busy\_Set(t, 50, 51);$$
$$P_{\langle retry,t=50\rangle} = send \cdot busy\_Set(t, 50, 51) \cdot w.$$

Postamble :

$$R_{\langle idle,t\geq 0\rangle} = \epsilon \text{ (i.e., empty)}; \; R_{\langle ready,t\geq 0\rangle} = free\_Set(t, 25, 806, 807) \cdot w \cdot w \cdot w;$$
$$R_{\langle transmit,t\leq 24\rangle} = w \cdot w \cdot w; \; R_{\langle transmit,25\leq t\leq 805\rangle} = w \cdot w; \; R_{\langle transmit,t=806\rangle} = w;$$
$$R_{\langle retry,t\leq 49\rangle} = w \cdot Set(t, 25, 806, 807) \cdot w \cdot w \cdot w;$$
$$R_{\langle retry,t=50\rangle} = Set(t, 25, 806, 807) \cdot w \cdot w \cdot w.$$

Since here each W_q is a singleton, a single test case is sufficient for checking a transition. Let us mention twelve specific test cases which allow to check each of the twelve transitions of the io-FSA. For each transition Tr : "$q \xrightarrow{\sigma/\rho} r$", the corresponding test case is therefore defined by : P_q, σ, W_r and R_s, where s is the state reached after the execution of W_r from r (see Sect. 6.2). For lack of space, we do not present here the twelve test cases, but note that they can be easily deduced because all parameters they depend on are presented.

7. CONCLUSION AND FUTURE WORK

7.1. Contributions

This study deals with testing a protocol entity with timing constraints. If we consider previous works in this area, we think that those presented in [17, 6] are among the most interesting. Contrary to the latters who use a continuous time, in the present study we use a discrete time. In comparison with [17, 6], our main contribution is that we propose a test generation method and a test architecture which avoid *state explosion*. For that purpose, we have used two special events called *Set* and *Exp* which are generated by the test system and which represent the setting and expiration of timers, respectively.

7.2. Future Work

In the near future, we intend to investigate the following issues :

- To determine several rules which can be used to remove *impossible* sequences of *Exp*. In the present study, we have used a single rule (see Point 3 in Step 3 of Sect. 4.2).

- To study certain problems of test *execution*. Several aspects need to be considered such as :

    1 When a io-FSA B is Exp-Undeterministic, we propose in Sect. 6.4 to replace B by several deterministic io-FSAs $B_1, B_2, \cdots$ and then to generate test cases from these deterministic io-FSAs. But during

a test execution, the tester does not know in advance which sequence of Exp will be executed. In Sect. 6.4 we propose an idea to deal with this problem, but we intend to study more thoroughly this problem in order to propose a complete procedure of test execution.

2 If certain impossible sequences of Exp remain in the io-FSA, the test generation procedure will generate test cases for checking impossible (i.e., unreachable) transitions. Therefore during a test execution, the tester will try to check unreachable transitions. An idea to deal with this problem is : *if after a given number N of attempts the tester does not succeed to execute a transition Tr, then it may consider that Tr is unreachable.* The question which arises is : *how to select N ?* We intend to study more thoroughly this problem.

- To study the test of *distributed* systems.
- To apply and possibly adapt the proposed method for testing multimedia applications with timing constraints.

References

[1] R. Alur, C. Courcoubetis, D. Dill, and H. Wong-Toi. Minimization of timed transitions systems. In *CONCUR*, pages 341–354. Springer-Verlag LNCS 630, 1992.

[2] R. Alur and D. Dill. A theory of timed automata. *Theoretical Computer Science*, 126:183–235, 1994.

[3] D. Clarke and I. Lee. Automatic generation of tests for timing constraints from requirements. In *Third International Workshop on Object-Oriented Real-Time Dependable Systems*, Newport Beach, California, February 1997.

[4] D. L. Dill. Timing assumptions and verification of finite-state concurrent systems. In *Automatic Verification Methods for Finite State Systems*, pages 197–212. Springer-Verlag LNCS 407, 1989.

[5] A. En-Nouaary, R. Dssouli, and A. Elqortobi. Génération de tests temporisés. In *6th Colloque Francophone de l'Ingénierie des Protocoles*. HERMES, 1997.

[6] A. En-Nouaary, R. Dssouli, F. Khendek, and A. Elqortobi. Timed test generation based on state characterization technique. In *19th IEEE Real-Time Systems Symposium (RTSS)*, Madrid, Spain, December 1998.

[7] A. En-Nouaary, F. Khendek, and R. Dssouli. Fault coverage in testing real-time systems. In *6th Intern. Conf. on Real-Time Computing Systems and Applications (RTCSA)*, Hong-Kong, December 1999.

[8] S. Fujiwara, G. v. Bochmann, F. Khendek, M. AAmalou, and A. Ghedamsi. Test selection based on finite state models. *IEEE Transactions on Software Engineering*, 17(6):591–603, 1991.

[9] A. Khoumsi, M. Akalay, A. En-Nouaary, R. Dssouli, and L. Granger. An approach for testing real time protocol entities. Technical Report 1173, Université de Montréal. Département IRO, Montréal, Québec, Canada, May 2000.

[10] A. Khoumsi and K. Saleh. Two formal methods for the synthesis of discrete event systems. *Computer Networks and ISDN*, 29:759–780, 1997.

[11] M. Lawford and W.M. Wonham. Equivalence preserving transformations of timed transitions models. *IEEE Transactions on Automatic Control*, 40:1167–1179, July 1995.

[12] F. Liu. Test generation based on an FSM model with timers and counters. Master's thesis, University of Montreal, Department IRO, 1993.

[13] G. Luo, R. Dssouli, G.v. Bochmann, P. Venkataram, and A. Ghedamsi. Test generation with respect to distributed interfaces. *Computer Standards and Interfaces*, 16:119–132, 1994.

[14] D. Mandrioli, S. Morasca, and A. Morzenti. Generating test cases for real-time systems from logic specifications. *ACM Transactions on Computer Systems*, 13(4):365–398, November 1995.

[15] J.S. Ostroff and W.M. Wonham. A framework for real-time discrete event control. *IEEE Transactions on Automatic Control*, 35(4):386–397, 1990.

[16] P. Ramanathan and K.G. Shin. Fault tolerant clock synchronization in distributed systems. *IEEE Computer*, 23:33–42, October 1990.

[17] J. Springintveld, F. Vaadranger, and P. Dargenio. Testing timed automata. Technical Report CTIT97-17, University of Twente, Amsterdam, The Netherlands, 1997.

[18] Q. M. Tan, A. Petrenko, and G. v. Bochmann. A test generation tool for specifications in the form of state machines. In *International Communications Conference (ICC)*, pages 225–229, Dallas, Texas, June 1996.

[19] A. Tanenbaum. *Computer Networks*. Prentice-Hall (3rd edition), 1996.

19

TEST GENERATION IN THE PRESENCE OF CONFLICTING TIMERS

Mariusz A. Fecko, Paul D. Amer
Computer and Information Sciences Department
University of Delaware, Newark, DE
[Fecko,Amer]@cis.udel.edu

M. Ümit Uyar, Ali Y. Duale
Electrical Engineering Department
The City College of the City University of New York, NY
[Umit,Duale]@ee-mail.engr.ccny.cuny.edu

Abstract   The UD's and CCNY's ongoing research to generate conformance tests for the Army network protocol MIL-STD 188-220 addressed test generation when multiple timers are running simultaneously. A test sequence may become unrealizable if there are conflicting conditions based on a protocol's timers. This problem is handled in the hitherto generated tests by manually expanding a protocol's extended FSM based on the set of conflicting timers, resulting in test sequences that are far from minimum-length. Similar inconsistencies, but based on arbitrary linear variables, are present in the extended FSMs modeling VHDL specifications. This paper presents an efficient solution to the conflicting timers problem that eliminates the redundancies of manual state expansion. CCNY's inconsistency removal algorithms are applied to a new model for testing protocols with multiple timers, in which complex timing dependencies are captured by simple linear expressions. This test generation technique is expected to significantly shorten the test sequences without compromising their fault coverage.

---

*Prepared through collaborative participation in the Advanced Telecommunication Information Distribution Research Program (ATIRP) Consortium sponsored by the U.S. Army Research Laboratory under the Federated Laboratory Program, Cooperative Agreement DAAL01-96-2-0002. The U.S. Government is authorized to reproduce and distribute reprints for Government purposes notwithstanding any copyright notation thereon.

Keywords: conformance testing; test case generation; timing constraints; timer testing

1.   INTRODUCTION

The on-going collaboration between the City College of the City University of New York (CCNY) and the University of Delaware (UD) [7] focuses on the generation of test cases automatically from Estelle specifications. Tests are being generated for the US Department of Defense (DoD)/Joint protocol— military standard developed in the US Army, Navy and Marine Corps systems for mobile combat network radios [7]. Within this effort, several theoretical problems have been investigated, including generation of test sequences uninterrupted by active timers [21], and the improvement of test coverage by using the semicontrollable interfaces [8].

This paper studies the problem of test case generation for network protocols with timers, where a test sequence may become unrealizable due to conflicting conditions based on a protocol's timers. This problem is termed the *conflicting timers problem.*

The research has been motivated by the ongoing effort to generate tests for MIL-STD 188-220. The protocol's Datalink Layer defines several timers that can run concurrently and affect the protocol's behavior. For example, *BUSY* and *ACK* timers may be running independently in *FRAME_BUFFERED* state. If either timer is running, a buffered frame cannot be transmitted. If *ACK* timer expires while *BUSY* timer is <u>not</u> running, a buffered frame is retransmitted. If, however, *ACK* timer expires while *BUSY* timer is running, no output is generated.

In the test cases delivered to the US Army Communications-Electronics Command (CECOM), such conflicts are handled by manually expanding EFSMs based on the set of conflicting timers. This procedure results in test sequences that are far from minimum-length [7]. Similar conflicts, but based on arbitrary linear variables, are present in EFSM models of VHDL specifications [19]. Uyar and Duale present algorithms for detecting [19] and removing [5, 20] such inconsistencies in VHDL specifications. Current research at UD and CCNY focuses on adapting these algorithms to eliminate inconsistencies caused by a protocol's conflicting timers, with a view to applying the methodology to conformance test generation for MIL-STD 188-220.

This paper presents a new model for testing real-time protocols with multiple timers, which captures complex timing dependencies by using simple linear expressions. This modeling technique, combined with the CCNY's inconsistency removal algorithms, is expected to significantly shorten test sequences without compromising their fault coverage.

The proposed solution is expected to have a broader application due to a recent proliferation of protocols with real-time requirements [12, 15]. The functional errors in such protocols are usually caused by the unsatisfiability of time constraints and (possibly conflicting) conditions involving timers; therefore, significant research is required to develop efficient algorithms for test generation for such protocols. The results presented here are expected to contribute towards achieving this goal.

2. PROBLEM DEFINITION

Suppose that a protocol specification defines a set of timers $K = \{tm_1, \ldots, tm_{|K|}\}$, such that a timer tm_j may be started and stopped by arbitrary transitions defined in the specification. Each timer tm_j can be associated with a boolean variable T_j whose value is true if tm_j is running, and false if tm_j is not running. Let ϕ be a time formula obtained from variables $T_1, \ldots, T_k$ by using logical operands $\wedge$, $\vee$, and $\neg$. Suppose that a specification contains transitions with time conditions of a form "if ϕ" for some time formula ϕ. It is clear that there may exist infeasible paths in an FSM modeling a protocol, if two or more edges in a path have inconsistent conditions. For example, for transitions e_1: if (T_j) then $\{\varphi_1\}$ and e_2: if $(\neg T_j)$ then $\{\varphi_2\}$, a path (e_1, e_2) is inconsistent unless the action of φ_1 in e_1 sets T_j to false (which happens when timer tm_j expires in transition e_1). The solution to the above problem is expected to allow generating low-cost tests free of such conflicts.

The conflicting timers problem is a special case of the feasibility problem of test sequences, which is an open research problem for the general case [9, 18]. However, there are two simplifying features of the conflicting timers problem: (1) time variables are linear, and (2) time-keeping variable values implicitly increase with time. By considering these features, we expect to find an efficient solution to this special case.

2.1 General Approach

The goal of the presented technique is to achieve at least the following fault coverage: *cover every state transition at least once.* During the testing of a system with multiple timers, when a node v_p is visited, an efficient test sequence should either (1) traverse as many self-loops as possible before a timeout or (2) leave v_p immediately through a non-timeout transition. Once the maximum allowable number of self-loops are traversed, a test sequence may leave v_p through any outgoing transition. Such an approach does not let perform full reachability analysis; however, it can be easily proven that considering only the above two cases is sufficient to include at least one feasible path for each transition (if such a feasible path is not prohibited by the original specification).

In general, the goal of an optimization is to generate a low-cost test sequence that follows the above guidelines, satisfies time conditions of all composite edges and is not disrupted by timeout events during traversal (i.e., contains only feasible transitions). In Section 3., a model will be introduced that allows the generation of test sequences satisfying the above criteria.

2.2   Related Work

Conformance test generation is an active research area [1, 3, 9, 11, 14, 17, 18]. The related work on testing systems with timing dependencies focuses on testing the so-called Timed Automata (TA) [2, 16], which are a formalism primarily used in system verification. However, there is relatively little work reported in the literature on successful application of timed automata to conformance testing. (Other FDTs, such as ET-LOTOS [13], can also be used to describe timed systems.)

Springintveld et al. [16] present the first published theoretical framework for testing timed automata. En-Nouaary et al. [6] introduce a method based on the state characterization technique using a timed extension of the Wp-method [9]. Higashino et al. [10] define several kinds of test sequence executability for real-time systems and present an algorithm for verifying if a test sequence is executable. Cardell-Oliver and Glover [4] propose a method based on the model of Timed Transition Systems (TTSs) [2].

A major goal of these methods is to limit the number of tests, which otherwise may become prohibitively large; hence, each technique offers a means to reduce the test suite size. The reader may consult the relevant papers [4, 6, 10, 16] for more details.

The new model presented in this paper offers several advantages over the TA-based modeling:

- it is tailored-designed only for testing purposes, which does not require to perform full reachability analysis;
- it allows more intuitive modeling of an IUT and testing procedure (each input/output exchange is assigned certain time to realize; there are no instantaneous transitions as in TA);
- it makes it possible to define a timer length as a constant or variable rather than a fixed value as in TA, with which many properties such as service delivery, proper timeout settings, etc. can be modeled and tested.

3.   NOVEL MODEL FOR TESTING SYSTEMS WITH TIMERS

A protocol can be modeled as a deterministic, completely specified FSM (Mealy) represented by a directed graph $G(V, E)$ and a set of timers $K = \{tm_1, \ldots, tm_{|K|}\}$. As part of this model, we also introduce a set of constants and the set of variables $\mathcal{V} = \{T_1, f_1, \ldots, T_{|K|}, f_{|K|}, L_1, \ldots, L_{|V|}, t^s_{1,1}, \ldots, t_{|V|,M_{|V|}}\}$, as defined below.

For each timer tm_j, we introduce the following parameters:

- $T_j \in \{0, 1\}$—boolean variable indicating if the timer is running. $T_j = 1$ if tm_j is running; $T_j = 0$ otherwise
- $D_j \in \mathcal{R}^+$—the timeout value (i.e., timer length) for tm_j
- $f_j \in \mathcal{R}^+ \cup \{0\} \cup \{-\infty\}$—time-keeping variable denoting the current time of tm_j. If $0 \leq f_j < D_j$, then tm_j is running; if $f_j \geq D_j$ or $f_j = -\infty$, tm_j is not running. (It is expired or stopped). f_j is set to 0 when tm_j is started; it is set to $-\infty$ when tm_j is stopped or has expired.

Let us define $EX(T_1, \ldots, T_{|K|})$ as the set of all boolean expressions on $T_1, \ldots, T_{|K|}$. Let a time formula ϕ be defined as an element of EX.

A transition $e_i \in E$ is associated with the following parameters:

- $c_i \in \mathcal{R}^+$—the time needed to traverse e_i
- time condition $\langle \phi_i \rangle$—e_i can trigger only if its associated time formula ϕ_i is satisfied; if no time formula is associated with e_i, its time condition is defined as $\langle 1 \rangle$. For example, if e_i's time condition involves $\phi_i = T_1 \wedge \neg T_3$, the transition can trigger only if tm_1 is running and tm_3 is not running, regardless of the state of other timers
- action list $\{\varphi_{i,1}, \varphi_{i,2}, \ldots\}$—each action $\varphi_{i,k}$ is an ordered pair $(x \in V, \text{update}(x) \in EX(V, \mathcal{R}, \{+, -, *, /\})$, where $\text{update}(x)$ belongs to the set of all linear expressions involving V, the set of real numbers $\mathcal{R}$, and arithmetic operands. Expression $\text{update}(x)$ is used to update x's value, e.g., the two actions of $\{T_1 = 1; f_2 = f_2 + 5\}$ start timer tm_1 and increment the value of the time-keeping variable associated with timer tm_2 by 5 units

The following parameters are defined for each state $v_p \in V$:

- $c_p^s \in \mathcal{R}^+$—the time needed to traverse a self-loop of v_p. The majority of self-loops are inopportune transitions, whose traversal times are therefore comparable and can be approximated with one value c_p^s
- $N_{p,l}^s$—a set of merged non-timeout self-loops of v_p sharing the same time condition $\langle \phi_{p,l} \rangle$, where $1 \leq l \leq M_p$. (A self-loop that starts/stops a timer cannot be merged with others)
- M_p—the number of sets of $N_{p,l}^s$ for node v_p
- $t_{p,l}^s$—the number of untested self-loops in $N_{p,l}^s$. $t_{p,l}^s$ is initialized to $|N_{p,l}^s|$
- $L_p \in \{0, 1, 2\}$—the `exit' condition for state v_p. If $L_p = 0$, no transition outgoing from v_p and no timeout transition in v_p may be traversed; if $L_p = 1$, a test sequence may leave v_p through an outgoing non-timeout transition; if $L_p = 2$, any outgoing transitions (including timeouts) may be traversed

In the next three sections, the time-related behavior of the IUT will be modeled by defining proper time constraints and actions for various types of transitions defined in the specification.

3.1   Types of Transitions

In general, the model distinguishes four types of transitions:
- *Type 1:* timeout transition $e_i^j(v_p, v_q)$, defined for each timer tm_j (e_i^j may be a self-loop, i.e., $p = q$)
- *Type 2:* non-timeout non-self-loop transition $e_i(v_p, v_q)$, where $p \neq q$
- *Type 3:* merged self-loop transition $e_{p,l}(v_p, v_p)$, defined for each node v_p and each set $N_{p,l}^s$
- *Type 4:* merged self-loop transition $e_{p,l}^j(v_p, v_p)$, defined for each node v_p, each set $N_{p,l}^s$ that contains more than one self-loop, and each timer tm_j

While visiting v_p, if there is enough time to test all self-loops of $N_{p,l}^s$ before any timer expires, $e_{p,l}$ (*Type 3*) will be traversed; otherwise, $e_{p,l}^j$ (*Type 4*) will be traversed with tm_j expiring before all self-loops of $N_{p,l}^s$ can be tested.

3.2   Conditions

A number of timing constraints must be appended to the time conditions for all transitions, as defined below.

For each timeout transition $e_i^j(v_p, v_q)$ (*Type 1*), the following condition holds for each timer $tm_{k \neq j}$: `exit' condition for timeouts in v_p true <u>AND</u> timer tm_j running <u>AND</u> (timer tm_k not running <u>OR</u> tm_j expires before tm_k), which is formalized as:*

$$\langle (L_p == 2) \wedge (T_j == 1) \wedge ((T_j == 0) \vee (D_j - f_j < D_k - f_k)) \rangle \equiv$$
$$\langle (L_p == 2) \wedge (T_j == 1) \wedge (D_j - f_j < D_k - f_k) \rangle$$

For each non-timeout non-self-loop $e_i(v_p, v_q)$ (*Type 2*), the following condition holds for each timer tm_k: `exit' condition for v_p true <u>AND</u> (timer tm_k not running <u>OR</u> there is time left to tm_k's timeout). Formally, this condition is:*

$$\langle (L_p > 0) \wedge ((T_j == 0) \vee (f_k < D_k)) \rangle \equiv \langle (L_p > 0) \wedge (f_k < D_k) \rangle$$

For each merged self-loop transition $e_{p,l}$ (*Type 3*), the following condition holds for each timer tm_k: *there are untested self-loops in $N_{p,l}^s$ <u>AND</u> (timer tm_k not running <u>OR</u> all untested self-loops of $N_{p,l}^s$ can be tested before tm_k expires).* For each $e_{p,l}$, all self-loops $N_{p,l}^s$ can be tested by traversing $e_{p,l}$. This condition can be formalized as:

$$\langle (t_{p,l}^s > 0) \wedge ((T_j == 0) \vee (t_{p,l}^s * c_p^s < D_k - f_k)) \rangle \equiv$$
$$\langle (t_{p,l}^s > 0) \wedge (t_{p,l}^s * c_p^s < D_k - f_k) \rangle$$

For each merged self-loop transition $e_{p,l}^j$ (*Type 4*), the following condition holds for each timer $tm_{k \neq j}$: *there are untested self-loops in $N_{p,l}^s$ <u>AND</u> (timer* *tm$_j$ running <u>AND</u> there is enough time left before tm$_j$ expires to test at least one but not all untested self-loops in N$_{p,l}^s$) <u>AND</u> (timer tm$_k$ not running <u>OR</u> tm$_j$ expires before tm$_k$).* In other words, only some of the self-loops of $N_{p,l}^s$ can be tested by traversing $e_{p,l}^j$. Formally, this condition is:

$$\langle (t_{p,l}^s > 0) \wedge ((T_j == 1) \wedge (c_p^s < D_j - f_j < t_{p,l}^s * c_p^s)) \wedge ((T_k == 0) \vee$$
$$(D_j - f_j < D_k - f_k))\rangle \equiv \langle (t_{p,l}^s > 0) \wedge (T_j == 1) \wedge$$
$$(c_p^s < D_j - f_j < t_{p,l}^s * c_p^s) \wedge (D_j - f_j < D_k - f_k)\rangle$$

3.3   Actions

A number of actions must be appended to the action lists for all transitions, as defined below.

For each timeout transition $e_i^j(v_p, v_q)$ *(Type 1)*, for each $k \neq j$:
- set variable T_j to 0 indicating timer expiry: $T_j = 0$
- increment tm_k's current time by the sum of e_i's traversal time and the amount of time left until tm_j's timeout: $f_k = f_k + c_i + \max(0, D_j - f_j)$
- set tm_j's time-keeping variable: $f_j = -\infty$

Since `max' is not a linear action, to utilize any test generation technique that allows only linear actions (as in [20]), e_i^j should be split into $e_{i,1}^j$ and $e_{i,2}^j$ as follows:

$$e_{i,1}^j : \langle (L_p == 2) \wedge (T_j == 1) \wedge (f_j \geq D_j) \wedge (D_j - f_j < D_k - f_k)\rangle$$
$$\{T_j = 0; f_k = f_k + c_i; f_j = -\infty\}$$
$$e_{i,2}^j : \langle (L_p == 2) \wedge (T_j == 1) \wedge (f_j < D_j) \wedge (D_j - f_j < D_k - f_k)\rangle$$
$$\{T_j = 0; f_k = f_k + c_i + D_j - f_j; f_j = -\infty\}$$

The above concept is illustrated in Figure 1. Timer tm_j is started at time $f_j = 0$. After f_j reaches a value of f_j^0, the two feasible transitions are e_1 and e_2. Consider the case where e_1 triggers and f_j is advanced to a value of $f_j^1 = f_j^0 + c_1 < D_j$. In this case, tm_j's timeout corresponds to traversing $e_{i,2}^j$, which advances all timers by $c_i + D_j - f_j^1$. In the case where e_2 triggers, f_j is advanced to a value of $f_j^2 = f_j^0 + c_2 > D_j$, with tm_j's timeout modeled by $e_{i,1}^j$. All timers will be advanced by $e_{i,1}^j$ only by its execution time c_i, because timer tm_j expired while e_2 was being traversed.

In addition, a non-self-loop e_i^j should set the `exit' condition for its end state v_q to 1 by the appended action of $\{L_q = 1\}$.

For each non-timeout non-self-loop $e_i(v_p, v_q)$ *(Type 2)*:
- set the `exit' condition for e_i's end state v_q to true: $L_q = 1$
- for each k, increment tm_k's current time by e_i's traversal time: $f_k = f_k + c_i$

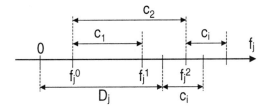

Figure 1.   Time dependencies in timeout transition e_i^j.

For each merged self-loop transition $e_{p,l}$ (*Type 3*):
- set the `exit' condition for state v_p to false: $L_p = 0$
- for each k, increment tm_k's current time by the time needed to traverse all untested self-loops in $N_{p,l}^s$: $f_k = f_k + t_{p,l}^s * c_p^s$
- set the number of untested self-loops in $N_{p,l}^s$ to 0: $t_{p,l}^s = 0$

If no self-loops can be traversed (i.e., there are no untested self-loops of v_p whose time condition is satisfied), L_p should be set to 2 (from either 0 or 1), enabling timeouts and all outgoing transitions in v_p. In this case, L_p will be set to 2 by a so-called *observer self-loop transition* s_p, with the the following condition:

$$\text{for each } l: \quad \langle (L_p < 2) \wedge (t_{p,l}^s == 0 \vee (t_{p,l}^s > 0 \wedge \neg\phi_{p,l})) \rangle \qquad (1)$$

and an action $\{L_p = 2\}$. Condition (1) is satisfied when all self-loops of v_p whose time condition is satisfied are tested (if there are no self-loops defined for v_p, the condition is trivially true).

For each merged self-loop transition $e_{p,l}^j$ (*Type 4*):
- set the `exit' condition for state v_p to true: $L_p = 2$
- for each k, increment tm_k's current time by the time needed to traverse all of the untested self-loops in $N_{p,l}^s$ that can be tested before tm_j expires: $f_k = f_k + c_p^s * \lfloor (D_j - f_j)/c_p^s \rfloor$
- decrement the number of untested self-loops: $t_{p,l}^s = t_{p,l}^s - \lfloor (D_j - f_j)/c_p^s \rfloor$

The `exit' condition L_p works as follows. A test sequence comes to state v_p through an incoming edge, which sets L_p to 1. Then the test sequence may leave immediately through an outgoing non-timeout edge, or take *Type 3* and/or *Type 4* edges. Once *Type 3* edge is taken, it sets L_p to 0, preventing a test sequence from leaving a state and timeouts from occurring. Then the test sequence may include either a *Type 4* edge (which sets L_p to 2, thus enabling timeouts and outgoing edges) or traverse further *Type 3* edges. If neither *Type 3* nor *Type 4* edges can be traversed (this includes the case where none are defined for v_p), the observer edge s_p sets L_p to 2.

Condition (1) results in 2^{M_p} parallel edges due to the presence of M_p number of "OR" statements. Clearly, the technique does not scale well. To prevent the exponential growth of the number of parallel edges, s_p will be replaced with the set of vertices and edges as depicted in Figure 2. The appended conditions and actions of the edges in Figure 2 are derived from (1) as follows:

$$s'_{p,l} : \langle L_p < 2 \wedge t^s_{p,l} == 0 \rangle \{\}, \qquad \check{s}_p : \langle 1 \rangle \{L_p = 2\} \tag{2}$$
$$s''_{p,l} : \langle L_p < 2 \wedge (t^s_{p,l} > 0 \wedge \neg \phi_{p,l}) \rangle \{\} \tag{3}$$

Condition (1) is satisfied when a feasible path exists from v_p to w_{p,M_p}. Since the edges of $s'_{p,k}$ and $s''_{p,k}$ are mutually exclusive, only one such a path is possible. The outgoing edge of w_{p,M_p}, i.e., $\check{s}_p$, sets the `exit' condition to true.

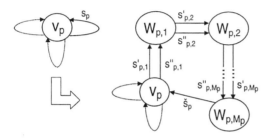

Figure 2. Graph extension to make an observer transition s_p scalable.

4. MODEL REFINEMENT

4.1 Modifying Nonlinear Actions

As can be seen, *Type 4* actions are non-linear, since the number of self-loop traversals before a timeout is computed in $e^j_{p,l}$'s actions by rounding down a fractional value to an integer $z = \lfloor (D_j - f_j)/c^s_p \rfloor$. Since VHDL inconsistencies removal algorithms are applicable only to linear actions, this nonlinearity will be removed by avoiding the computation of z. Instead, a test sequence is forced to traverse one of a number of extra edges with the index of the traversed extra edge equal to z.

To employ this idea, the following steps are taken. Let us first note that $Z_{p,l}$, the number of self-loops of $N_{p,l}$ that can be traversed in any *Type 4* transition, is upper bounded by the cardinality of $N^s_{p,l}$ and the maximum number of self-loop traversals allowed by timers, as defined in (4). The maximum number of self-loop traversals at any time during the execution of a test sequence is therefore obtained by (4).

$$Z = \max_{p,l} Z^s_{p,l}, \quad \text{where} \quad Z^s_{p,l} = \min(|N^s_{p,l}| - 1, \max_{k \leq |K|} \lfloor D_k/c^s_p \rfloor) \tag{4}$$

Having computed the value of Z, we define additional variables D, c^s, z, and r, and extend graph $G(V, E)$ with two vertices u_1 and u_2, as depicted in Figure 3. Next, each $e_{p,l}^j(v_p, v_p)$ is replaced with $\hat{e}_{p,l}^j(v_p, u_1)$, with the unchanged conditions, and the following actions:

- memorize which set of $N_{p,l}$ is represented: $r = p.l$
- set the cost of a self-loop traversal: $c^s = c_p^s$
- set the time remaining until timeout: $D = D_j - f_j$

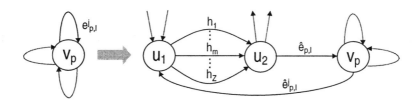

Figure 3.   Graph extension to remove nonlinear actions.

Vertices u_1 and u_2 are connected with extra edges $h_1, \ldots, h_Z$, with the following condition for h_m: *there is enough time left before a timer expires to test m but not $m + 1$ untested self-loops.* Using the additional variables, this condition can be formalized as $\langle m + 1 > D/c^s \geq m \rangle$. The action of h_m sets the value of z to h_m's index m: $\{z = m\}$. In this way, since the conditions of $h_1, \ldots, h_Z$ are mutually exclusive, only one transition h_m indicating the proper number m of allowed self-loop traversals will be traversed (setting z to a value of m).

Finally, M_p edges of $\hat{e}_{p,l}(u_2, v_p)$ are added from u_2 to each v_p with *Type 4* edges, with the condition of $\langle r == p.l \rangle$, which allows a test sequence that left v_p through $\hat{e}_{p,l}^j$ to return to the same v_p, and decrement the proper $t_{p,l}^s$. The <u>linear</u> actions of $\hat{e}_{p,l}$ replace the nonlinear actions of $e_{p,l}^j$ as follows:

- set `exit' condition for v_p to true: $L_p = 2$
- for each k, increment tm_k's current time by the time needed to traverse all of the untested self-loops in $N_{p,l}^s$ that can be tested before timeout: $f_k = f_k + c^s * z$
- decrement the number of untested self-loops: $t_{p,l}^s = t_{p,l}^s - z$

4.2   Delaying Start of Timers

Every transition e_i has the appended conditions and actions as defined in Section 3.. In addition, if e_i stops timer tm_j, the actions of $\{T_j = 0; f_j = -\infty\}$ must be appended to e_i's action list. If e_i starts timer tm_j, the two actions of $\{T_j = 1; f_j = 0\}$ must be appended to e_i's action list.

To have good test coverage, a test sequence should traverse all feasible transitions of an IUT. Some edges in the IUT graph are reachable only if a transition(s) that starts a timer is *delayed* in the test sequence by certain amount of time. The action of delaying such transitions allows us to explore various ordering of timers' expirations by causing certain timers to expire before others.

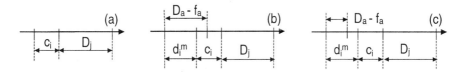

Figure 4. Delaying transition e_i: (a) all timers inactive, no delay; (b) tm_a to expire first, delay less than $D_a - f_a$; (c) tm_a to expire first, delay greater than $D_a - f_a$ cannot be applied due to tm_a's timeout.

Suppose that $e_i = (v_p, v_q)$ starts timer tm_j. Before e_i is traversed, one of the timers—say tm_a—is to expire first. Let d_i^m be the amount of time by which e_i is delayed in this case. It is clear that if e_i is to be traversed instead of tm_a's timeout, d_i^m must be less than $D_a - f_a$ (Figure 4 (b)). In the case where none of the timers are running before traversing e_i (Figure 4 (a)), d_i^m may be set to 0 because time passage does not affect system behavior if all timers are inactive.

Based on the above observations, each e_i will be replaced by two sets of transitions. The first one, which handles the case with d_i^m set to 0 where all timers are inactive before traversing e_i, contains transition e_i^0. Transition e_i^0 has the following appended condition for each timer tm_k: *timer tm_k not running.* Formally, this condition is $\langle T_k == 0 \rangle$.

The second set, which handles the case where d_i^m is upper bounded by a running timer tm_a with the shortest time to expire, contains transitions e_i^a, defined for each $a : 1 \leq a < |K|$. The transitions e_i^a have the following appended condition that holds for each timer $tm_{k \neq a}$: *timer tm_a running AND timer tm_a is to expire before tm_k.* Formally, this condition is:

$$\langle (T_a == 1) \wedge (D_a - f_a < D_k - f_k) \rangle$$

Each e_i^a also has the following appended action:
- for each k, increment tm_k's current time by the introduced delay: $f_k = f_k + d_i^m$, where $0 \leq d_i^m < D_a - f_a$

In the above conversion, e_i is replaced with $|K| + 1$ transitions, out of which only one has a consistent condition, i.e., e_i^0 if no timer is running, or e_i^a for a particular tm_a that is to expire first.

The delay of d_i^m is involved in actions as a parameter with lower (0) and upper ($D_a - f_a$) bounds. During an application of the inconsistencies removal algorithm, the two inequalities of $d_i^m \geq 0$ and $d_i^m < D_a - f_a$ must be included in the consistency check of conditions involving d_i^m. The actual instantiation of d_i^m, i.e., assigning a particular value from between d_i^m's bounds, takes place after generating a test sequence.

5.   APPLICATION TO EXAMPLE FSM

The FSM in Figure 5 consists of three states v_0 (the initial state), v_1, and v_2, and eight transitions e_1 through e_8. Transition e_3 takes 3sec and the remaining transitions each take 1sec to traverse. There are two timers defined for the FSM: tm_1 (started by e_2) with the length of $D_1 = 5.5$ and the timeout transition e_8, and tm_2 (started by e_4 and stopped by e_2) with the length of $D_2 = 3.7$ and the timeout transition e_7. Transition e_1 is associated with time condition $\langle T_1 == 0 \wedge T_2 == 1 \rangle$, transitions e_5 and e_6 are associated with time condition $\langle T_1 == 1 \wedge T_2 == 1 \rangle$, and, for simplicity, the remaining transitions have the time condition $\langle 1 \rangle$.

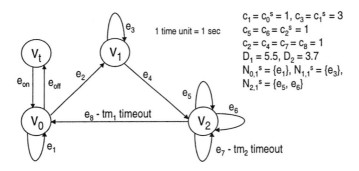

$$c_1 = c_0^s = 1, c_3 = c_1^s = 3$$
$$c_5 = c_6 = c_2^s = 1$$
$$c_2 = c_4 = c_7 = c_8 = 1$$
$$D_1 = 5.5, D_2 = 3.7$$
$$N_{0,1}^s = \{e_1\}, N_{1,1}^s = \{e_3\},$$
$$N_{2,1}^s = \{e_5, e_6\}$$

Figure 5.   FSM with conflicting timers tm_1 and tm_2.

State v_t is introduced as the system initialization state, where a test sequence originates and terminates. A test sequence would start in state v_t with edge e_{on} : $\langle 1 \rangle \{T_1 = 0; T_2 = 0; f_1 = -\infty; f_2 = -\infty; t_{0,1} = 1; t_{1,1} = 1; t_{2,1} = 2; L_0 = 1\}$, which initializes all timers and the variables of $t_{p,l}$. A test sequence would terminate when, after arriving at v_0, edge e_{off} : $\langle T_1 == 0 \wedge T_2 == 0 \rangle \{\}$ is traversed, bringing the IUT back to state v_t. The time condition of e_{off} ensures that all timers are inactive when the test sequence is terminated. Note that, unlike the regular states v_0 through $v_{|V|}$, v_t is not split by the inconsistencies removal algorithm—the final inconsistency-free graph contains only one copy of v_t.

One can give examples of invalid test sequences for the FSM of Figure 5. A test sequence beginning with $(e_{on}, e_1, e_2, \ldots)$ does not satisfy the time condition for e_1: $\langle T_1 == 0 \wedge T_2 == 1 \rangle$, since after traversing e_{on} (initial power-up), neither timer is running. Similarly, any test sequence containing

$(\ldots, e_4, e_7, e_5, \ldots)$ is invalid, because e_5's time condition requires that both timers be running, which does not hold after tm_2 expires in e_7.

Let us first consider transitions of *Type 1* (e_7, e_8). Transition e_7 has the following appended conditions and actions (the conditions and actions for e_8 are analogous):

$$e_{7,1}^2 : \langle L_2 == 2 \wedge f_2 \geq 3.7 \wedge (3.7 - f_2 < 5.5 - f_1) \wedge T_2 == 1 \rangle$$
$$\{L_2 = 1; T_2 = 0; f_1 = f_1 + 1; f_2 = -\infty\}$$
$$e_{7,2}^2 : \langle L_2 == 2 \wedge f_2 < 3.7 \wedge (3.7 - f_2 < 5.5 - f_1) \wedge T_2 == 1 \rangle$$
$$\{L_2 = 1; T_2 = 0; f_1 = f_1 - f_2 + 4.7; f_2 = -\infty\}$$

For transitions of *Type 2* (e_2, e_4), the appended conditions and actions are as follows:

$$e_2 : \langle L_0 > 0 \wedge f_1 < 5.5 \wedge f_2 < 3.7 \rangle$$
$$\{f_1 = f_1 + 1; f_2 = f_2 + 1; L_1 = 1; T_1 = 1; f_1 = 0; T_2 = 0; f_2 = -\infty\}$$
$$e_4 : \langle L_1 > 0 \wedge f_1 < 5.5 \wedge f_2 < 3.7 \rangle$$
$$\{L_2 = 1; f_1 = f_1 + 1; f_2 = f_2 + 1; T_2 = 1; f_2 = 0\}$$

Vertex v_2 has two merged self-loops in $N_{2,1}^s = \{e_5, e_6\}$. Therefore, transitions of both *Type 3* $(e_{2,1})$ and *Type 4* $(e_{2,1}^1, e_{2,1}^2)$ are defined in v_2.

The value of Z is obtained from (19.4) as $Z = Z_{2,1} = \min(1, \max(3, 5)) = 1$. The only extra edge h_1 has the condition of $\langle 2 * c^s > D \geq c^s \rangle$, and the action of $\{z = 1\}$. Edge $\hat{e}_{2,1}$ has the condition of $\langle r == 2.1 \rangle$, and the actions of $\{L_p = 2; f_k = f_k + c^s * z; t_{2,1}^s = t_{2,1}^s - z\}$. Edge $e_{2,1}^1$ is replaced with $\hat{e}_{2,1}^1$, with the unchanged conditions and the following actions: $\{r = 2.1; c^s = 1; D = 5.5 - f_1\}$. Similarly, the actions of $\hat{e}_{2,1}^2$ replacing $e_{2,1}^2$ are $\{r = 2.1; c^s = 1; D = 3.7 - f_2\}$.

In this example, the above augmentation is unnecessary, since $z = 1$ implies that, in any *Type 4* in v_2, $\lfloor 5.5 - f_1 \rfloor = 1$ and $\lfloor 3.7 - f_2 \rfloor = 1$. Therefore, the appended conditions and actions are as follows:

$$e_{2,1} : \langle t_{2,1}^s > 0 \wedge (T_1 == 1 \wedge T_2 == 1) \wedge$$
$$(t_{2,1}^s < 5.5 - f_1) \wedge (t_{2,1}^s < 3.7 - f_2) \rangle$$
$$\{L_2 = 0; f_1 = f_1 + t_{2,1}^s; f_2 = f_2 + t_{2,1}^s; t_{2,1}^s = 0\}$$
$$e_{2,1}^1 : \langle (T_1 == 1 \wedge (1 < 5.5 - f_1 < t_{2,1}^s)) \wedge t_{2,1}^s > 0 \wedge$$
$$(5.5 - f_1 < 3.7 - f_2) \wedge (T_1 == 1 \wedge T_2 == 1) \rangle$$
$$\{L_2 = 2; f_1 = f_1 + 1; f_2 = f_2 + 1; t_{2,1}^s = t_{2,1}^s - 1\}$$
$$e_{2,1}^2 : \langle (T_2 == 1 \wedge (1 < 3.7 - f_2 < t_{2,1}^s)) \wedge t_{2,1}^s > 0 \wedge$$
$$(3.7 - f_2 < 5.5 - f_1) \wedge (T_1 == 1 \wedge T_2 == 1) \rangle$$
$$\{L_2 = 2; f_1 = f_1 + 1; f_2 = f_2 + 1; t_{2,1}^s = t_{2,1}^s - 1\}$$

Since only a single self-loop is defined in vertices v_0 and v_1, both vertices will have merged self-loop transitions of *Type 3* only. For v_0 and v_1, merged self-loop transitions $e_{0,1}$ and $e_{1,1}$ are defined for the sets of $N_{0,1}^s = \{e_1\}$ and $N_{1,1}^s = \{e_3\}$, respectively, with the appended conditions and actions derived as for $e_{2,1}$.

Consider the test sequence for the FSM in Figure 5 (Table 1). While the test sequence is being executed, the values of timer-related variables of the model change with the progress of time.

Table 1.   Valid test sequence for the FSM of Figure 5.

Test step	Edge name	Edge cost	T_1	T_2	f_1	f_2
(1)	e_{on}	0	0	0	$-\infty$	$-\infty$
(2)	e_2	1	1	0	0	$-\infty$
(3)	e_3	3	1	0	3	$-\infty$
(4)	e_4	1	1	1	4	0
(5)	e_5	1	1	1	5	1
(6)	e_8	1	0	1	$-\infty$	2.5
(7)	e_1	1	0	1	$-\infty$	3.5
(8)	e_2	1	1	0	0	$-\infty$
(9)	e_4	1	1	1	1	0
(10)	e_6	1	1	1	2	1
(11)	e_7	1	1	0	5.7	$-\infty$
(12)	e_8	1	0	0	$-\infty$	$-\infty$
(13)	e_{off}	0	0	0	$-\infty$	$-\infty$

Let us now trace the execution of the test sequence. After system initialization by transition e_{on}, transition e_2 starts timer tm_1. After arriving at state v_1, there are 5.5sec left until tm_1's timeout; so, transition $e_{1,1}$ can be tested, which takes 3sec. After leaving v_1, tm_1 has 2.5sec left until timeout. In transition e_4, timer tm_2 is started and the time-keeping variable for tm_1 reaches $f_1 = 4$. After the test sequence arrives at state v_2, tm_1 and tm_2 have 1.5sec and 3.7sec left until timeout, respectively—tm_1 will therefore expire first. There is not enough time to traverse $e_{2,1}$ (i.e., to test both e_5 and e_6); therefore, $e_{2,1}^1$ is traversed (e_5 is tested). In fact, traversing $e_{2,1}^1$ is equivalent to traversing a sequence of edges $(\hat{e}_{2,1}^1, h_1, \hat{e}_2)$, which contain only linear actions. This step leaves 0.5sec and 2.7sec until timeouts for tm_1 and tm_2, respectively. After tm_1 expires, the time-keeping variable for tm_2 is advanced to $f_2 = 2.5$, which gives enough time (1.2 sec) to traverse $e_{0,1}$. Traversing $e_{0,1}$ is equivalent to testing e_1 with the time condition of $\langle T_1 == 0 \wedge T_2 == 1 \rangle$. Since at this point tm_1 has expired and tm_2 is running, e_1's time condition is satisfied and the transition is tested.

Afterwards, e_2 are e_4 are traversed consecutively without spending time on already tested e_3. The test sequence arrives again at state v_2, with 4.5sec and 3.7sec left until timeouts for tm_1 and tm_2, respectively. Now tm_2 is to expire first, leaving sufficient time to traverse $e_{2,1}$ (test e_6). Then, tm_2 expires and the time-keeping variable for tm_1 is advanced to $f_1 = 5.7$, exceeding tm_1's length by 0.2. Therefore, e_8 is traversed immediately, since tm_1 expired while e_7 was being traversed. Now the IUT is back in its initial state v_0 with both timers inactive and all transitions tested, so the test sequence returns to the system initialization state v_t through transition e_{off}.

The test sequence shown in Table 1 satisfies all timing constraints imposed by the two timers tm_1 and tm_2. In addition, the time conditions for all transitions in the FSM are satisfied at any time during the test sequence traversal. Section 6. presents an algorithmic technique to obtain low-cost test sequences satisfying the above criteria.

6.   INCONSISTENCIES REMOVAL

The interdependence among the variables used in the actions and conditions of an EFSM, or an FSM with time variables, may cause various *inconsistencies* among the actions and conditions of the model. For example, in Figure 5, the actions of e_7 set T_2 to 0. Since the time condition of e_5 requires that $\langle T_2 == 1 \rangle$, e_7's action causes inconsistency with e_5's condition. Similarly, a test sequence that includes both e_1 and e_5 contains condition inconsistency—e_1 requires that $\langle T_1 == 0 \rangle$ and e_5 that $\langle T_1 == 1 \rangle$. Both test sequences are therefore infeasible.

Feasible test sequences can be generated from the EFSM models if the inconsistencies are eliminated. The algorithms by Uyar and Duale [5, 19, 20] eliminate inconsistencies from an EFSM in two phases. First, action inconsistencies are detected and eliminated. Next, the algorithms proceed with the detection and elimination of condition inconsistencies by employing linear programming techniques.

In these algorithms, both edge actions and conditions are represented by sets of matrices to analyze their interdependence. In addition, the actions and conditions accumulated along the paths in the graph are represented by sets of *Action Update Matrix (AUM)* pairs and *Accumulated Condition Matrix (ACM)* triplets [5], respectively. While traversing the EFSM graph in a modified breadth-first (MBF) and a depth-first (DF) manner, inconsistencies are eliminated by splitting the nodes and edges of the EFSM graph. During this split, unnecessary growth of the number of states and transitions is avoided. Only the edges with feasible conditions and the nodes that can be reached from the initial node are selected from the split nodes and edges to be included in the resulting FSM. (See paper [5] in these proceedings for a detailed presentation of the inconsistencies removal algorithms.)

In the methodology presented in this paper, the inconsistency removal algorithms are adapted for handling the conflicts caused by multiple timers, and are incorporated in the proposed technique as follows:

Step (1)—Graph augmentation:
Augment an original graph with vertex v_t, edges of e_{on} and e_{off}, and a number of observer edges as described in Section 3. (see Figure 6 for an example). Mark and queue vertex v_0 as $v_{0.0}$.

Step (2)—Inconsistencies removal:
Unqueue vertex $v_{0.k}$ (copy of the initial state state v_0). Apply VHDL inconsistencies removal algorithms in MBF and DF manners starting from $v_{0.k}$ until $v_{0.k}$ is reached again through a set of edges denoted by $E_{0.k}$ (the set of incoming edges of $v_{0.k}$).

Step (3)—Initial state splitting:
Split vertex $v_{0.k}$ into a set of vertices $V_{0.k} \cup \{v_{0.k}\}$; $V_{0.k}$'s cardinality is equal to the number of distinct AUMs associated with edges in $E_{0.k}$ (note: $v_{0.k}$ may belong to $V_{0.k}$). The set of $V_{0.k}$ is further divided into $V_{0.k}^{\text{inc}}$, which contains vertices associated with AUMs corresponding to all timers inactive, and $V_{0.k}^{\text{act}}$, containing the remaining vertices in $V_{0.k}$. The set of edges $E_{0.k}$ is divided accordingly into $E_{0.k}^{\text{inc}}$ and $E_{0.k}^{\text{act}}$.

Edge e_{on}, whose traversal is mandatory in the test sequence, is incoming only to vertex v_0; an edge e_{off} is outgoing from each vertex in $V_{0.k}^{\text{inc}}$. All copies of e_{off} are optional to traverse—they will be included in the test sequence only when necessary.

Step (4)—Redundant paths pruning:
Remove from the graph edges in $E_{0.k}^{\text{inc}}$ using the following two-phase heuristic procedure. First, any edge $e_i \in E_{0.k}^{\text{inc}}$ is deleted if $\exists\, e_j \in E_{0.k}^{\text{inc}}$ such that:

- AUM_j includes AUM_i. Since all timers are inactive in $V_{0.k}^{\text{inc}}$, a sufficient condition for AUM_j to include AUM_i is as follows: $(\forall_{p \leq |V|} \forall_{l \leq M_p}) t_{p,l}^{s\,(j)} \leq t_{p,l}^{s\,(i)}$. This means that AUM_j allows testing more self-loops than AUM_i.
- All edges in the paths from $v_{0.k}$ to vertices in $V_{0.k}^{\text{inc}}$ associated with AUM_i have their copies in the paths from $v_{0.k}$ to vertices in $V_{0.k}^{\text{inc}}$ associated with AUM_j.

Second, any edge $e_i \in E_{0.k}^{\text{inc}}$ is deleted if neither of the following conditions is true:

- A new edge can be traversed by keeping e_i in the graph, i.e., the paths from $v_{0.k}$ to vertices in $V_{0.k}^{\text{inc}}$ associated with AUM_i should contain an edge that has not been traversed before unqueuing $v_{0.k}$.
- Some untested self-loops can be traversed by keeping e_i in the graph, i.e., $(\exists_{p \leq |V|} \exists_{l \leq M_p}) t_{p,l}^{s\,(j)} < t_{p,l}^{s\,(0.k)}$.

Step (5)—Queueing and marking copies of the initial state:
Queue all unmarked vertices in $V_{0.k}^{act}$ and unmarked vertices in $V_{0.k}^{inc}$ with at least one undeleted edge in $E_{0.k}^{inc}$. Mark queued vertices.

If the queue is empty, terminate; otherwise, go back to Step (2). $\square$

Typically, a test sequence is divided into a number of subtours—subsequences of a full test sequence that start and stop in v_0. Each subtour may or may not be preceded by a system power-down/power-up; therefore, when an IUT starts executing, not only should it be brought to state v_0, in addition, all timers must be inactive. To ensure this behavior, each v_0's copy corresponding to an AUM with all timers inactive (i.e., any vertex in $V_{0.k}^{inc}$) may be considered the start state of a subtour.

Let us now apply the above algorithm to the FSM of Figure 5. First, the FSM is augmented with the auxiliary edges of e_{on} and e_{off}, and a number of observer edges as shown in Figure 6. The conditions and actions of the observer edges are defined based on (2)–(3) as follows:

$$s_{0,1}' : \langle L_0 < 2 \wedge t_{0,1}^s == 0 \rangle \; \{\}, \quad s_{1,1}' : \langle L_1 < 2 \wedge t_{1,1}^s == 0 \rangle \; \{\}$$
$$s_{2,1}' : \langle L_2 < 2 \wedge t_{2,1}^s == 0 \rangle \; \{\}$$
$$s_{0,1}'' : \langle L_0 < 2 \wedge (t_{0,1}^s > 0 \wedge (T_1 == 1 \vee T_2 == 0))) \rangle \; \{\}$$
$$s_{2,1}'' : \langle L_2 < 2 \wedge (t_{2,1}^s > 0 \wedge (T_1 == 0 \vee T_2 == 0))) \rangle \; \{\}$$
$$\check{s}_0 : \langle 1 \rangle \{L_0 = 2\}, \quad \check{s}_1 : \langle 1 \rangle \{L_1 = 2\}, \quad \check{s}_2 : \langle 1 \rangle \{L_2 = 2\}$$

An application of the algorithm described in this section to the graph of Figure 6 produces the final graph shown in Figure 7. A minimum-cost test sequence, given by (5)–(7), can be derived as a solution to the Rural Chinese Postman Problem [1] on this final graph. The test sequence of (5)–(7) consists of three subtours containing the edges defined in the original graph (Figure 5) and the auxiliary edges of e_{on} and e_{off}; the observer edges are dropped. All edges defined in the graph of Figure 5 are included without the explicit delaying of timers tm_1 and tm_2; therefore, the technique presented in Section 4. need not be applied in this case. Note that the test sequence of Table 1, which was derived manually, corresponds to *Subtour 1* of (5).

$$1 : \quad e_{on}, e_2, e_3, e_4, e_5, e_8, e_1, e_2, e_4, e_6, e_7, e_8, e_{off} \tag{5}$$
$$2 : \quad e_{on}, e_2, e_4, e_5, e_6, e_7, e_8 \tag{6}$$
$$3 : \quad e_2, e_3, e_4, e_8, e_1, e_2, e_4, e_7, e_8, e_{off} \tag{7}$$

7. CONCLUSION

As a recent result of on-going collaboration between UD and CCNY, this paper presents the study of conformance test generation when multiple timers

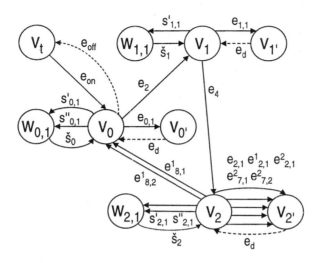

Figure 6.   Augmented graph for the FSM of Figure 5.

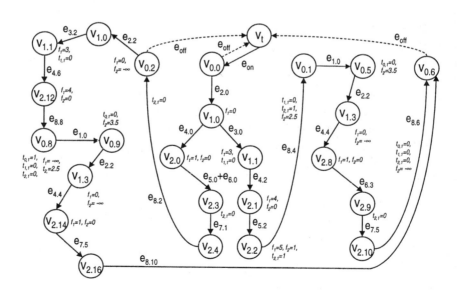

Figure 7.   The final graph for the FSM of Figure 5.

are running simultaneously. CCNY's inconsistency removal algorithms are applied to a new model for testing real-time protocols with multiple timers. As introduced in this paper, the new model captures complex timing dependencies by using simple linear expressions. This modeling technique, combined with the inconsistency removal algorithms, is expected to significantly shorten the test sequences without compromising their fault coverage. Currently, a software tool applying inconsistency removal algorithms to EFSMs models is being implemented at CCNY. Completion of this software project will enable the application of the presented methodology to MIL-STD 188-220.

The methodology presented in this paper is expected to detect transfer and output faults, where an IUT moves into a wrong state (a state other than the one specified) or generates a wrong output (an output other than the one specified) to a given input. As future work, fault detection issues will be pursued further. In particular, a fault model taking into account specific faults caused by the violation of timing constraints and time conditions should be considered. Computing the fault coverage of the presented methodology also needs to be investigated. [1]

Notes

1. The views and conclusions contained in this document are those of the authors and should not be interpreted as representing the official policies, either expressed or implied, of the Army Research Laboratory or the U.S. Government.

References

[1] A. V. Aho, A. T. Dahbura, D. Lee, and M. U. Uyar. An optimization technique for protocol conformance test generation based on UIO sequences and rural Chinese postman tours. *IEEE Trans. Commun.*, 39(11):1604–1615, Nov. 1991.

[2] R. Alur and D. L. Dill. A theory of timed automata. *Theoret. Comput. Sci.*, 126:183–235, 1994.

[3] L. Cacciari and O. Rafiq. Controllability and observability in distributed testing. In K. Saleh and R. Robert, eds, *Communications Software Engineering*, vol. 41(11-12) of *Inform. Softw. Techn.*, pp. 767–780. (special issue), Sept. 1999.

[4] R. Cardell-Oliver and T. Glover. A practical and complete algorithm for testing real-time systems. In *Proc. Int'l Symp. Formal Techn. Real-Time Fault-Toler. Syst. (FTRTFT)*, vol. 1486 of *LCNS*, pp. 251–261, Lyngby, Denmark, Sept. 1998. Springer-Verlag.

[5] A. Y. Duale and M. U. Uyar. Generation of feasible test sequences for EFSM models. In H. Ural, R. L. Probert, and G. v. Bochmann, eds, *Proc. IFIP Int'l Conf. Test. Communicat. Syst. (TestCom)*, Ottawa, Canada, Sept. 2000.

[6] A. En-Nouaary, R. Dssouli, F. Khendek, and A. Elqortobi. Timed test cases generation based on state characterisation technique. In *Proc. IEEE Real-Time Syst. Symp. (RTSS)*, pp. 220–229, Madrid, Spain, Dec. 1998.

[7] M. A. Fecko, M. U. Uyar, P. D. Amer, A. S. Sethi, T. J. Dzik, R. Menell, and M. McMahon. A success story of formal description techniques: Estelle specification and test generation

for MIL-STD 188-220. In R. Lai, ed, *FDTs in Practice*, vol. 23 of *Comput. Commun.* (special issue), Summer 2000. (in press).

[8]   M. A. Fecko, M. U. Uyar, A. S. Sethi, and P. D. Amer. Conformance testing in systems with semicontrollable interfaces. In S. Budkowski and E. Najm, eds, *Protocol Engineering: Part 2*, vol. 55(1-2) of *Annals Telecommun.* (special issue), Jan.–Feb. 2000.

[9]   S. Fujiwara, G. v. Bochmann, F. Khendek, M. Amalou, and A. Ghedamsi. Test selection based on finite state models. *IEEE Trans. Softw. Eng.*, 17(6):591–603, June 1991.

[10] T. Higashino, A. Nakata, K. Taniguchi, and A. R. Cavalli. Generating test cases for a timed I/O automaton model. In G. Csopaki, S. Dibuz, and K. Tarnay, eds, *Proc. IFIP Int'l Wksp Test. Communicat. Syst. (IWTCS)*, pp. 197–214, Budapest, Hungary, Sept. 1999. Boston, MA: Kluwer Academic Publ.

[11] D. Hogrefe. On the development of a standard for conformance testing based on formal specifications. *Comput. Stand. Interf.*, 14(3):185–190, 1992.

[12] R. Lanphier, A. Rao, and H. Schulzrinne. Real time streaming protocol (RTSP). RFC 2326, Internet Eng. Task Force, Apr. 1998.

[13] L. Leonard and G. Leduc. An introduction to ET-LOTOS for the description of time-sensitive systems. *Comput. Networks ISDN Syst.*, 29(3):271–290, 1997.

[14] G. Luo, G. v. Bochmann, and A. F. Petrenko. Test selection based on communicating nondeterministic finite state machines using a generalized Wp-method. *IEEE Trans. Softw. Eng.*, 20(2):149–162, 1994.

[15] H. Schulzrinne, S. Casner, R. Frederick, and V. Jacobson. RTP: A transport protocol for real-time applications. RFC 1889, Internet Eng. Task Force, Jan. 1996.

[16] J. Springintveld, F. Vaandrager, and P. R. D' Argenio. Testing timed automata. Technical Report CTIT-97-17, Univ. of Twente, the Netherlands, 1997. (Invited talk at TAPSOFT' 97, Lille, France, Apr 1997).

[17] J. Tretmans. Conformance testing with labelled transitions systems: Implementation relations and test generation. *Comput. Networks ISDN Syst.*, 29(1):49–79, 1996.

[18] H. Ural. Formal methods for test sequence generation. *Comput. Commun.*, 15(5):311–325, June 1992.

[19] M. U. Uyar and A. Y. Duale. Modeling VHDL specifications as consistent EFSMs. In *Proc. IEEE Military Commun. Conf. (MILCOM)*, Monterey, CA, Nov. 1997.

[20] M. U. Uyar and A. Y. Duale. Resolving inconsistencies in EFSM-modeled specifications. In *Proc. IEEE Military Commun. Conf. (MILCOM)*, Atlantic City, NJ, Nov. 1999.

[21] M. U. Uyar, M. A. Fecko, A. S. Sethi, and P. D. Amer. Testing protocols modeled as FSMs with timing parameters. *Comput. Networks*, 31(18):1967–1988, Sept. 1999.

Author Index